Biodegradable Metals

Biodegradable Metals

Special Issue Editor

Eli Aghion

MDPI • Basel • Beijing • Wuhan • Barcelona • Belgrade

MDPI

Special Issue Editor
Eli Aghion
Ben-Gurion University of the Negev
Israel

Editorial Office
MDPI
St. Alban-Anlage 66
4052 Basel, Switzerland

This is a reprint of articles from the Special Issue published online in the open access journal *Metals* (ISSN 2075-4701) from 2017 to 2018 (available at: https://www.mdpi.com/journal/metals/special_issues/biodegradable_metals)

For citation purposes, cite each article independently as indicated on the article page online and as indicated below:

LastName, A.A.; LastName, B.B.; LastName, C.C. Article Title. *Journal Name* **Year**, *Article Number*, Page Range.

ISBN 978-3-03897-386-7 (Pbk)
ISBN 978-3-03897-387-4 (PDF)

Cover image courtesy of Eli Aghion.

Contents

About the Special Issue Editor

Eli Aghion is a Professor in the Department of Materials Engineering and holder of the Stephen and Edith Berger Chair in Physical Metallurgy at Ben-Gurion University of the Negev, Beer-Sheva, Israel. He received his Doctor of Science from the Technion, Israel Institute of Technology and served as Vice-president and Director of the Magnesium Research Institute at Dead Sea Magnesium, a company affiliated with Israel Chemicals, Ltd. for over a decade. His research activities are mainly related to (i) the development of biodegradable metal implants, (ii) assessing the mechano-chemical behavior and corrosion performance of metals and alloys produced by Additive Manufacturing (AM) technology (3D printing) and (iii) describing the environmental behavior and properties of light metals (Mg, Al, Ti) with nano/sub-micron structures.

Preface to "Biodegradable Metals"

Extensive scientific efforts have been dedicated to the development of biodegradable metal implants, mainly for orthopedic and cardiovascular applications. This was largely due to the enhanced mechanical properties of metals, as compared to biodegradable polymers. Obviously, the main focus of such efforts was directed at metals naturally presenting excellent biocompatibility, such as Mg, Fe and Zn. However, it soon became evident that these metals possess inherent limitations when used as structural implants in in vivo conditions. Magnesium-based implants showed accelerated corrosion rates accompanied by hydrogen gas evolution that can lead to premature loss of mechanical integrity, separation of tissues and gas embolism. Iron-based implants with acceptable corrosion rates showed accumulation of voluminous corrosion products that repel neighboring cells and that did not appear to be metabolized or excreted at appreciable rates. Zinc-based implants have relatively low mechanical properties and reduced corrosion rates that can provoke fibrous encapsulation processes which critically limit their biodegradable capabilities. In view of these limitations, this special issue presents the latest scientific advances in developing biodegradable metal-based systems with special attention to their safe performance in in vivo conditions.

Eli Aghion
Special Issue Editor

metals

MDPI

Editorial

Biodegradable Metals

Eli Aghion

Department of Materials Engineering, Ben-Gurion University of the Negev, Beer-Sheva, Israel; egyon@bgu.ac.il

Received: 25 September 2018; Accepted: 2 October 2018; Published: 8 October 2018

1. Introduction and Scope

Over the last two decades, significant scientific efforts have been devoted to developing biodegradable metal implants for orthopedic and cardiovascular applications, mainly due to their improved mechanical properties compared to those of biodegradable polymers. Naturally, the main focus of such efforts was directed at structural metals with the best biocompatibility characteristics, namely magnesium, iron, and zinc, as well as their alloys. However, it soon became evident that the use of such metal systems in vivo resulted in major problems, limiting their capabilities to act as acceptable structural materials for biodegradable implants in practical applications. These hurdles are exemplified by the accelerated corrosion rate of Mg, accompanied by hydrogen gas evolution that can lead to premature loss of mechanical integrity of the implant, separation of tissues, and, in extreme situations, gas embolism [1,2]. In the case of iron-based implants, although iron corrodes at a reasonable rate, it accumulates a voluminous corrosion product that repels neighboring cells and biological matrices and does not appear to be either metabolized or excreted at an appreciable rate [3]. As for zinc-based implants, they possess relatively low mechanical properties in vivo, as well as reduced corrosion rates, which can provoke fibrous encapsulation processes, which limit their prospects as practical biodegradable implants [4,5]. As such, efforts have been directed at overcoming these hurdles.

This Special Issue introduces the latest scientific advances in developing innovative biodegradable metal implants for various applications and examines the safety of such materials. The collection of high-quality papers comprising this Special Issue is divided into three parts according to the major metal matrix component involved, namely Mg-, Fe-, and Zn-based implants.

2. Contributions

The first part of this Special Issue addresses scientific progress made with magnesium-based implants. The paper by Lumei Liu et al. [6] evaluates the safety of biodegradable Mg-based implants by assessing their impacts at the cellular/molecular level, including in terms of cell adhesion, signaling, immune response, and tissue growth, during the degradation process. In addition, they also evaluate the effect of Mg-based implants on gene expression/protein biosynthesis at the site of implantation, as well as throughout the body. The outcomes of their study serve as the basis for an innovative prediction method to assess the safety of magnesium-based implants. Muhammad Imran Rahim et al. [7] concentrated their efforts on bacterial biofilm infections, as well as bone growth stimulation, given the mechanical forces imposed by magnesium corrosion products. Their novel model for examining implant-derived infections suggests that host cell adhesion to implants is important to prevent bacterial invasion of the exposed host tissue surface, and not, as previously thought, to prevent bacterial adhesion to the implant. According to their model, they predict that passive antibacterial implant-coating strategies would not be efficacious in vivo. Jakub Tkacz et al. [8] examined the effects of solution composition and material surface finish on the corrosion degradation behavior of AZ31 and AZ61 alloys by analyzing the corrosion products created at the implant surface. Their results reveal differences in the response of the Mg alloys to the commonly used cell culture medium Hank's balanced salt solution (HBSS), lacking Mg and Ca ions, and enriched HBSS (HBSS+), containing these ions. Although both alloys exhibited better corrosion resistance in the enriched HBSS+ solution, AZ61 presented higher values of polarization resistance than did AZ31 in both corrosion solutions. As for surface finish (i.e., ground vs. polished), no significant effects were observed during EIS measurements of AZ31 alloy in both HBSS solutions. By contrast, ground samples of AZ61 in HBSS+ solution displayed R_p values which are higher than those obtained with polished samples. The effect of processing parameters on mechanical properties and corrosion degradation of pure Mg in terms of powder metallurgy technology was evaluated by Matej Brezina et al. [9]. They found that cold-compacted samples were quite brittle, with reduced strength (up to 50 MPa) and accelerated corrosion degradation, as compared to hot-pressed samples that yielded much higher strength (up to 250 MPa) and significantly improved corrosion resistance. Overall, applying temperatures treatments of 300 and 400 °C and high pressures of 300–500 MPa had a significantly positive influence on material bonding, and mechanical and electrochemical corrosion properties. A higher compaction temperature of 500 °C had a detrimental effect on material consolidation processes at compacting pressures above 200 MPa. Apart from using Mg-based implants as a supporting device for orthopedic and cardiovascular applications, Da-Jun Lin et al. [10] evaluated the possibility of using Mg-5Zn-0.5Zr alloy for dental-guided bone regeneration. They managed to develop and integrate an optimized solution of heat-treatment and surface fluoride coating to produce an Mg-based regeneration membrane. Their heat-treated Mg regeneration membrane ARRm-H380 was able to provide a proper concentration of Mg ions to accelerate early stage bone growth, encouraging the formation of more than 80% of the new bone.

The second part of the Special Issue is devoted to innovative efforts aimed at using iron as a structural material for biodegradable implants. These activities were carried out mainly due to the inherent limitation of iron, manifested by a reduced corrosion rate in vivo. Although different methods have been proposed for improving the corrosion rate of iron, including modification of alloying elements, poly (lactic-co-glycolic acid) infiltration, and coating, Reza Alavi et al. [11] examined the possibility of using iron-foams as an alternative. The objective of their study was to investigate the mechanical behavior of iron foams with different cell sizes in various compression tests under dry and wet conditions and after being subjected to degradation in HBSS. In general, they found that a wet environment did not significantly change the mechanical behavior of the iron foams, while degradation processes resulted in reductions of elastic modulus, yield strength, and energy absorption. Another attempt to overcome the reduced corrosion degradation of iron is introduced in the review from Mohammad Asgari et al. [12], which notes that sandblasting treatment can increase the degradation rate of pure Fe in simulated body fluid (SBF). The main reasons for the increased corrosion rate were changes in surface composition, and the high roughness and density of dislocations.

The third and final section of the Special Issue is dedicated to the prospects of zinc as a promising alternative to magnesium and iron to overcome the critical limitations of these metals in terms of their suitability for clinical applications. The review paper by Galit Levy Katarivas et al. [13] indicates that Zn^{2+}, the main byproduct of zinc metal corrosion, is highly regulated within physiological systems and plays a critical role in numerous fundamental cellular processes. However, the use of pure Zn as a biodegradable metal for most medical device applications is limited due to its insufficient strength, plasticity, and hardness, as well as its reduced corrosion rate. Although a number of zinc alloys with relatively improved mechanical properties have been developed, such as Zn–Mg and Zn–Al, they still do not present satisfactory properties for practical biodegradable implants. Innovative efforts to develop a new zinc-based alloy were carried out by Alon Kafri et al. [14]. Their research aimed at evaluating the possibility of using Fe as a relatively cathodic biocompatible alloying element in zinc that can tune the implant degradation rate via microgalvanic effects. The selected Zn–1.3wt %Fe alloy composition produced by gravity casting was examined both in vitro and in vivo. The absence of undesirable systemic effects in terms of gain, subject well-being, and hematological characteristics (i.e., red blood cell, hemoglobin, and white blood cell levels) of rats during 14 weeks of implantation, as well as adequate histology results in subcutaneous tissues close to the tested implants, suggests that the Zn–1.3%Fe alloy can be considered as a potential candidate for biodegradable implants. Another attempt carried out by Michaela Krystynova et al. [15] aimed at developing zinc implants via powder metallurgy technology. Their study focused on consolidating zinc powders with two different particle sizes, 7.5 µm and 150 µm, by cold-pressing, followed by sintering and hot-pressing. The obtained results showed that the mechanical properties of samples made from 150 µm particle size powder were better than those prepared with 7.5 µm particle size powder.

In summary, the articles comprising this Special Issue clearly demonstrate the potential of magnesium-, iron-, and zinc-based systems to be a suitable options as structural materials for biodegradable metals implants. However, looking ahead, there are still many challenges to be overcome before these options can be realized. These include: (i) Precisely controlling the degradation kinetics of biocompatible metals and their corrosion products according to in vivo absorption capabilities, while maintaining mechanical integrity prior to significant degradation processes; (ii) correlating the mechanical properties of metal implants according to the required properties of the designated medical device application; and (iii) conducting long-term clinical trials to obtain full biocompatibility responses. Considering the importance of realizing these goals, it is reasonable to expect solutions sooner, rather than later.

References

1. Aghion, E.; Levy, G. The effect of Ca on the in vitro corrosion performance of biodegradable Mg-Nd-Y-Zr alloy. *J. Mater. Sci.* **2010**, *45*, 3096–3101. [CrossRef]
2. Aghion, E.; Levy, G.; Ovadia, S. In vivo behavior of biodegradable Mg-Nd-Y-Zr-Ca alloy. *J. Mater. Sci. Mater. Med.* **2012**, *23*, 805–812. [CrossRef] [PubMed]
3. Bowen, P.K.; Shearier, E.R.; Zhao, S.; Guillory, R.J., II; Zhao, F.; Goldman, J.; Drelich, J.W. Biodegradable Metals for Cardiovascular Stents: From Clinical Concerns to Recent Zn-Alloys. *Adv. Healthc. Mater.* **2016**, *5*, 1121–1140. [CrossRef] [PubMed]
4. Bakhsheshi-Rad, H.R.; Hamzah, E.; Low, H.T.; Kasiri-Asgarani, M.; Farahany, S.; Akbari, E.; Cho, M.H. Fabrication of biodegradable Zn-AlMg alloy: mechanical properties, corrosion behavior, cytotoxicity and antibacterial activities. *Mater Sci Eng C.* **2017**, *73*, 215. [CrossRef] [PubMed]
5. Guillory, R.J.; Bowen, P.K.; Hopkins, S.P.; Shearier, E.R.; Earley, E.J.; Gillette, A.A.; Aghion, E.; Bocks, M.; Drelich, J.W.; Goldman, J. Corrosion Characteristics Dictate the Long-Term Inflammatory Profile of Degradable Zinc Arterial Implants. *ACS Biomater. Sci. Eng.* **2016**, *2*, 2355–2364. [CrossRef]
6. Liu, L.; Wang, J.; Russell, T.; Sankar, J.; Yun, Y. The Biological Responses to Magnesium-Based Biodegradable Medical Devices. *Metals* **2017**, *7*, 514. [CrossRef]
7. Rahim, M.; Ullah, S.; Mueller, P. Advances and Challenges of Biodegradable Implant Materials with a Focus on Magnesium-Alloys and Bacterial Infections. *Metals* **2018**, *8*, 532. [CrossRef]
8. Tkacz, J.; Slouková, K.; Minda, J.; Drábiková, J.; Fintová, S.; Doležal, P.; Wasserbauer, J. Influence of the Composition of the Hank's Balanced Salt Solution on the Corrosion Behavior of AZ31 and AZ61 Magnesium Alloys. *Metals* **2017**, *7*, 465. [CrossRef]
9. Březina, M.; Minda, J.; Doležal, P.; Krystýnová, M.; Fintová, S.; Zapletal, J.; Wasserbauer, J.; Ptáček, P. Characterization of Powder Metallurgy Processed Pure Magnesium Materials for Biomedical Applications. *Metals* **2017**, *7*, 461. [CrossRef]
10. Lin, D.; Hung, F.; Lee, H.; Yeh, M. Development of a Novel Degradation-Controlled Magnesium-Based Regeneration Membrane for Future Guided Bone Regeneration (GBR) Therapy. *Metals* **2017**, *7*, 481. [CrossRef]
11. Alavi, R.; Trenggono, A.; Champagne, S.; Hermawan, H. Investigation on Mechanical Behavior of Biodegradable Iron Foams under Different Compression Test Conditions. *Metals* **2017**, *7*, 202. [CrossRef]
12. Asgari, M.; Hang, R.; Wang, C.; Yu, Z.; Li, Z.; Xiao, Y. Biodegradable Metallic Wires in Dental and Orthopedic Applications: A Review. *Metals* **2018**, *8*, 212. [CrossRef]
13. Katarivas Levy, G.; Goldman, J.; Aghion, E. The Prospects of Zinc as a Structural Material for Biodegradable Implants—A Review Paper. *Metals* **2017**, *7*, 402. [CrossRef]

14. Kafri, A.; Ovadia, S.; Goldman, J.; Drelich, J.; Aghion, E. The Suitability of Zn–1.3%Fe Alloy as a Biodegradable Implant Material. *Metals* **2018**, *8*, 153. [CrossRef]
15. Krystýnová, M.; Doležal, P.; Fintová, S.; Březina, M.; Zapletal, J.; Wasserbauer, J. Preparation and Characterization of Zinc Materials Prepared by Powder Metallurgy. *Metals* **2017**, *7*, 396. [CrossRef]

metals

MDPI

Review

The Biological Responses to Magnesium-Based Biodegradable Medical Devices

Lumei Liu [1,2], Juan Wang [3], Teal Russell [1,2], Jagannathan Sankar [1] and Yeoheung Yun [1,2,*]

[1]   National Science Foundation-Engineering Research Center for Revolutionizing Metallic Biomaterials,
     North Carolina Agricultural and Technical State University, Greensboro, NC 27401, USA;
     llumei@aggies.ncat.edu (L.L); tarussel@ncat.edu (T.R.); sankar@ncat.edu (J.S.)
[2]   FIT BEST Laboratory, Department of Chemical, Biological, and Bioengineering,
     North Carolina Agricultural and Technical State University, Greensboro, NC 27401, USA
[3]   Department of Anesthesiology, School of Medicine, Yale University, New Haven, CT 06519, USA;
     juan.wang@yale.edu
*   Correspondence: yyun@ncat.edu; Tel.: +1-336-285-3226

Received: 20 September 2017; Accepted: 18 November 2017; Published: 21 November 2017

Abstract: The biocompatibility of Magnesium-based materials (MBMs) is critical to the safety of biodegradable medical devices. As a promising metallic biomaterial for medical devices, the issue of greatest concern is devices' safety as degrading products are possibly interacting with local tissue during complete degradation. The aim of this review is to summarize the biological responses to MBMs at the cellular/molecular level, including cell adhesion, transportation signaling, immune response, and tissue growth during the complex degradation process. We review the influence of MBMs on gene/protein biosynthesis and expression at the site of implantation, as well as throughout the body. This paper provides a systematic review of the cellular/molecular behavior of local tissue on the response to Mg degradation, which may facilitate a better prediction of long-term degradation and the safe use of magnesium-based implants through metal innovation.

Keywords: biological responses; biocompatibility; biodegradable; magnesium-based materials

1. Introduction

The degradability of magnesium-based materials (MBMs) makes these biomaterials a great choice for clinical devices, especially for orthopedic and cardiovascular applications. The biocompatibility of MBM refers to their ability to interact with the body organic tissues without causing an unacceptable degree of harm. From a biological perspective, human tissue can not only tolerate, but can even benefit from the interaction with MBM implants by proper responses. On the other hand, the interaction between MBMs and organic tissue in vivo has also been shown to cause phenomena that are not observed in vitro. In an aqueous environment, whether that be organic tissue or in vitro cell culture, Mg reacts with water, generating magnesium hydroxide ($Mg(OH)_2$) and molecular hydrogen (H_2). The biological responses of Mg-based materials have been studied both in vivo and in vitro [1–5]. In vivo, MBM implantation results in the formation of gas pockets in tissue containing different concentrations of H_2, O_2, CO_2, and/or N_2; a high deposition of calcium phosphate (Ca-P), which acts as a mineral layer between tissue and MBM implants; and an increase in the local pH of body fluid [2,6–8]. In contrast, there is no formation of gas pockets in vitro since it is freely released, while in vivo, the gas pockets are trapped by local tissue. Instead, molecular hydrogen escapes to the atmosphere, and cell-adhesion behavior on the surface of MBM implants indicates biocompatibility [1,3]. As a product of MBM corrosion, H_2 was also found to be a potential antioxidant that is involved in cell signaling and has a novel role in preventive and therapeutic applications [9–11]. Furthermore, the Ca-P mineral layer that is associated with magnesium can promote osteoinductivity and osteoconductivity, which aids in the biocompatibility of magnesium alloys as a bone regenerative material [12]. The increase in pH has a positive correlation in hemoglobin picking up oxygen in the blood based on the Bohr effect and a negative correlation in cell-mediated bone resorption by rat osteoclasts in vitro [13,14]. To better understand the biological response to MBMs both in vivo and in vitro, the mechanism of these phenomena should be investigated on the molecular/cellular level.

Like many non-degradable biomaterials, the surface of MBMs is adhered to via protein integrins (heterodimeric receptors in the cell membrane) from the extracellular matrix, within nanoseconds after contact with tissue. Integrins are also involved in intracellular signaling and thus participate in a diverse range of cell functions [15–17]. For cardiovascular applications, MBMs are subject to the Vroman effect, which is exhibited by the absorption of blood serum proteins to the biomaterial surface [18]. However, unlike non-degradable biomaterials, at the time of protein adhesion, Mg reacts with the aqueous environment to generate hydrogen gas (H_2) and $Mg(OH)_2$, thus increasing the concentration of Mg^{2+}. It is known that the physiologically active form of Mg^{2+} serves as a catalyst for over 300 enzymes, including those for ATP synthesis, as well as those that use other nucleotides to synthesize DNA and RNA [19]. Both MBMs and permanent biomaterials, such as Titanium-based alloys, are mixed with biocompatible elements (e.g., rear earth, Nb) [20,21]. However, the biological responses to the added elements and the molecular mechanisms that need to be determined by in vitro and in vivo cytotoxicity evaluation.

When MBMs are implanted into the lesion area, a layer of proteins rapidly adsorbs from the blood (or serum). These proteins effectively translate the structure and composition of the foreign surface into biological signals. The signals that are generated by the recognition of the foreign MBM implant are then transmitted from the extracellular environment to the interior of the cell to regulate gene and protein expression; thus, initiating and mediating cellular behaviors like migration, proliferation, differentiation, and apoptosis in different cell types [22–24]; in addition to stimulating constructive responses that favor wound healing and tissue integration. This layer of proteins determines the activation of the coagulation cascade, complements system, platelets, and immune cells, and guides their interplay, which results in the formation of a transient provisional matrix and the onset of an inflammatory response from the immune system [25,26]. Further research should be done on this protein layer and its expression profile to better understand its involvement in the biological response to MBMs.

Finally, the immune response leads to an encapsulation of the implants, which also indicates the growth of tissue. The regular foreign body reaction process of encapsulation includes inflammation, granulation and regeneration, and fibrosis. It has been shown that Mg^{2+} on bioceramic surfaces substantially affects the phenotype of osteogenic cells in vivo and in vitro [27–30]. A number of studies have demonstrated that Mg^{2+} plays a critical role in bone remodeling and skeletal development [31]. The mechanism of these phenomena is not yet known, but the function of Mg^{2+} in protein synthesis and molecular regulation is a possible explanation. Knowing which genes and proteins are expressed differently due to the influence of MBM implants and how these molecules are affected will not only give further insight into the biocompatibility of MBMs, but will also indicate whether MBMs influence other biological functions involving these proteins. This is of great importance to modern MBM implant design, which should make full use of these differentially expressed molecules to improve implant integration [32]. In this article, these molecules from local molecular/cellular response to the degradation of Mg-based alloys are categorized and reviewed based on their involvement in four functions: cell adhesion, transportation signaling, immune response, and tissue growth.

2. Degradation of Mg-Based Alloys

The degradation behavior of MBMs has been studied and reviewed [33–36]. The mechanism of MBMs degradation involves the reaction of magnesium with its aqueous environment, which produces magnesium hydroxide ($Mg(OH)_2$) and hydrogen gas (H_2). A general summary of the corrosion reaction kinetics that takes place is given below [34,37]:

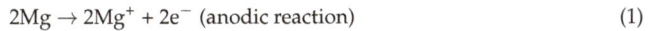

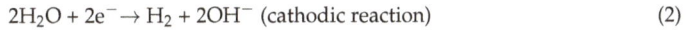

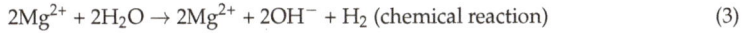

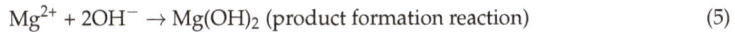

$$2Mg \rightarrow 2Mg^+ + 2e^- \text{ (anodic reaction)} \tag{1}$$

$$2H_2O + 2e^- \rightarrow H_2 + 2OH^- \text{ (cathodic reaction)} \tag{2}$$

$$2Mg^{2+} + 2H_2O \rightarrow 2Mg^{2+} + 2OH^- + H_2 \text{ (chemical reaction)} \tag{3}$$

$$Mg + 2H_2O \rightarrow Mg^{2+} + 2OH^- + H_2 \text{ (overall reaction)} \tag{4}$$

$$Mg^{2+} + 2OH^- \rightarrow Mg(OH)_2 \text{ (product formation reaction)} \tag{5}$$

$$Mg(OH)_2 \rightarrow Mg^{2+} + 2OH^- \text{ (product dissolution reaction)} \tag{6}$$

Mg degradation is a dynamic process, including (1) degradation initiation, (2) degradation rate, (3) degradation product formation, (4) the composition of degradation products, (5) removal of the product from flow-induced shear stress, and (6) localized pitting with hydrogen evolution. This complex process is constantly interacting with local tissue, which involves a typical foreign body reaction composed of macrophages and foreign body giant cells formation [38]. A local physiological environment, such as loading and flow affects Mg degradation and finding the most important factors that influence degradation is the key. These dynamic reactions not only produce corrosion products, such as solid $Mg(OH)_2$ and H_2 gas, but also generate charged molecules that might affect cellular and molecular responses. For example, it has been studied that responding to different concentrations of Mg^{2+}, osteosarcoma (U2OS) cells have different gene expression related to cell growth, apoptosis, inflammation, and migration [39]. While Mg degrades in the body, the neighboring tissue is expected to regenerate and sustain normal functions. The active interface between degrading MBMs' surface and regenerating local tissue should be monitored and controlled to address the medical concern of biocompatibility [40].

3. Protein-Mediated Cell Adhesion

The MBM implants enhance the adhesion of surrounding cells that are mediated by proteins in the extracellular matrix. It is known that cell adhesion and morphology influence their proliferation and differentiation [41]. The ability of biomaterials to adsorb the proteins from serum in a favorable conformation determines their ability to support cell adhesion and spreading [42]. The MBMs have this ability, indicating an important aspect of their relative biocompatibility with adjustable biodegradation [43]. For example, $\alpha 5\beta 1$- and $\beta 1$-integrin were found to mediate cell adhesion to biomaterial surfaces. The expression of $\alpha 5\beta 1$-integrin receptor was increased in human bone-derived cells (HBDC) responding to Mg^{2+}-enriched substrates [44]. It has also been shown that the presence of Mg in bioceramics can significantly increase the expression of $\beta 1$-, $\alpha 5\beta 1$-, and $\alpha 3\beta 1$-integrins that are vital for osteoblast activity [44,45]. Mg^{2+} promotes cell adhesion via $5\beta 1$- and $\beta 1$-integrin-associated signal transduction pathways, which are involved in the enhanced activation of the key signaling adaptor protein Shc (Src homology collagen), resulting in the enhanced gene expression of extracellular matrix proteins [46,47]. In our recent studies, we found that platelets have a different adhesion rate on different MBMs surfaces in dynamic conditions [5]. The major platelet integrin $\alpha IIb\beta 3$ in relation to MBMs has not been studied. This integrin is required for platelet interactions with proteins in plasma and the extracellular matrices (ECM) that are essential for platelet adhesion and aggregation during hemostasis and arterial thrombosis [48].

Surface chemistry modification with Mg^{2+} also plays an important role in focal adhesion kinase (FAK; $pp125^{FAK}$)-mediated signal transduction via cell surface integrin-ECM interaction [44]. It has been shown that FAK expression is enhanced in osteoblasts growing on Al_2O_3-Mg^{2+}, suggesting that tyrosine phosphorylation of signaling proteins was enhanced by binding to Mg^{2+}-supplemented bioceramics [44]. In addition to Shc and FAK, other key proteins, such as collagen type 1, vitronectin, and fibronectin, are also highly expressed by osteoblast cells in the presence of Mg [47]. In vitro, osteoblastic cells and other cell types have been shown to depend primarily on adsorbed vitronectin or fibronectin for initial adhesion and spreading on various materials, including tissue culture polystyrene, titanium, stainless steel, and hydroxyapatite [49–51]. Furthermore, vitronectin and/or fibronectin have been detected among the proteins adsorbed from whole blood and plasma in vitro and in vivo by implanted surfaces [52–55]. According to the Vroman effect, under stagnant conditions, initial protein deposition takes place in this sequence: albumin, globulin, fibrinogen, fibronectin, factor XII, and HMWK [18]. It has been studied that Mg^{2+} improves smooth muscle cells adhesion at 10 mM with certain interaction time. This study revealed some genes that related the influence of Mg^{2+} to cell adhesion (SERPINE 1) and inflammation (HMOX1, IL-1β) functions [56]. One exception to the adhesion-promotion effects of Mg^{2+} is the rapid formation of hydrogen bubbles that accumulated next to the MBM surfaces [57], which physically occupy the position for cell attachment [5]. However, this effect can be moderated by the Ca-P mineral layer coating the surface of MBM implants, which has been shown to enhance cell attachment and spreading [58]. It has also been demonstrated that pH-related proteins near isoelectric pH adsorb more on uncharged biomaterial surfaces [59–61]. Thus, the increasing pH of the surroundings and surface ion change caused by MBM corrosion might decrease cell adhesion.

4. Transportation Signaling

MBM implants increase the concentration of Mg^{2+}, which may modify its transportation signaling pathway between intracellular and extracellular space. Intracellular Mg^{2+} concentration incorporating with Mg^{2+} channels is related to cell growth [62–65]. Mg^{2+}-related functions in the nucleus and mitochondria, such as ATP synthesis, will change due to the increased amount of Mg^{2+} transported by cell membrane magnesium transporters (Figure 1): transient receptor potential melastatin (TRPM) 6 and 7, SLC41A1, CNNM2, and Claudin-16 and 19 (CLDN 16/19). Calcium homeostasis may also be altered (Figure 2).

TRPM6 and TRPM7 were characterized as magnesium "gatekeepers" on the cell membrane that monitor cellular magnesium homeostasis [66]. TRPM7 is responsible for intracellular Mg ion homeostasis in osteoblast cells and plays an important role in osteoblast proliferation and survival [67]. Thus, tight regulation of magnesium homeostasis is crucial for bone health. Another Mg^{2+} transporter is SLC41A1, which was found to be expressed in all of the human tissues tested, but at varying levels, with the heart and testis having the highest expression of the gene [68]. No explanation of the expression pattern has been given with regard to Mg^{2+}-related physiology, though it has been suggested that SLC41 proteins are likely to be the metazoan equivalent of the Mg transporter E (MgtE) that is found in bacteria [68]. This will need to be verified using one of the now standard experimental systems for examining transport, especially in terms of the interface between tissue and MBM implants. Ancient conserved domain protein 2 (ACDP2) is encoded by CNNM2 and regulates physiological magnesium homeostasis in humans [69]. It belongs to the ACDP family and is widely expressed in human tissues, with the highest levels of expression in the brain, kidney, and placenta [70]. Furthermore, studies provide evidence for its involvement in magnesium transport [71,72]. Claudins allow for Mg^{2+} transport via the paracellular pathway; that is, that they mediate the transport of Mg ions through the tight junctions between cells that form an epithelial cell layer. In the claudin family, Claudin-19, which is encoded by the CLDN19 gene, has been implicated in magnesium transport [73,74]. Claudin-16 allows the selective re-uptake of Mg^{2+} in the kidney [75]. Defects in CLDN16 and CLDN19 can cause primary hypomagnesemia, which is characterized by massive renal magnesium wasting and hypercalciuria, resulting in nephrocalcinosis and renal failure [76].

Federica I. Wolf et al. suggested that the magnesium-deficient condition led to the increased cells percentage in the G0/G1-phase and the decreased cells percentage in the S-phase of the cell cycle [77]. Hypermagnesemia is uncommonly reported because the kidney is very efficient in excreting excess magnesium, thus we believe that patients with renal dysfunctions may not be suitable candidates for MBM implants. Besides this, hypomagnesemia and increased pH also affect cell morphology. Echinocytes (red blood cells with a spike-like cell membrane) can be seen with mild hemolysis in hypomagnesemia and are caused by an increase in pH in vitro [78]. At the site of MBM implantation, the Mg^{2+} concentration and pH are increased, and it has not been clearly reported whether MBM implants will increase the number of echinocytes [79], thus causing acanthocytosis. It seems that host tissue has regulation on the magnesium transporters overcompensate for the increase in magnesium ion concentration during the corrosion. However, evidence, such as channels behaviors before, during, and after Mg-based alloys implantation need to be studied.

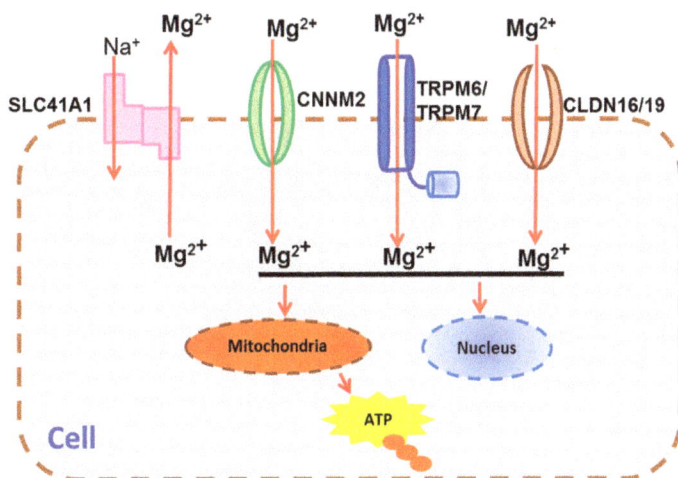

Figure 1. An illustration of the main magnesium transporters on the cell membrane.

The layer of Ca-P deposition formed between the host tissue and MBM implants indicates the transportation of Mg^{2+} has a tight connection with Ca^{2+} transportation. TRPM7 by itself appears to be a Ca^{2+} channel [80], but in the presence of TRPM6, the affinity series of transported cations places Mg^{2+} above Ca^{2+} [67,81]. It has been found that the intestinal absorption and the renal excretion of the two ions are interdependent [82]. Furthermore, the Ca-P layer is the direct cause of vascular calcification [83]. Studies have shown that magnesium reduces calcification in bovine vascular smooth muscle cells (BVSMC) in a dose-dependent manner. Higher magnesium levels prevented BVSMC calcification and inhibited the expression of osteogenic proteins, apoptosis induced by β-glycerophosphate (BGP), and further progression of already established calcification [84]. It has been demonstrated that Mg^{2+} interferes with calcium homeostasis and Ca-P deposition in vascular smooth muscle cells (Figure 2) in the following ways: (1) Mg^{2+} can stabilize the Ca-P complex and inhibit the apatite transformation from Ca-P, instead forming more soluble magnesium-substituted whitlockite [85–87]; (2) Mg^{2+} suppresses apoptosis resulting in the formation of fewer apoptotic bodies; (3) Mg^{2+} blocks the entry of Ca^{2+} into the cells by being transported into cells as a Ca^{2+}-channel antagonist [88], and then impedes the formation of Ca-P particles and Ca-P particle matrix vesicles; (4) Mg^{2+} enters cells through TRPM7 to balance the expression of calcification promotors and inhibitors by suppressing the negative effect of Pi (inorganic phosphate, transported by Pit-1 and Pit-2) on calcification inhibitors (MGP and BMP7) and regressing the activating effect of phosphate on calcification promotors (RUNX2 and BMP2) [89]; (5) Due to the effect of Mg^{2+} on these two calcification promotors, vascular smooth muscle cells are prevented from undergoing osteogenic differentiation and vascular calcification by the same pathway [84]; and, (6) Mg^{2+} activates calcium-sensing receptor (CaSR), which inhibits vascular smooth muscle cell calcification [90,91]. Theoretically, Mg^{2+} should prevent the formation of the Ca-P layer. In reality, however, a Ca-P layer is still formed between MBM implants and host tissue and its deposition to tissue depended on the Mg degradation rate [4].

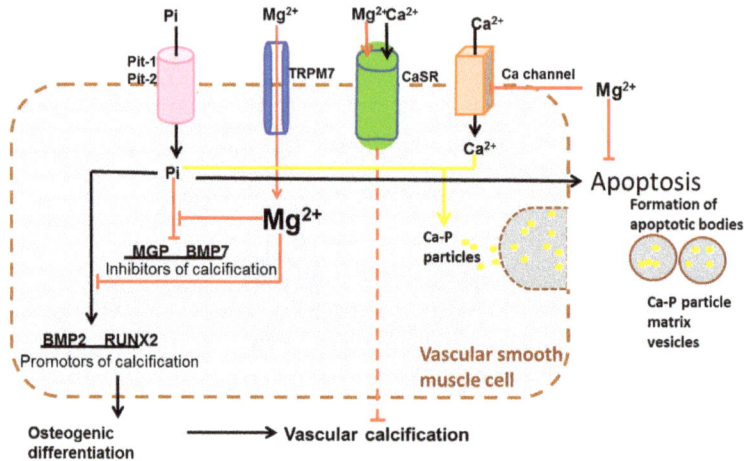

Figure 2. Mg^{2+} interferes with calcium homeostasis and Ca-P layer deposition. Abbreviations: Pit, inorganic phosphate transporter; MGP, matrix Gla protein; BMP, bone morphogenetic protein; RUNX2, runt-related transcription factor 2.

5. Immune Responses

As an implantation biomaterial, MBMs should induce injury, blood-material interactions, provisional matrix formation, inflammation, chronic inflammation, granulation tissue development, foreign body reaction, and fibrosis/fibrous capsule development [38,92–95]. Immune cytokines, such as IL-4 and IL-13, may be involved to induce monocytes adhesion on MBMs surface and monocytes/macrophage fusion to form foreign body giant cells [38]. However, because of the degradability of MBMs, the immune responses are affected by the corrosion products and surface changes of MBMs. Magnesium ions participate in immune responses in numerous ways: as a cofactor for immunoglobulin synthesis, C'3 convertase, immune cell adherence, antibody-dependent cytolysis, IgM-lymphocyte binding, macrophage response to lymphokines, T helper cell-B cell adherence, binding of substance P to lymphoblasts, and antigen binding to macrophage RNA [96]. As biocompatible materials, MBMs do not elicit a detrimental immune response. In fact, some of the immunological responses that are generated by MBMs reflect their beneficial properties.

In one in vitro study, the expression of inflammation-related genes (IL-8, PDGF, TGF-β1, Angio1, βFGF, VEGF, ET-1, CXCR-1, HIF-1α) was either increased or decreased with different magnesium ion concentrations [39]. In magnesium-deficient rodents, TNFα, IL-1, and IL-6 are increased in both the serum and bone marrow microenvironment [97]. Low extracellular magnesium increases endothelial secretion of growth factors and cytokines, such as interleukin 1 (IL-1), which perpetuates cell dysfunction and affects smooth muscle cell functions [98]. These factors have important roles in the immune system. For example, IL-1α and IL-1β are cytokines that participate in the regulation of immune responses, inflammatory reactions, and hematopoiesis [99]. Interleukin 6 (IL-6), also referred to as B-cell stimulatory factor-2 (BSF-2) and interferon beta-2, is a cytokine involved in a wide variety of immune functions, such as antibody secretion, acute phase reaction, and inflammation [100]. Interleukin 8 (IL-8), known as a neutrophil chemotactic factor, is a chemokine produced by macrophages and other cell types, including epithelial cells, airway smooth muscle cells [101], and endothelial cells.

The most significant aspect of MBMs that are related to the immune response is hydrogen gas production [102]. The expression of several pro-inflammatory factors can be decreased by molecular H_2, including TNF-α, IL-6, IL-1β, CCL2, IL-10, TNF-γ, IL-12, CAM-1 [103], HMGB-1 [104], PGE2 [105], and nuclear factor-κB (NF-κB) [106]. The design of MBM implants should make use of the immune response to improve implant integration while avoiding its perpetuation, leading to chronic inflammation and foreign body reactions, and thus loss of intended function [32].

6. Tissue Growth

MBMs implanted into living tissue initiate host immune responses that reflect the first step of tissue growth [107] and fibrous encapsulation [38]. There were concerns about tissue damage because of the evolved hydrogen bubbles and alkalization of solution that are caused by magnesium degradation [43,108]. In some cases, hydrogen bubbles from a degrading MBM surface can be accumulated next to the implant and separate tissues and tissue layers, which will delay the healing of the surgery region and lead to the necrosis of tissues [58]. However, promising studies of magnesium-based biodegradable materials in vivo have shown that they can enhance new bone formation in the vicinity of implantation, including the enhanced local formation of the periosteum and endosteum, two distinct membrane layers that cover the outer and inner surfaces of the bone [109]. MBMs have been shown to be non-toxic and can stimulate bone tissue healing because a high concentration of magnesium ions can lead to bone cell activation [12]. For cardiovascular tissue growth, we recently studied Magnesium implantation in arteries both ex vivo and in vivo. Though there are gas pockets in intima around the implanted Mg wire, the tissue showed normal morphology [4]. A complex signaling network of growth factors includes epidermal growth factor (EGF), fibroblast growth factor (FGF), granulocyte macrophage colony stimulating factor (GM-CSF), transform growth factor-β (TGF-β), vascular endothelial growth factor (VEGF), and platelet derived growth factor (PDGF). This signaling network controls adhesion, migration, proliferation, and differentiation of fibroblasts, keratinocytes, and endothelial cells in wound healing [110]. According to Vroman Effect [18], during the vascular wound healing process, blood proteins will deposit on MBMs surface in a provisional matrix manner, which provides structural, biochemical, and cellular components to processes wound healing [38].

Increased expression of collagen I extracellular matrix protein was found in human bone-derived cells (HBDC) responding to Mg^{2+}-enriched substrates [44], further suggesting that magnesium promotes bone growth. In addition to magnesium, studies have shown that the Ca-P layer that is generated by MBM implants can also promote tissue growth during the biodegradation process both in vivo and in vitro [12,111]. This layer has been proven to facilitate the differentiation and proliferation of osteoblastic cells in a Ca-P ratio-dependent manner, indicating that the Ca-P layer promotes bone formation [112]. There is also Ca-P layer formation due to blood-triggered corrosion of magnesium alloys [113]. The molecular mechanism of this effect has not been discovered yet; however, it might be related to the ability of the Ca-P layer to increase cell adhesion and spreading.

There are still some molecules that have not been related to MBM implants that are associated with tissue growth. For example, Damsky has suggested a role for the integrin molecules $\alpha5\alpha1$ and $\alpha3\alpha1$ in bone formation [114]. It has also been shown that inhibitor of κB kinase–nuclear factor-κB (IKK-NF-κB) inhibits osteoblastic bone formation by restricting the expression of Fos-related antigen-1 (Fra-1), an essential transcription factor that is involved in bone matrix formation in vitro and in vivo [115]. Therefore, targeting IKK–NF-κB, $\alpha5\alpha1$, and $\alpha3\alpha1$ may help to promote bone formation and treat bone resorption that occurs due to the inflammatory response after MBM implantation.

7. Systematic Integration

The biodegradation of Mg elicits an increase of Mg^{2+}, hydrogen gas, and other corrosion products to homeostasis. The molecules that have been proved or might be related to the responses to these corrosion products are converged in Table 1. The molecules generally function in cell adhesion, transportation signaling, immune responses, and tissue growth. The further study of key molecules that are involved in the in vivo and in vitro response to MBM implants, including their functions and pathway, are advanced approaches to understand the biocompatibility of MBMs.

Table 1. Molecular factors involved in or possibly related to the response to magnesium-based materials (MBM) implant corrosion products.

Biological Responses	Mg^{2+}	Ca-P	H$_2$
Cell Adhesion	α5β1-, α3β1-, β1-integrins [44], Shc [46], FAK [44], vitronectin and fibronectin [47,49], SERPINE 1 [56]		
Transportation Signaling	TRPM6/7 [67,81,82], SLC41A1 [68], CLDN16/19 [76], CNNM2 [69]	CaSR [90,91], BGP [84]	
Immune Response	IL-8, PDGF, TGF-β1, Angio1, βFGF, VEGF, ET-1, CXCR-1, HIF-1α [39]; HMOX1 [56], IL-1, TNFα, IL-6 [97]; IL-1 α and IL-1 β [100]; BSF-2 [100]		TNF-α, IL-6, IL-1β, CCL2 and IL-10, TNF-γ, IL-12, CAM-1 [103]; HMGB-1 [104]; PGE2 [105], NF-κB [106]
Tissue Growth	collagen I extracellular matrix protein [44]; EGF, FGF, GM-CSF, TGF-β, VEGF, PDGF [110]; IKK-NF-κB [110,115]; α5α1 and α3α1 [114]		

8. Conclusions

The biocompatibility and degradation properties of Mg alloys make them remarkable implant materials. The most significant problem with MBMs is the difference in their corrosion behavior between in vitro and in vivo studies, which reflects the difficulty in predicting the biological responses of MBMs in the in vitro studies. Another problem is the rapid corrosion of MBMs and the products generated as a result. Systematically understanding the cellular/molecular responses to MBMs implants in the aspect of cell adhesion, transportation signaling, immune responses, and tissue growth are innovative strategies to evaluate their long-term safety for clinical use.

Acknowledgments: This work was supported by NIH NIGMS grant (ISC3GM113728), National Science Foundation (NSF) EAGER grant (1649243), and Engineering Research Center (ERC) for Revolutionizing Metallic Biomaterials (NSF-0812348) at North Carolina A & T State University.

Author Contributions: Lumei Liu conceptualized, wrote and edited the paper; Juan Wang, Teal Russell and Yeoheung Yun revised the paper; Juan Wang gave idea and advice; Jagannathan Sankar and Yeoheung Yun gave instructions and directions.

Conflicts of Interest: The authors declare no conflict of interest.

References

1. Kuhlmann, J.; Bartsch, I.; Willbold, E.; Schuchardt, S.; Holz, O.; Hort, N.; Höche, D.; Heineman, W.R.; Witte, F. Fast escape of hydrogen from gas cavities around corroding magnesium implants. *Acta Biomater.* **2013**, *9*, 8714–8721. [CrossRef] [PubMed]
2. McBride, E.D. Absorbable metal in bone surgery: A further report on the use of magnesium alloys. *J. Am. Med. Assoc.* **1938**, *111*, 2464–2467. [CrossRef]
3. Zhang, S.; Li, J.; Song, Y.; Zhao, C.; Zhang, X.; Xie, C.; Zhang, Y.; Tao, H.; He, Y.; Jiang, Y. In vitro degradation, hemolysis and mc3t3-e1 cell adhesion of biodegradable Mg–Zn alloy. *Mater. Sci. Eng. C* **2009**, *29*, 1907–1912. [CrossRef]
4. Wang, J.; Liu, L.; Wu, Y.; Maitz, M.F.; Wang, Z.; Koo, Y.; Zhao, A.; Sankar, J.; Kong, D.; Huang, N. Ex vivo blood vessel bioreactor for analysis of the biodegradation of magnesium stent models with and without vessel wall integration. *Acta Biomater.* **2017**, *50*, 546–555. [CrossRef] [PubMed]

5. Liu, L.; Koo, Y.; Collins, B.; Xu, Z.; Sankar, J.; Yun, Y. Biodegradability and platelets adhesion assessment of magnesium-based alloys using a microfluidic system. *PLoS ONE* **2017**, *12*, e0182914. [CrossRef] [PubMed]
6. McCord, C.P.; Prendergast, J.J.; Meek, S.F.; Harrold, G.C. Chemical gas gangrene from metallic magnesium. *Ind. Med.* **1942**, *11*, 71–75.
7. Edwards, J.D. *Application of the Interferometer to Gas Analysis*; Government Publishing Office: Washington, DC, USA, 1919.
8. Song, G.; Song, S.-Z. A possible biodegradable magnesium implant material. *Adv. Eng. Mater.* **2007**, *9*, 298–302. [CrossRef]
9. Ohsawa, I.; Ishikawa, M.; Takahashi, K.; Watanabe, M.; Nishimaki, K.; Yamagata, K.; Katsura, K.-I.; Katayama, Y.; Asoh, S.; Ohta, S. Hydrogen acts as a therapeutic antioxidant by selectively reducing cytotoxic oxygen radicals. *Nat. Med.* **2007**, *13*, 688–694. [CrossRef] [PubMed]
10. Salganik, R.I. The benefits and hazards of antioxidants: Controlling apoptosis and other protective mechanisms in cancer patients and the human population. *J. Am. Coll. Nutr.* **2001**, *20*, 464S–472S. [CrossRef] [PubMed]
11. Liu, H.; Colavitti, R.; Rovira, I.I.; Finkel, T. Redox-dependent transcriptional regulation. *Circ. Res.* **2005**, *97*, 967–974. [CrossRef] [PubMed]
12. Witte, F.; Kaese, V.; Haferkamp, H.; Switzer, E.; Meyer-Lindenberg, A.; Wirth, C.J.; Windhagen, H. In vivo corrosion of four magnesium alloys and the associated bone response. *Biomaterials* **2005**, *26*, 3557–3563. [CrossRef] [PubMed]
13. Arnett, T.R.; Dempster, D.W. Effect of ph on bone resorption by rat osteoclasts in vitro. *Endocrinology* **1986**, *119*, 119–124. [CrossRef] [PubMed]
14. Bohr, C.; Hasselbalch, K.; Krogh, A. Concerning a biologically important relationship—The influence of the carbon dioxide content of blood on its oxygen binding. *Skand. Arch. Physiol.* **1904**, *16*, 402. [CrossRef]
15. Schwartz, M.A.; Schaller, M.D.; Ginsberg, M.H. Integrins: Emerging paradigms of signal transduction. *Annu. Rev. Cell Dev. Biol.* **1995**, *11*, 549–599. [CrossRef] [PubMed]
16. Grzesik, W.J. Integrins and bone—Cell adhesion and beyond. *Arch. Immunol. Ther. Exp.* **1996**, *45*, 271–275.
17. Damsky, C.H.; Ilić, D. Integrin signaling: It's where the action is. *Curr. Opin. Cell Biol.* **2002**, *14*, 594–602. [CrossRef]
18. Vroman, L.; Adams, A.; Fischer, G.; Munoz, P. Interaction of high molecular weight kininogen, factor XII, and fibrinogen in plasma at interfaces. *Blood* **1980**, *55*, 156–159. [PubMed]
19. Coté, C.J.; Lerman, J.; Todres, I.D. *A Practice of Anesthesia for Infants and Children E-Book: Expert Consult: Online and Print*; Elsevier Health Sciences: Amsterdam, The Netherlands, 2012.
20. Ehtemam-Haghighi, S.; Liu, Y.; Cao, G.; Zhang, L.-C. Phase transition, microstructural evolution and mechanical properties of Ti-Nb-Fe alloys induced by fe addition. *Mater. Des.* **2016**, *97*, 279–286. [CrossRef]
21. Ehtemam-Haghighi, S.; Prashanth, K.; Attar, H.; Chaubey, A.K.; Cao, G.; Zhang, L. Evaluation of mechanical and wear properties of Ti-xNb-7Fe alloys designed for biomedical applications. *Mater. Des.* **2016**, *111*, 592–599. [CrossRef]
22. Hynes, R.O. Integrins: Versatility, modulation and signaling. *Cell* **1992**, *69*, 11–25. [CrossRef]
23. Zhu, X.; Ohtsubo, M.; Böhmer, R.M.; Roberts, J.M.; Assoian, R.K. Adhesion-dependent cell cycle progression linked to the expression of cyclin D1, activation of cyclin E-cdk2, and phosphorylation of the retinoblastoma protein. *J. Cell Biol.* **1996**, *133*, 391–403. [CrossRef] [PubMed]
24. Van der Flier, A.; Sonnenberg, A. Structural and functional aspects of filamins. *Biochim. Biophys. Acta (BBA) Mol. Cell Res.* **2001**, *1538*, 99–117. [CrossRef]
25. Wilson, C.J.; Clegg, R.E.; Leavesley, D.I.; Pearcy, M.J. Mediation of biomaterial-cell interactions by adsorbed proteins: A review. *Tissue Eng.* **2005**, *11*, 1–18. [CrossRef] [PubMed]
26. Gorbet, M.B.; Sefton, M.V. Biomaterial-associated thrombosis: Roles of coagulation factors, complement, platelets and leukocytes. *Biomaterials* **2004**, *25*, 5681–5703. [CrossRef] [PubMed]
27. Howlett, C.R.; Zreiqat, H.; Wu, Y.; McFall, D.W.; McKenzie, D.R. Effect of ion modification of commonly used orthopedic materials on the attachment of human bone-derived cells. *J. Biomed. Mater. Res.* **1999**, *45*, 345–354. [CrossRef]
28. Zreiqat, H.; Evans, P.; Howlett, C.R. Effect of surface chemical modification of bioceramic on phenotype of human bone-derived cells. *J. Biomed. Mater. Res.* **1999**, *44*, 389–396. [CrossRef]

29. Bilek, M.; Evans, P.; Mckenzie, D.; McCulloch, D.; Zreiqat, H.; Howlett, C. Metal ion implantation using a filtered cathodic vacuum arc. *J. Appl. Phys.* **2000**, *87*, 4198–4204. [CrossRef]

30. Kokubo, T. *Bioceramics and Their Clinical Applications*; Elsevier: Amsterdam, The Netherlands, 2008.

31. Anast, C.S.; Mohs, J.M.; Kaplan, S.L.; Burns, T.W. Evidence for parathyroid failure in magnesium deficiency. *Science* **1972**, *177*, 606–608. [CrossRef] [PubMed]

32. Williams, D.F. On the mechanisms of biocompatibility. *Biomaterials* **2008**, *29*, 2941–2953. [CrossRef] [PubMed]

33. Neil, W.; Forsyth, M.; Howlett, P.; Hutchinson, C.; Hinton, B. Corrosion of magnesium alloy ZE41—The role of microstructural features. *Corros. Sci.* **2009**, *51*, 387–394. [CrossRef]

34. Song, G.L.; Atrens, A. Corrosion mechanisms of magnesium alloys. *Adv. Eng. Mater.* **1999**, *1*, 11–33. [CrossRef]

35. Smith, C.E.; Xu, Z.; Waterman, J.; Sankar, J. Cytocompatibility assessment of mgznca alloys. *Emerg. Mater. Res.* **2013**, *2*, 283–290. [CrossRef]

36. Atrens, A.; Liu, M.; Abidin, N.I.Z. Corrosion mechanism applicable to biodegradable magnesium implants. *Mater. Sci. Eng. B* **2011**, *176*, 1609–1636. [CrossRef]

37. Witte, F.; Hort, N.; Vogt, C.; Cohen, S.; Kainer, K.U.; Willumeit, R.; Feyerabend, F. Degradable biomaterials based on magnesium corrosion. *Curr. Opin. Solid State Mater. Sci.* **2008**, *12*, 63–72. [CrossRef]

38. Anderson, J.M.; Rodriguez, A.; Chang, D.T. Foreign body reaction to biomaterials. In *Seminars in Immunology*; Elsevier: Amsterdam, The Netherlands, 2008; pp. 86–100.

39. Yun, Y.; Dong, Z.; Tan, Z.; Schulz, M.J. Development of an electrode cell impedance method to measure osteoblast cell activity in magnesium-conditioned media. *Anal. Bioanal. Chem.* **2010**, *396*, 3009–3015. [CrossRef] [PubMed]

40. Yun, Y.; Dong, Z.; Lee, N.; Liu, Y.; Xue, D.; Guo, X.; Kuhlmann, J.; Doepke, A.; Halsall, H.B.; Heineman, W. Revolutionizing biodegradable metals. *Mater. Today* **2009**, *12*, 22–32. [CrossRef]

41. Ingber, D.E. Tensegrity ii. How structural networks influence cellular information processing networks. *J. Cell Sci.* **2003**, *116*, 1397–1408. [CrossRef] [PubMed]

42. Steele, J.G.; Dalton, B.A.; Johnson, G.; Underwood, P.A. Adsorption of fibronectin and vitronectin onto primaria™ and tissue culture polystyrene and relationship to the mechanism of initial attachment of human vein endothelial cells and BHK-21 fibroblasts. *Biomaterials* **1995**, *16*, 1057–1067. [CrossRef]

43. Song, G. Control of biodegradation of biocompatable magnesium alloys. *Corros. Sci.* **2007**, *49*, 1696–1701. [CrossRef]

44. Zreiqat, H.; Howlett, C.R.; Zannettino, A.; Evans, P.; Schulze-Tanzil, G.; Knabe, C.; Shakibaei, M. Mechanisms of magnesium-stimulated adhesion of osteoblastic cells to commonly used orthopaedic implants. *J. Biomed. Mater. Res.* **2002**, *62*, 175–184. [CrossRef] [PubMed]

45. Gronthos, S.; Stewart, K.; Graves, S.E.; Hay, S.; Simmons, P.J. Integrin expression and function on human osteoblast-like cells. *J. Bone Min. Res.* **1997**, *12*, 1189–1197. [CrossRef] [PubMed]

46. Shakibaei, M.; Schulze-Tanzil, G.; de Souza, P.; John, T.; Rahmanzadeh, M.; Rahmanzadeh, R.; Merker, H.-J. Inhibition of mitogen-activated protein kinase kinase induces apoptosis of human chondrocytes. *J. Biol. Chem.* **2001**, *276*, 13289–13294. [CrossRef] [PubMed]

47. Schlaepfer, D.D.; Hanks, S.K.; Hunter, T.; van der Geer, P. Integrin-mediated signal transduction linked to ras pathway by GRB2 binding to focal adhesion kinase. *Nature* **1994**, *372*, 786–791. [CrossRef] [PubMed]

48. Shattil, S.J.; Newman, P.J. Integrins: Dynamic scaffolds for adhesion and signaling in platelets. *Blood* **2004**, *104*, 1606–1615. [CrossRef] [PubMed]

49. Steele, J.G.; McFarland, C.; Dalton, B.A.; Johnson, G.; Evans, M.D.; Rolfe Howlett, C.; Underwood, P.A. Attachment of human bone cells to tissue culture polystyrene and to unmodified polystyrene: The effect of surface chemistry upon initial cell attachment. *J. Biomater. Sci. Polym. Ed.* **1994**, *5*, 245–257. [CrossRef]

50. Howlett, C.R.; Evans, M.D.; Walsh, W.R.; Johnson, G.; Steele, J.G. Mechanism of initial attachment of cells derived from human bone to commonly used prosthetic materials during cell culture. *Biomaterials* **1994**, *15*, 213–222. [CrossRef]

51. Kilpadi, K.L.; Chang, P.L.; Bellis, S.L. Hydroxylapatite binds more serum proteins, purified integrins, and osteoblast precursor cells than titanium or steel. *J. Biomed. Mater. Res.* **2001**, *57*, 258–267. [CrossRef]

52. Bale, M.D.; Wohlfahrt, L.A.; Mosher, D.F.; Tomasini, B.; Sutton, R.C. Identification of vitronectin as a major plasma protein adsorbed on polymer surfaces of different copolymer composition. *Blood* **1989**, *74*, 2698–2706. [PubMed]

53. Fabrizius-Homan, D.J.; Cooper, S.L. A comparison of the adsorption of three adhesive proteins to biomaterial surfaces. *J. Biomater. Sci. Polym. Ed.* **1992**, *3*, 27–47. [CrossRef]

54. Babensee, J.E.; Cornelius, R.M.; Brash, J.L.; Sefton, M.V. Immunoblot analysis of proteins associated with hema-mma microcapsules: Human serum proteins in vitro and rat proteins following implantation. *Biomaterials* **1998**, *19*, 839–849. [CrossRef]

55. Rosengren, Å.; Pavlovic, E.; Oscarsson, S.; Krajewski, A.; Ravaglioli, A.; Piancastelli, A. Plasma protein adsorption pattern on characterized ceramic biomaterials. *Biomaterials* **2002**, *23*, 1237–1247. [CrossRef]

56. Ma, J.; Zhao, N.; Zhu, D. Biphasic responses of human vascular smooth muscle cells to magnesium ion. *J. Biomed. Mater. Res. Part A* **2016**, *104*, 347–356. [CrossRef] [PubMed]

57. Meyer-Lindenberg, A.; Windhugen, H.; Witte, F. Medical Implant for the Human or Animal Body. U.S. Patents US20,040,241,036 A1, 2 December 2004.

58. Maxian, S.H.; Zawadsky, J.P.; Dunn, M.G. Effect of Ca/P coating resorption and surgical fit on the bone/implant interface. *J. Biomed. Mater. Res.* **1994**, *28*, 1311–1319. [CrossRef] [PubMed]

59. Norde, W. Driving forces for protein adsorption at solid surfaces. In *Macromolecular Symposia*; Wiley Online Library: New York, NY, USA, 1996; pp. 5–18.

60. Ohno, Y.; Maehashi, K.; Yamashiro, Y.; Matsumoto, K. Electrolyte-gated graphene field-effect transistors for detecting pH and protein adsorption. *Nano Lett.* **2009**, *9*, 3318–3322. [CrossRef] [PubMed]

61. Dee, K.C.; Puleo, D.A.; Bizios, R. *An Introduction to Tissue-Biomaterial Interactions*; John Wiley & Sons: Hoboken, NJ, USA, 2003.

62. Maier, J.A.; Bernardini, D.; Rayssiguier, Y.; Mazur, A. High concentrations of magnesium modulate vascular endothelial cell behaviour in vitro. *Biochim. Biophys. Acta (BBA) Mol. Basis Dis.* **2004**, *1689*, 6–12. [CrossRef] [PubMed]

63. Moomaw, A.S.; Maguire, M.E. The unique nature of Mg^{2+} channels. *Physiology* **2008**, *23*, 275–285. [CrossRef] [PubMed]

64. Günther, T. Concentration, compartmentation and metabolic function of intracellular free Mg^{2+}. *Magnes. Res.* **2006**, *19*, 225–236. [PubMed]

65. Bo, S.; Pisu, E. Role of dietary magnesium in cardiovascular disease prevention, insulin sensitivity and diabetes. *Curr. Opin. Lipidol.* **2008**, *19*, 50–56. [CrossRef] [PubMed]

66. Schlingmann, K.P.; Waldegger, S.; Konrad, M.; Chubanov, V.; Gudermann, T. TRPM6 and TRPM7—Gatekeepers of human magnesium metabolism. *Biochim. Biophys. Acta (BBA) Mol. Basis Dis.* **2007**, *1772*, 813–821. [CrossRef] [PubMed]

67. Nadler, M.J.; Hermosura, M.C.; Inabe, K.; Perraud, A.-L.; Zhu, Q.; Stokes, A.J.; Kurosaki, T.; Kinet, J.-P.; Penner, R.; Scharenberg, A.M. Ltrpc7 is a Mg·ATP-regulated divalent cation channel required for cell viability. *Nature* **2001**, *411*, 590–595. [CrossRef] [PubMed]

68. Wabakken, T.; Rian, E.; Kveine, M.; Aasheim, H.-C. The human solute carrier SLC41A1 belongs to a novel eukaryotic subfamily with homology to prokaryotic MgtE Mg^{2+} transporters. *Biochem. Biophys. Res. Commun.* **2003**, *306*, 718–724. [CrossRef]

69. Meyer, T.E.; Verwoert, G.C.; Hwang, S.-J.; Glazer, N.L.; Smith, A.V.; van Rooij, F.J.; Ehret, G.B.; Boerwinkle, E.; Felix, J.F.; Leak, T.S.; et al. Genome-wide association studies of serum magnesium, potassium, and sodium concentrations identify six loci influencing serum magnesium levels. *PLoS Genet.* **2010**, *6*, e1001045. [CrossRef] [PubMed]

70. Wang, C.-Y.; Shi, J.-D.; Yang, P.; Kumar, P.G.; Li, Q.-Z.; Run, Q.-G.; Su, Y.-C.; Scott, H.S.; Kao, K.-J.; She, J.-X. Molecular cloning and characterization of a novel gene family of four ancient conserved domain proteins (ACDP). *Gene* **2003**, *306*, 37–44. [CrossRef]

71. Goytain, A.; Quamme, G.A. Functional characterization of ACDP2 (ancient conserved domain protein), a divalent metal transporter. *Physiol. Genom.* **2005**, *22*, 382–389. [CrossRef] [PubMed]

72. Will, C.; Breiderhoff, T.; Thumfart, J.; Stuiver, M.; Kopplin, K.; Sommer, K.; Günzel, D.; Querfeld, U.; Meij, I.C.; Shan, Q.; et al. Targeted deletion of murine cldn16 identifies extra-and intrarenal compensatory mechanisms of Ca^{2+} and Mg^{2+} wasting. *Am. J. Physiol.-Ren. Physiol.* **2010**, *298*, F1152–F1161. [CrossRef] [PubMed]

73. Naeem, M.; Hussain, S.; Akhtar, N. Mutation in the tight-junction gene claudin 19 (CLDN19) and familial hypomagnesemia, hypercalciuria, nephrocalcinosis (FHHNC) and severe ocular disease. *Am. J. Nephrol.* **2011**, *34*, 241–248. [CrossRef] [PubMed]

74. Konrad, M.; Schaller, A.; Seelow, D.; Pandey, A.V.; Waldegger, S.; Lesslauer, A.; Vitzthum, H.; Suzuki, Y.; Luk, J.M.; Becker, C. Mutations in the tight-junction gene claudin 19 (CLDN19) are associated with renal magnesium wasting, renal failure, and severe ocular involvement. *Am. J. Hum. Genet.* **2006**, *79*, 949–957. [CrossRef] [PubMed]

75. Hou, J.; Renigunta, A.; Konrad, M.; Gomes, A.S.; Schneeberger, E.E.; Paul, D.L.; Waldegger, S.; Goodenough, D.A. Claudin-16 and claudin-19 interact and form a cation-selective tight junction complex. *J. Clin. Investig.* **2008**, *118*, 619. [CrossRef] [PubMed]

76. Hou, J.; Renigunta, A.; Gomes, A.S.; Hou, M.; Paul, D.L.; Waldegger, S.; Goodenough, D.A. Claudin-16 and claudin-19 interaction is required for their assembly into tight junctions and for renal reabsorption of magnesium. *Proc. Natl. Acad. Sci. USA* **2009**, *106*, 15350–15355. [CrossRef] [PubMed]

77. Wolf, F.I.; Cittadini, A. Magnesium in cell proliferation and differentiation. *Front. Biosci.* **1999**, *4*, D607–D617. [CrossRef] [PubMed]

78. Hoffman, R.; Benz, E.J., Jr.; Silberstein, L.E.; Heslop, H.; Weitz, J.; Anastasi, J. *Hematology: Basic Principles and Practice, Expert Consult Premium Edition-Enhanced Online Features*; Elsevier Health Sciences: Amsterdam, The Netherland, 2012.

79. Wang, J.; He, Y.; Maitz, M.F.; Collins, B.; Xiong, K.; Guo, L.; Yun, Y.; Wan, G.; Huang, N. A surface-eroding poly (1,3-trimethylene carbonate) coating for fully biodegradable magnesium-based stent applications: Toward better biofunction, biodegradation and biocompatibility. *Acta Biomater.* **2013**, *9*, 8678–8689. [CrossRef] [PubMed]

80. Runnels, L.W.; Yue, L.; Clapham, D.E. The TRPM7 channel is inactivated by PIP2 hydrolysis. *Nat. Cell Biol.* **2002**, *4*, 329–336. [CrossRef] [PubMed]

81. Monteilh-Zoller, M.K.; Hermosura, M.C.; Nadler, M.J.; Scharenberg, A.M.; Penner, R.; Fleig, A. TRPM7 provides an ion channel mechanism for cellular entry of trace metal ions. *J. Gen. Physiol.* **2003**, *121*, 49–60. [CrossRef] [PubMed]

82. Paunier, L. Effect of magnesium on phosphorus and calcium metabolism. *Monatsschr. Kinderheilkd. Organ Deutsch. Ges. Kinderheilkd.* **1992**, *140*, S17–S20.

83. Massy, Z.A.; Drüeke, T.B. Magnesium and cardiovascular complications of chronic kidney disease. *Nat. Rev. Nephrol.* **2015**, *11*, 432–442. [CrossRef] [PubMed]

84. Kircelli, F.; Peter, M.E.; Ok, E.S.; Celenk, F.G.; Yilmaz, M.; Steppan, S.; Asci, G.; Ok, E.; Passlick-Deetjen, J. Magnesium reduces calcification in bovine vascular smooth muscle cells in a dose-dependent manner. *Nephrol. Dial. Transplant.* **2011**, *27*, 514–521. [CrossRef] [PubMed]

85. LeGeros, R. Formation and transformation of calcium phosphates: Relevance to vascular calcification. *Z. Kardiol.* **2001**, *90*, 116–124. [CrossRef] [PubMed]

86. Cheng, P.-T.; Grabher, J.; LeGeros, R. Effects of magnesium on calcium phosphate formation. *Magnesium* **1987**, *7*, 123–132.

87. Peters, F.; Epple, M. Simulating arterial wall calcification in vitro: Biomimetic crystallization of calcium phosphates under controlled conditions. *Z. Kardiol.* **2001**, *90*, 81–85. [CrossRef] [PubMed]

88. Altura, B.; Altura, B.; Carella, A.; Gebrewold, A.; Murakawa, T.; Nishio, A. Mg^{2+}-Ca^{2+} interaction in contractility of vascular smooth muscle: Mg^{2+} versus organic calcium channel blockers on myogenic tone and agonist-induced responsiveness of blood vessels. *Can. J. Physiol. Pharmacol.* **1987**, *65*, 729–745. [CrossRef] [PubMed]

89. Montezano, A.C.; Zimmerman, D.; Yusuf, H.; Burger, D.; Chignalia, A.Z.; Wadhera, V.; van Leeuwen, F.N.; Touyz, R.M. Vascular smooth muscle cell differentiation to an osteogenic phenotype involves TRPM7 modulation by magnesium. *Hypertension* **2010**, *56*, 453–462. [CrossRef] [PubMed]

90. Brown, E.M.; Gamba, G.; Riccardi, D.; Lombardi, M.; Butters, R.; Kifor, O.; Sun, A.; Hediger, M.A.; Lytton, J.; Hebert, S.C. Cloning and characterization of an extracellular Ca^{2+}-sensing receptor from bovine parathyroid. *Nature* **1993**, *366*, 575–580. [CrossRef] [PubMed]

91. Ivanovski, O.; Nikolov, I.G.; Joki, N.; Caudrillier, A.; Phan, O.; Mentaverri, R.; Maizel, J.; Hamada, Y.; Nguyen-Khoa, T.; Fukagawa, M. The calcimimetic R-568 retards uremia-enhanced vascular calcification and atherosclerosis in apolipoprotein E deficient (apoE−/−) mice. *Atherosclerosis* **2009**, *205*, 55–62. [CrossRef] [PubMed]

92. Anderson, J.M. Biological responses to materials. *Annu. Rev. Mater. Res.* **2001**, *31*, 81–110. [CrossRef]

93. Anderson, J.M. Multinucleated giant cells. *Curr. Opin. Hematol.* **2000**, *7*, 40–47. [CrossRef] [PubMed]

94. Gretzer, C.; Emanuelsson, L.; Liljensten, E.; Thomsen, P. The inflammatory cell influx and cytokines changes during transition from acute inflammation to fibrous repair around implanted materials. *J. Biomater. Sci. Polym. Ed.* **2006**, *17*, 669–687. [CrossRef] [PubMed]

95. Luttikhuizen, D.T.; Harmsen, M.C.; Luyn, M.J.V. Cellular and molecular dynamics in the foreign body reaction. *Tissue Eng.* **2006**, *12*, 1955–1970. [CrossRef] [PubMed]

96. Galland, L. Magnesium and immune function: An overview. *Magnesium* **1988**, *7*, 290–299. [PubMed]

97. Mazur, A.; Maier, J.A.M.; Rock, E.; Gueux, E.; Nowacki, W.; Rayssiguier, Y. Magnesium and the inflammatory response: Potential physiopathological implications. *Arch. Biochem. Biophys.* **2007**, *458*, 48–56. [CrossRef] [PubMed]

98. Maier, J.A.; Malpuech-Brugère, C.; Zimowska, W.; Rayssiguier, Y.; Mazur, A. Low magnesium promotes endothelial cell dysfunction: Implications for atherosclerosis, inflammation and thrombosis. *Biochim. Biophys. Acta (BBA) Mol. Basis Dis.* **2004**, *1689*, 13–21. [CrossRef] [PubMed]

99. Sims, J.E.; March, C.J.; Cosman, D.; Widmer, M.B.; MacDonald, H.R.; McMahan, C.J.; Grubin, C.E.; Wignall, J.M.; Jackson, J.L.; Call, S.M.; et al. cDNA expression cloning of the IL-1 receptor, a member of the immunoglobulin superfamily. *Science* **1988**, *241*, 585–589. [CrossRef] [PubMed]

100. Hirano, T.; Yasukawa, K.; Harada, H.; Taga, T.; Watanabe, Y.; Matsuda, T.; Kashiwamura, S.-I.; Nakajima, K.; Koyama, K.; Iwamatsu, A.; et al. Complementary DNA for a novel human interleukin (BSF-2) that induces b lymphocytes to produce immunoglobulin. *Nature* **1986**, *324*, 73–76. [CrossRef] [PubMed]

101. Hedges, J.C.; Singer, C.A.; Gerthoffer, W.T. Mitogen-activated protein kinases regulate cytokine gene expression in human airway myocytes. *Am. J. Respir. Cell Mol. Biol.* **2000**, *23*, 86–94. [CrossRef] [PubMed]

102. Seitz, J.M.; Eifler, R.; Bach, F.; Maier, H.J. Magnesium degradation products: Effects on tissue and human metabolism. *J. Biomed. Mater. Res. Part A* **2014**, *102*, 3744–3753. [CrossRef] [PubMed]

103. Buchholz, B.M.; Kaczorowski, D.J.; Sugimoto, R.; Yang, R.; Wang, Y.; Billiar, T.R.; McCurry, K.R.; Bauer, A.J.; Nakao, A. Hydrogen inhalation ameliorates oxidative stress in transplantation induced intestinal graft injury. *Am. J. Transplant.* **2008**, *8*, 2015–2024. [CrossRef] [PubMed]

104. Xie, K.; Yu, Y.; Pei, Y.; Hou, L.; Chen, S.; Xiong, L.; Wang, G. Protective effects of hydrogen gas on murine polymicrobial sepsis via reducing oxidative stress and hmgb1 release. *Shock* **2010**, *34*, 90–97. [CrossRef] [PubMed]

105. Kawasaki, H.; Guan, J.; Tamama, K. Hydrogen gas treatment prolongs replicative lifespan of bone marrow multipotential stromal cells in vitro while preserving differentiation and paracrine potentials. *Biochem. Biophys. Res. Commun.* **2010**, *397*, 608–613. [CrossRef] [PubMed]

106. Chen, H.; Sun, Y.P.; Li, Y.; Liu, W.W.; Xiang, H.G.; Fan, L.Y.; Sun, Q.; Xu, X.Y.; Cai, J.M.; Ruan, C.P.; et al. Hydrogen-rich saline ameliorates the severity of L-arginine-induced acute pancreatitis in rats. *Biochem. Biophys. Res. Commun.* **2010**, *393*, 308–313. [CrossRef] [PubMed]

107. Franz, S.; Rammelt, S.; Scharnweber, D.; Simon, J.C. Immune responses to implants—A review of the implications for the design of immunomodulatory biomaterials. *Biomaterials* **2011**, *32*, 6692–6709. [CrossRef] [PubMed]

108. Song, G. Recent progress in corrosion and protection of magnesium alloys. *Adv. Eng. Mater.* **2005**, *7*, 563–586. [CrossRef]

109. Zhang, E.; Xu, L.; Yu, G.; Pan, F.; Yang, K. In vivo evaluation of biodegradable magnesium alloy bone implant in the first 6 months implantation. *J. Biomed. Mater. Res. Part A* **2009**, *90*, 882–893. [CrossRef] [PubMed]

110. Barrientos, S.; Stojadinovic, O.; Golinko, M.S.; Brem, H.; Tomic-Canic, M. Growth factors and cytokines in wound healing. *Wound Repair Regen.* **2008**, *16*, 585–601. [CrossRef] [PubMed]

111. Li, L.; Gao, J.; Wang, Y. Evaluation of cyto-toxicity and corrosion behavior of alkali-heat-treated magnesium in simulated body fluid. *Surf. Coat. Technol.* **2004**, *185*, 92–98. [CrossRef]

112. Hulshoff, J.; Van Dijk, K.; De Ruijter, J.; Rietveld, F.; Ginsel, L.; Jansen, J. Interfacial phenomena: An in vitro study of the effect of calcium phosphate (Ca-P) ceramic on bone formation. *J. Biomed. Mater. Res.* **1998**, *40*, 464–474. [CrossRef]

113. Geis-Gerstorfer, J.; Schille, C.; Schweizer, E.; Rupp, F.; Scheideler, L.; Reichel, H.P.; Hort, N.; Nolte, A.; Wendel, H.P. Blood triggered corrosion of magnesium alloys. *Mater. Sci. Eng. B* **2011**, *176*, 1761–1766. [CrossRef]

114. Damsky, C.H. Extracellular matrix-integrin interactions in osteoblast function and tissue remodeling. *Bone* **1999**, *25*, 95–96. [CrossRef]
115. Chang, J.; Wang, Z.; Tang, E.; Fan, Z.; McCauley, L.; Franceschi, R.; Guan, K.; Krebsbach, P.H.; Wang, C.-Y. Inhibition of osteoblastic bone formation by nuclear factor-κB. *Nat. Med.* **2009**, *15*, 682–689. [CrossRef] [PubMed]

metals

MDPI

Review

Advances and Challenges of Biodegradable Implant Materials with a Focus on Magnesium-Alloys and Bacterial Infections

Muhammad Imran Rahim [1], Sami Ullah [2] and Peter P. Mueller [3],*

[1] Department of Prosthetic Dentistry and Biomedical Materials Science, Lower Saxony Centre for Biomedical Engineering, Implant Research and Development, Hannover Medical School, Carl-Neuberg-Straße 1, 30625 Hannover, Germany; M.Imran.Rahim@outlook.com

[2] Department of MSYS, Helmholtz Center for Infection Research, Inhoffenstrasse 7, 38124 Braunschweig, Germany; Sami.Ullah@helmholtz-hzi.de

[3] Department of Chemical Biology, Helmholtz Center for Infection Research, Inhoffenstrasse 7, 38124 Braunschweig, Germany

* Correspondence: pmu@gbf.de; Tel.: +49-531-6181-5070

Received: 12 June 2018; Accepted: 4 July 2018; Published: 10 July 2018

Abstract: Medical implants made of biodegradable materials could be advantageous for temporary applications, such as mechanical support during bone-healing or as vascular stents to keep blood vessels open. After completion of the healing process, the implant would disappear, avoiding long-term side effects or the need for surgical removal. Various corrodible metal alloys based on magnesium, iron or zinc have been proposed as sturdier and potentially less inflammatory alternatives to degradable organic polymers, in particular for load-bearing applications. Despite the recent introduction of magnesium-based screws, the remaining hurdles to routine clinical applications are still challenging. These include limitations such as mechanical material characteristics or unsuitable corrosion characteristics. In this article, the salient features and clinical prospects of currently-investigated biodegradable implant materials are summarized, with a main focus on magnesium alloys. A mechanism of action for the stimulation of bone growth due to the exertion of mechanical force by magnesium corrosion products is discussed. To explain divergent in vitro and in vivo effects of magnesium, a novel model for bacterial biofilm infections is proposed which predicts crucial consequences for antibacterial implant strategies.

Keywords: bioresorbable implants; corrosion layer; vascular stents; orthopedic implants; microbial infections

---

1. Introduction

For metallic implants, industrially-developed, inert and long-lasting materials, such as titanium (Ti) alloys, stainless steel (SS) and cobalt–chromium (CoCr) alloys, are most frequently used [1–4]. The duration of the healing process is highly variable depending on the extent of injury, disease state, age and treatment. In general, healing time may range from a brief one month period, up to a six month period in more complex cases. Permanent implants are frequently removed after the completion of the healing process to avoid diverse side effects. Long-term disadvantages of this practice include the failure to adapt to rapid growth in young children, bone degradation by stress shielding, microbial implant infections, excessive fibrosis or persistent inflammation. Novel bioresorbable metal implants could provide support during the healing process, and then disappear to avoid long-term side effects without requiring surgical removal [5,6]. Conventionally, initial material tests are done most economically in vitro under precisely defined technical conditions. Subsequent assays are performed under increasingly complex and more costly cell culture conditions, followed by small animal experiments and eventually tested in large animals or in clinical trials. However, corrosion results obtained under simple technical conditions cannot be extrapolated to clinical circumstances. In one study, the corrosion rate for magnesium alloys was reported to differ four orders of magnitude between in vitro and in vivo conditions [7]. Due to inherent limitations in reproducing the complexities of living tissue in vitro, this review preferentially refers to animal models or clinical trials if available [8,9]. For molecular genetic and economic reasons, small animal studies are most popular. However, for load bearing applications in particular, size is an important parameter that must be kept in mind, and eventually large animal experiments and clinical data are essential. Even though intense research efforts recently culminated in clinical reports, degradable metallic implants are not yet routinely applied (see Section 7 for details). Compared to organic polymers, biodegradable metals can achieve higher strengths and ductility and would therefore would be preferential for load bearing applications such as bone plates, screws or coronary stents [10]. Three types of metal alloys have been commonly-investigated as degradable implants for biomedical applications based on magnesium, iron or zinc. The main purpose of this article is a brief and easily understandable overview of virtues and clinical hurdles of self-degrading implants as screws, plates or intramedullary rods for load-bearing orthopedic (musculoskeletal) applications or as vascular stents. In addition, a novel model for implant infections is proposed to explain divergent effects of magnesium on bacteria in vitro and in vivo.

2. Material Requirements for Fully-Bioresorbable Vascular Stents

Clinical requirements provide the basis for the required implant material characteristics. Age-related vascular malfunctions such as vessels clogged by a blood clot are of growing importance [11,12]. One of the earliest effective treatments was antithrombotic therapy, but this required some time until the clot was dissolved. Then, vascular stents, used to keep blood vessels open, were shown to be superior, despite the fact that treatment had to be delayed to prevent patient deaths. Balloon angioplasty is routinely applied and requires minimal invasive surgery. A small folded stent on a balloon at the tip of a catheter is maneuvered through blood vessels until it is located at the site of the restriction. The position within the body can be monitored with the help of an X-ray camera. For this reason, an X-ray dense stent material is an advantage. Once positioned, the balloon is inflated to unfold the stent. Stents are overextended to a limited degree to allow a firm anchoring in the vessel wall to prevent migration, and to compensate for the inherent elastic recoil of the stent after the balloon had deflated. The stent material must be sturdy enough to allow for thin struts, minimize the recoil, and withstand the pressure of the tissue and the forces during movements of the body [13]. It is clinically well established that thin, yet robust and highly ductile, stainless steel or shape memory alloy stents fulfill these requirements. Nevertheless, initial stent overextension and subsequent persistent mechanical stress, due to interactions with pulsing blood vessel walls, stimulates smooth muscle cell proliferation in the vessel walls. In a process, termed restenosis, a growing mass of proliferating vascular smooth muscle-related cells eventually obstruct the stented vessel again. The blood flow may be reestablished by inserting a second stent or through surgical bypass, leading to additional patient discomfort, risks and costs [14,15]. In clinical applications restenosis has been successfully curbed by drug-eluting stents [16]. Thereby, unwanted cell growth is suppressed locally by clinically well-established drug-loaded polymer coated stents that gradually release immune suppressive agents like sirolimus or the antiproliferative-acting drug paclitaxel. Even though these drugs could reduce the incidence of restenosis, serious side effects include a delayed healing response, inflammation and persistent thrombosis risks [17–20]. This results in the requirement of costly regular and prolonged antiplatelet treatments, and non-complying patients drastically increase the thrombosis hazard [21]. Therefore, the application of drug-eluting stents must be carefully considered for each patient individually depending on the restenosis risks and the treatment-associated bleeding vulnerability. As an alternative, long-term side effects could be avoided by using fully biodegradable stents which provide the essential support for a few weeks during the healing process, and then completely disappear [22–25]. Key degradable stent material requirements are appropriate corrosion characteristics, biocompatibility, high elasticity to allow for small folded stents and sufficient strength to resist collapsing.

3. Material Requirements for Degradable Orthopedic Implants

To allow healing, broken bones must be firmly stabilized to avoid even micro-movements under the influence of considerable forces. Since inflammation may antagonize bone repair, the implant must be highly biocompatible. Clinically, all requirements are met by sturdy plates, screws or intramedullary nails made of titanium alloys or stainless steel. Nevertheless, after completion of the healing process stress shielding implants are mostly removed since their prolonged presence can lead to bone degradation [26]. Strong, tissue friendly self-degrading implants with bone-like mechanical parameters, to minimize stress-shielding, and suitable degradation characteristics could reduce such side effects. Furthermore, they allow patients to avoid a second surgery for implant removal, making them a highly attractive option. Whereas conventional permanent implant materials are sturdy and biologically inert, resorbable polymeric materials, as well as corrodible metals, have distinct biological characteristics (Table 1). In the following table, the cardinal properties of the most intensively investigated prospective biodegradable implant materials for load-bearing applications are described.

Table 1. Basic properties of degradable implant materials.

Implant Material	Degradation Speed	Physical and Corrosion Characteristics	Biological Effects	References
Organic polymers	Adjustable	Potentially flexible but mostly too weak for load-bearing applications; Implant swelling in moist environments; X-ray transparent	Inflammatory acidic hydrolysis products	[27,28]
Iron	Very slow, complete degradation may require several years	Sturdy but irregular corrosion characteristics	Accumulation of inflammatory iron hydroxide particles in various tissues	[29–31]
Zinc-based	Slow, life-time by far exceeds expected healing periods	Suboptimal strength	Non-inflammatory	[32,33]
Magnesium-based	Rapid, danger of mechanical implant failure before the healing process is completed	Alloys with sufficient strength available; compliance can be adjusted; irregular pitting corrosion; corrosion coat formation due to slowly dissolving solid precipitates resulting in reduction of initial corrosion rates	Non-inflammatory; gas accumulation in the tissue; accumulating solid corrosion products or gaseous hydrogen may exert pressure on non-yielding bony tissue	[34–36]
Surgical steel	inert	Sturdy, suitable for load-bearing applications, allows for ductile thin vascular stent struts	Non-inflammatory, inert	[1]
Titanium	inert	Sturdy, highly suitable for load-bearing applications	Non-inflammatory, bone-friendly surface oxide layer	[1]

4. Polymeric Vascular Stents

Even though they may act somewhat inflammatory compared to metals, biodegradable polymeric implants have been routinely employed as suture material and to temporarily fix tendons to bones until they eventually adhere by themselves [37,38]. Popular hydrolysable polymers used for bioresorbable scaffolds are poly(lactic-co-glycolic) acid (PLGA), polylactic acid (PLA) or polyglycolic acid (PGA) [39]. A main research focus has been polymeric stents, resulting in data that revealed several features that had to be optimized. Since polymeric materials are generally less sturdy than metals, thicker struts are required. This results in stents that are more difficult to direct through small vessels. Moreover, they are X-ray transparent and therefore harder to localize in the patient. Polymers also tend to swell in aqueous environments and acidic hydrolysis products can act inflammatory [27]. In experimental animal models degrading polymer stents resulted in increased restenosis rates [40,41]. One of the first commercially available fully absorbable polylactic acid stent (Absorb, Abbot) that was FDA approved in 2016 dissolved in two to three years, but despite promising short-term results long-term side effects were negative and sales were terminated by 2017 [42,43]. In clinical trials these polymer stents were more difficult to insert due to the increased efforts required for imaging, and over a two-year period induced higher rates of in-stent thrombosis than drug eluting metal stents [44]. In summary, the presently investigated resorbable polymer stents were deemed inferior to established metal stents.

5. Iron as a Prospective Stent Material

Pure iron and iron alloys were proposed in 2001 for corrodible stent materials [45]. Despite appropriate mechanical properties, iron implants take years to disappear. The corrosion rate is an order of magnitude too small for the implant to disappear without long-term side effects [30,46,47]. The immediate oxidation products (Fe^{2+}) and ferrous (Fe^{3+}) ions are essential for life and presumably non-toxic at the expected concentrations [48–52]. In pioneering animal experiments iron implants analysis revealed insoluble iron hydroxide precipitates that accumulated mainly at the site of implantation [45,53]. Further analyses, in a mouse model, revealed iron precipitates engulfed by local cells. After a few weeks these iron laden cells could be detected in various organs throughout the body [54]. In war veterans, corroding iron fragments from grenade splinters have been shown to migrate in the body and to cause chronic inflammation [55–57]. Overall, the slow degradation rate prolonged possible side effects after completion of the healing process, and inflammatory precipitates impede clinical applications of iron implants.

6. Zinc Alloy Stents

Corrodible zinc-based implants have been introduced relatively recently in 2013 (reviewed in [58]). Even though the mechanical properties can be adjusted according to the requirements, zinc alloys, with a reported yield strength up to 300 MPa, are not as strong as titanium or stainless steel [59]. When tested, zinc alloys corroded with favorable kinetics, faster than iron, but less rapidly than magnesium alloys. Zinc alloy degradation products were considered sufficiently biocompatible [60]. In a rat model, zinc stents were still structurally intact after four months in the abdominal aorta. The implant and the relevant degradation product Zn^{2+} appeared non-toxic and even anti-inflammatory [61]. One year after the implantation of a pure zinc stent in a rabbit aorta, an examination revealed artery remodeling and tissue healing without signs of inflammation, platelet aggregation or thrombosis [33]. It was therefore concluded that selected zinc alloys had promising strength and excellent biocompatibility for prospective bio-corrodible stent applications [62]. Nevertheless, it remains to be demonstrated in clinical trials that zinc alloys provide advantages over clinically established permanent metal alloys.

7. Characteristics of Magnesium-Based Implants

The first reported medical application of degradable magnesium alloys in humans, as ligature wire, was investigated in 1878 [63]. Side effects included the occurrence of gas pockets in the tissue, and rapid, irregular pitting corrosion leading to premature implant failure. In part, pure magnesium has been experimentally used to simplify the interpretation of biological responses. In general, alloy metals such as aluminum, calcium, lithium, zirconium and rare earth elements have been used to adjust mechanical properties such as the same stiffness as bony tissue or to reduce the degradation rate. In addition, grain refinement, protective surface coatings, and metallic glasses obtained by ultrafast cooling techniques resulted in improved degradation characteristics, increased material strength and bone-compatible elastic moduli [64–75].

In biological environments magnesium reacts with water molecules in a pitting type corrosion with kinetics that depend on the surrounding tissue [76–78]. In addition, irregular corrosion could lead to premature mechanical implant failure [79,80]. The primary magnesium corrosion products—soluble magnesium ions (Mg^{2+}), hydroxide ions (OH^-), and hydrogen gas (H_2)—are well tolerated by the body. Mg^{2+} ions are essential for living cells, by complexing with the energy carrier adenosine triphosphate and numerous enzymatic processes, and excess Mg^{2+} can be excreted in the urine [81,82]. Soluble hydroxide ions could in principle lead to toxic pH increases [76]. However, in biological environments magnesium implants appear highly biocompatible presumably due to an adequate buffering capacity of the tissue. In addition, magnesium and hydroxide ions combine in a pH neutral way, and, together with carbonic acid, phosphates and other components present in surrounding body fluids, precipitate to form a corrosion-retarding and highly biocompatible implant-tissue interface [83,84]. However, perhaps initially surprisingly, during corrosion these precipitates can transiently lead to increases of the overall implant mass and volume. This is particularly critical for implants in non-yielding bony tissue. Stimulation of new bone growth and calcium phosphate deposition has also been observed. This may be due to magnesium hydroxide deposition, calcium phosphate precipitation at the tissue interface and the exertion of mechanical stress by the resulting volume increase [85–87]. One gram of Mg can generate around one liter of hydrogen gas. Hydrogen gas is non-toxic and easily diffusible, but excessive corrosion can nevertheless lead to formation of undesirable gas bubbles (emphysema) in surrounding soft tissue. Excessive corrosion may also lead to a buildup of pressure in bone enclosed cavities and may, therefore, stimulate bone growth in appropriate setups [88,89].

In orthopedic applications selected magnesium alloys could achieve mechanical properties more similar to human bone than titanium or steel. This could be favorable as it would reduce implant-associated stress shielding and bone degradation [90,91]. Magnesium-based screws have been used in bone healing clinical trials without notable side effects reported by patients [92,93]. The first commercial magnesium screws (Magnezix, Syntellix, Hannover, Germany) were available in 2013, and completely disappeared one to two years after implantation [94]. Furthermore, recently an additional interference screw, made of an MgYREZr-alloy, has been introduced to the market (Milagro; DePuy Mitek, Leeds, United Kingdom) [95]. A transient appearance of radio translucent areas around magnesium implants was reported [96]. In fact, such a phenomenon would be expected from the above proposed mechanism; an initial magnesium implant size expansion by the deposition and the subsequent resorption of solid corrosion products, leaving a temporary void space to be filled by bony tissue.

Vascular magnesium alloy stents with reduced corrosion rates have been shown to be mechanically stable for up to 6 months in animal experiments and were eventually evaluated in clinical trials [97–103]. Polymer-coated drug-eluting magnesium stents (Magmaris and DREAMS; Biotronik AG, 231, Bülach, Switzerland) were commercially offered and claimed to be resorbed to 95% within a year in clinical trials. Thus, they may thereby overcome long-term side effects [104–106]. Both orthopedic and vascular magnesium implants appear promising but, with the exception of small orthopedic implants like pins or screws, the development of these options is still in its infancy, and a broader clinical applicability needs to be demonstrated [107].

8. Magnesium Implant Infection Susceptibility Mechanism: Race for the Surface versus Susceptible Tissue Surface Model

Bacterial implant infections are difficult problem to treat in orthopedics, particularly in non-sterile environments like the oral cavity [108]. Bacteria can form recalcitrant biofilms on implant surfaces that are resistant to conventional antibiotic treatments. As a last resort, the entire implant may have to be removed to allow an efficacious antibiotic treatment before the implant can be replaced. Corroding magnesium has been shown to act as an antibacterial in vitro due to the generation of hydroxide ions and pH increases [109–112]. In animal studies, an enhanced susceptibility to bacterial infections has been observed [113,114]. The reasons that could enhance the susceptibility to infection in vivo are not understood, and difficult to explain. Any model must take into account that the corrosion effects are no different in vitro, where there is no such enhanced susceptibility. The proposed model is an attempt to explain this observation. Conventionally, exposed implant surfaces are thought to be susceptible to bacterial adherence in competition with host tissue adhesion [115]. To allow bacterial adhesion and survival on the freshly implanted magnesium, toxic pH increases directly at the interface would have to be prevented in vivo. Unfortunately, experimental observation of the initial steps of bacterial invasion has not been accomplished so far. However, this scenario appears unlikely if a freshly implanted magnesium surface does act bactericidal. Importantly, despite systemic antibiotic treatment, bacterial biofilms on magnesium were observed. Not only were they observed on the implant surface but, also in the adjacent tissue (Figure 1), suggesting that bacterial adhesion to the implant may actually not be essential for biofilm formation [113].

Alternatively, similar to burn wound infections or keratitis, initial bacterial invasion could occur via the wound liquid to susceptible wounded tissue surfaces (Figure 2) [116,117]. If true for implanted materials other than magnesium, this scenario would predict dire consequences for implant infection prevention strategies.

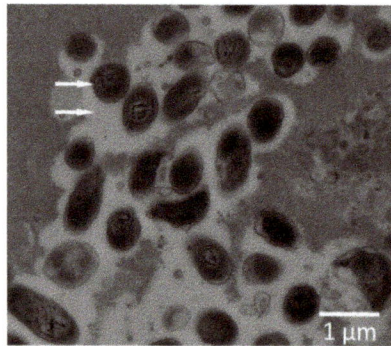

Figure 1. Bacterial biofilm in tissue pockets at a distance from the implant surface. Magnesium discs subcutaneously implanted into standard BALB/c mice were immediately infected with Pseudomonas aeruginosa. After one week, tissue adjacent to the implants was subjected to scanning transmission electron microscopic analysis (for a more detailed description see Reference [113]). Bacteria (upper arrow) surrounded by clear areas (lower arrow), indicating the presence of exopolysaccharide matrix material, a typical biofilm component. Reproduced with permission from *J. Biomed. Mater. Res.;* Published by Wiley 2016.

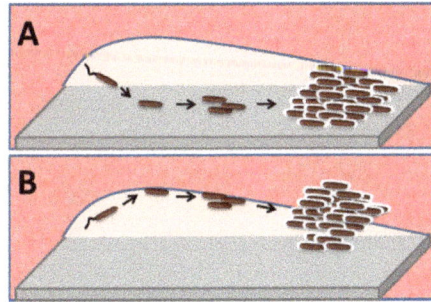

Figure 2. Model proposing tissue infection as initial key step of bacterial implant infections. (**A**) Conventional model. Consecutive steps of biofilm infections are shown from left to right. Planktonic bacteria (brown) enter the wound-liquid-filled interspace (colorless) between implant (grey) and tissue (pink). As a crucial step towards biofilm formation, bacteria first adhere to the implant surface and form micro-colonies. After reaching a critical density, bacteria switch to the biofilm mode and secrete extracellular matrix compounds. Biofilm features, such as the encapsulation in the matrix, nutrient restriction and slow growth, render the associated bacteria highly resistant to the host immune defenses and to antibiotics. Subsequently, secreted exotoxins and proteases allow bacteria to invade the adjacent host tissue. Alternatively, adhesion of host tissue to the implant acts to protect the implant surface from bacterial attachment and subsequent biofilm formation. Based on the in vitro results, in this scenario magnesium implants would be expected to act bactericidal. (**B**) Tissue infection model. Under normal circumstances contiguous epithelial cell layers protect living tissue, whereas wounding renders tissue highly susceptible to bacterial infections. After implant insertion the essential initial bacterial attachment occurs primarily at the susceptible injured tissue surface. Bacterial colonies growing on the tissue surface eventually switch to the biofilm mode with analogous outcomes as in the conventional model. While bacterial adhesion to the implant may occur, it plays no essential role for the course of the infection. Adhesion of host tissue to the implant would still be important to antagonize infections but predominantly to protect the wound tissue surface, and not the implant, from bacterial colonization. Despite acting bactericidal upon close contact, the observed enhanced infection susceptibility of magnesium implants is explained by interference of corroding magnesium with host tissue adhesion. Factors that prolong the wound surface exposure to bacteria could be alkaline pH immediately after implantation, and hydrogen gas evolution or eroding solid corrosion layers thereafter.

9. Implications for the Design of Antibacterial Implants

A wide variety of anti-infective implant strategies have been investigated, mostly in vitro [118]. In the light of the proposed tissue invasion model, in order to be efficacious, antibacterial substances would need to be diffusible to reach bacteria in the vicinity of the implant. Therefore, implant nanostructures that act antiadhesive, or passive coatings that act bactericidal upon contact, would not be expected to curb infections in patients. In addition, implant features that affect tissue adhesion play an important, through different, role than previously thought. That is, to primarily prevent bacterial adhesion to the injured tissue rather than to the implant (Table 2). Even though magnesium implants could not curb bacterial infections in mice, clinical data is needed before a final conclusion can be drawn. In addition, several alternative strategies are presently investigated, such as antibiotic-releasing coatings for magnesium-based implants or the addition of antibacterial acting alloy metals like silver, copper, or zinc that release cytotoxic ions [119–125]. The major challenge for such an approach is to maintain the balance of achieving efficacious bactericidal ion concentrations in vivo without damaging the host tissue.

Table 2. Implant features predicted by the tissue infection model to influence the susceptibility to infections in vivo.

Ineffective Coatings	Infection Risks	Favorable Measures
Surfaces that antagonize bacterial adhesion	Factors that hinder host tissue adhesion; convex or microporous surfaces	Surfaces favoring tissue integration; smooth, flat or concave forms;
Contact-dependent bactericidal surfaces	Relative movement of implants versus tissue	Antibacterial substance-releasing coatings

10. Conclusions

In long-term clinical trials biodegradable polymeric stents were inferior to conventional drug-eluting metal stents, while recently introduced biocorrodible magnesium-based bone screws were without noticeable side effects. However, since in vitro tests and even small animal studies cannot predict the outcome in human patients, long-term clinical confirmation of the expected benefits, with regard to potential risks, are needed. In addition, a novel model for implant infections suggests that host cell adhesion to implants is important to prevent bacterial invasion of the exposed host tissue surface, and not, as previously thought, to prevent bacterial adhesion to the implant. The model predicts that passive antibacterial implant coating strategies would not be efficacious in vivo.

Funding: We acknowledge the financial support by a joint grant to M.I.R. and to S.U. of the German Academic Exchange Service (DAAD), Germany, and the Higher Education Commission of Pakistan (HEC).

Conflicts of Interest: The authors declare that there is no conflict of interest regarding the publication of this paper.

References

1. Perren, S.M.; Regazzoni, P.; Fernandez, A.A. How to choose between the implant iaterials steel and titanium in orthopedic trauma surgery: Part 2—Biological aspects. *Acta Chir. Orthop. Traumatol. Cech.* **2017**, *84*, 85–90. [PubMed]
2. Osman, R.B.; Swain, M.V. A Critical Review of Dental Implant Materials with an Emphasis on Titanium versus Zirconia. *Materials (Basel)* **2015**, *8*, 932–958. [CrossRef] [PubMed]
3. Williams, D.F. On the nature of biomaterials. *Biomaterials* **2009**, *30*, 5897–5909. [CrossRef] [PubMed]
4. Elias, C.N.; Lima, J.H.C.; Valiev, R.; Meyers, M.A. Biomedical applications of titanium and its alloys. *JOM* **2008**, *60*, 46–49. [CrossRef]
5. Sheikh, Z.; Najeeb, S.; Khurshid, Z.; Verma, V.; Rashid, H.; Glogauer, M. Biodegradable Materials for Bone Repair and Tissue Engineering Applications. *Materials (Basel)* **2015**, *8*, 5744–5794. [CrossRef] [PubMed]
6. Middleton, J.C.; Tipton, A.J. Synthetic biodegradable polymers as orthopedic devices. *Biomaterials* **2000**, *21*, 2335–2346. [CrossRef]
7. Witte, F.; Fischer, J.; Nellesen, J.; Crostack, H.-A.; Kaese, V.; Pisch, A.; Beckmann, F.; Windhagen, H. In vitro and in vivo corrosion measurements of magnesium alloys. *Biomaterials* **2006**, *27*, 1013–1018. [CrossRef] [PubMed]
8. Martinez Sanchez, A.H.; Luthringer, B.J.; Feyerabend, F.; Willumeit, R. Mg and Mg alloys: How comparable are in vitro and in vivo corrosion rates? A review. *Acta Biomater.* **2015**, *13*, 16–31. [CrossRef] [PubMed]
9. Luthringer, B.J.; Feyerabend, F.; Willumeit-Romer, R. Magnesium-based implants: A mini-review. *Magnes. Res.* **2014**, *27*, 142–154. [PubMed]
10. Li, H.; Zheng, Y.; Qin, L. Progress of biodegradable metals. *Prog. Nat. Sci.* **2014**, *24*, 414–422. [CrossRef]
11. Miller, A.P.; Huff, C.M.; Roubin, G.S. Vascular disease in the older adult. *J. Geriatr. Cardiol. JGC* **2016**, *13*, 727–732. [PubMed]
12. Martens, A.; Beckmann, E.; Kaufeld, T.; Umminger, J.; Fleissner, F.; Koigeldiyev, N.; Krueger, H.; Puntigam, J.; Haverich, A.; Shrestha, M. Total aortic arch repair: Risk factor analysis and follow-up in 199 patients. *Eur. J. Cardiothorac. Surg.* **2016**, *50*, 940–948. [CrossRef] [PubMed]

13. Rittersma, S.Z.; de Winter, R.J.; Koch, K.T.; Bax, M.; Schotborgh, C.E.; Mulder, K.J.; Tijssen, J.G.; Piek, J.J. Impact of strut thickness on late luminal loss after coronary artery stent placement. *Am. J. Cardiol.* **2004**, *93*, 477–480. [CrossRef] [PubMed]

14. Onche, I.I.; Osagie, O.E.; INuhu, S. Removal of orthopaedic implants: Indications, outcome and economic implications. *J. West Afr. Coll. Surg.* **2011**, *1*, 101–112. [PubMed]

15. Levy, J.A.; Podeszwa, D.A.; Lebus, G.; Ho, C.A.; Wimberly, R.L. Acute complications associated with removal of flexible intramedullary femoral rods placed for pediatric femoral shaft fractures. *J. Pediatr. Orthop.* **2013**, *33*, 43–47. [CrossRef] [PubMed]

16. Buchanan, K.; Steinvil, A.; Waksman, R. Does the new generation of drug-eluting stents render bare metal stents obsolete? *Cardiovasc. Revasc. Med.* **2017**, *18*, 456–461. [CrossRef] [PubMed]

17. Waksman, R. A new generation of drug-eluting stents: Indications and outcomes of bioresorbable vascular scaffolds. *Cleve. Clin. J. Med.* **2017**, *84* (Suppl. S4), e20–e24. [CrossRef] [PubMed]

18. Pleva, L.; Kukla, P.; Hlinomaz, O. Treatment of coronary in-stent restenosis: A systematic review. *J. Geriatr. Cardiol.* **2018**, *15*, 173–184. [PubMed]

19. Artang, R.; Dieter, R.S. Analysis of 36 reported cases of late thrombosis in drug-eluting stents placed in coronary arteries. *Am. J. Cardiol.* **2007**, *99*, 1039–1043. [CrossRef] [PubMed]

20. Mauri, L.; Hsieh, W.H.; Massaro, J.M.; Ho, K.K.; D'Agostino, R.; Cutlip, D.E. Stent thrombosis in randomized clinical trials of drug-eluting stents. *N. Engl. J. Med.* **2007**, *356*, 1020–1029. [CrossRef] [PubMed]

21. Filion, K.B.; Roy, A.M.; Baboushkin, T.; Rinfret, S.; Eisenberg, M.J. Cost-effectiveness of drug-eluting stents including the economic impact of late stent thrombosis. *Am. J. Cardiol.* **2009**, *103*, 338–344. [CrossRef] [PubMed]

22. Erne, P.; Schier, M.; Resink, T.J. The road to bioabsorbable stents: Reaching clinical reality? *Cardiovasc. Intervent. Radiol.* **2006**, *29*, 11–16. [CrossRef] [PubMed]

23. Tan, L.; Yu, X.; Wan, P.; Yang, K. Biodegradable Materials for Bone Repairs: A Review. *J. Mater. Sci. Technol.* **2013**, *29*, 503–513. [CrossRef]

24. Zheng, Y.F.; Gu, X.N.; Witte, F. Biodegradable metals. *Mater. Sci. Eng. R Rep.* **2014**, *77*, 1–34. [CrossRef]

25. Waksman, R.; Pakala, R. Biodegradable and bioabsorbable stents. *Curr. Pharm. Des.* **2010**, *16*, 4041–4051. [CrossRef] [PubMed]

26. Sumner, D.R. Long-term implant fixation and stress-shielding in total hip replacement. *J. Biomech.* **2015**, *48*, 797–800. [CrossRef] [PubMed]

27. Ceonzo, K.; Gaynor, A.; Shaffer, L.; Kojima, K.; Vacanti, C.A.; Stahl, G.L. Polyglycolic acid-induced inflammation: Role of hydrolysis and resulting complement activation. *Tissue Eng.* **2006**, *12*, 301–308. [CrossRef] [PubMed]

28. Athanasiou, K.A.; Niederauer, G.G.; Agrawal, C.M. Sterilization, toxicity, biocompatibility and clinical applications of polylactic acid/polyglycolic acid copolymers. *Biomaterials* **1996**, *17*, 93–102. [CrossRef]

29. Pierson, D.; Edick, J.; Tauscher, A.; Pokorney, E.; Bowen, P.; Gelbaugh, J.; Stinson, J.; Getty, H.; Lee, C.H.; Drelich, J.; et al. A simplified in vivo approach for evaluating the bioabsorbable behavior of candidate stent materials. *J. Biomed. Mater. Res. B Appl. Biomater.* **2012**, *100*, 58–67. [CrossRef] [PubMed]

30. Bowen, P.K.; Drelich, J.; Buxbaum, R.E.; Rajachar, R.M.; Goldman, J. New approaches in evaluating metallic candidates for bioabsorbable stents. *Emerg. Mater. Res.* **2012**, *1*, 237–255. [CrossRef]

31. Hermawan, H.; Purnama, A.; Dube, D.; Couet, J.; Mantovani, D. Fe-Mn alloys for metallic biodegradable stents: Degradation and cell viability studies. *Acta Biomater.* **2010**, *6*, 1852–1860. [CrossRef] [PubMed]

32. Vojtech, D.; Kubasek, J.; Serak, J.; Novak, P. Mechanical and corrosion properties of newly developed biodegradable Zn-based alloys for bone fixation. *Acta Biomater.* **2011**, *7*, 3515–3522. [CrossRef] [PubMed]

33. Yang, H.; Wang, C.; Liu, C.; Chen, H.; Wu, Y.; Han, J.; Jia, Z.; Lin, W.; Zhang, D.; Li, W.; et al. Evolution of the degradation mechanism of pure zinc stent in the one-year study of rabbit abdominal aorta model. *Biomaterials* **2017**, *145*, 92–105. [CrossRef] [PubMed]

34. Zhao, D.; Wang, T.; Nahan, K.; Guo, X.; Zhang, Z.; Dong, Z.; Chen, S.; Chou, D.T.; Hong, D.; Kumta, P.N.; et al. In vivo characterization of magnesium alloy biodegradation using electrochemical H2 monitoring, ICP-MS, and XPS. *Acta Biomater.* **2017**, *50*, 556–565. [CrossRef] [PubMed]

35. Zhao, D.; Witte, F.; Lu, F.; Wang, J.; Li, J.; Qin, L. Current status on clinical applications of magnesium-based orthopaedic implants: A review from clinical translational perspective. *Biomaterials* **2017**, *112*, 287–302. [CrossRef] [PubMed]

36. Staiger, M.P.; Pietak, A.M.; Huadmai, J.; Dias, G. Magnesium and its alloys as orthopedic biomaterials: A review. *Biomaterials* **2006**, *27*, 1728–1734. [CrossRef] [PubMed]

37. Debieux, P.; Franciozi, C.E.; Lenza, M.; Tamaoki, M.J.; Magnussen, R.A.; Faloppa, F.; Belloti, J.C. Bioabsorbable versus metallic interference screws for graft fixation in anterior cruciate ligament reconstruction. *Cochrane Database Syst. Rev.* **2016**, *7*, CD009772. [CrossRef] [PubMed]

38. Seitz, J.M.; Durisin, M.; Goldman, J.; Drelich, J.W. Recent Advances in Biodegradable Metals for Medical Sutures: A Critical Review. *Adv. Healthc. Mater.* **2015**, *4*, 1915–1936. [CrossRef] [PubMed]

39. Doppalapudi, S.; Jain, A.; Khan, W.; Domb, A.J. Biodegradable polymers—An overview. *Polym. Adv. Technol.* **2014**, *25*, 427–435. [CrossRef]

40. Van der Giessen, W.J.; Lincoff, A.M.; Schwartz, R.S.; van Beusekom, H.M.; Serruys, P.W.; Holmes, D.R., Jr.; Ellis, S.G.; Topol, E.J. Marked inflammatory sequelae to implantation of biodegradable and nonbiodegradable polymers in porcine coronary arteries. *Circulation* **1996**, *94*, 1690–1697. [CrossRef] [PubMed]

41. Bunger, C.M.; Grabow, N.; Sternberg, K.; Kroger, C.; Ketner, L.; Schmitz, K.P.; Kreutzer, H.J.; Ince, H.; Nienaber, C.A.; Klar, E.; et al. Sirolimus-eluting biodegradable poly-L-lactide stent for peripheral vascular application: A preliminary study in porcine carotid arteries. *J. Surg. Res.* **2007**, *139*, 77–82. [CrossRef] [PubMed]

42. Rizik, D.G.; Hermiller, J.B.; Kereiakes, D.J. Bioresorbable vascular scaffolds for the treatment of coronary artery disease: Clinical outcomes from randomized controlled trials. *Catheter. Cardiovasc. Interv.* **2016**, *88* (Suppl. S1), 21–30. [CrossRef] [PubMed]

43. Ali, Z.A.; Serruys, P.W.; Kimura, T.; Gao, R.; Ellis, S.G.; Kereiakes, D.J.; Onuma, Y.; Simonton, C.; Zhang, Z.; Stone, G.W. 2-year outcomes with the Absorb bioresorbable scaffold for treatment of coronary artery disease: A systematic review and meta-analysis of seven randomised trials with an individual patient data substudy. *Lancet* **2017**, *390*, 760–772. [CrossRef]

44. Ali, Z.A.; Gao, R.; Kimura, T.; Onuma, Y.; Kereiakes, D.J.; Ellis, S.G.; Chevalier, B.; Vu, M.T.; Zhang, Z.; Simonton, C.A.; et al. Three-Year Outcomes with the Absorb Bioresorbable Scaffold: Individual-Patient-Data Meta-Analysis from the ABSORB Randomized Trials. *Circulation* **2018**, *137*, 464–479. [CrossRef] [PubMed]

45. Peuster, M.; Wohlsein, P.; Brugmann, M.; Ehlerding, M.; Seidler, K.; Fink, C.; Brauer, H.; Fischer, A.; Hausdorf, G. A novel approach to temporary stenting: Degradable cardiovascular stents produced from corrodible metal-results 6–18 months after implantation into New Zealand white rabbits. *Heart* **2001**, *86*, 563–569. [CrossRef] [PubMed]

46. Francis, A.; Yang, Y.; Virtanen, S.; Boccaccini, A.R. Iron and iron-based alloys for temporary cardiovascular applications. *J. Mater. Sci. Mater. Med.* **2015**, *26*, 138. [CrossRef] [PubMed]

47. Peuster, M.; Hesse, C.; Schloo, T.; Fink, C.; Beerbaum, P.; von Schnakenburg, C. Long-term biocompatibility of a corrodible peripheral iron stent in the porcine descending aorta. *Biomaterials* **2006**, *27*, 4955–4962. [CrossRef] [PubMed]

48. Fagali, N.S.; Grillo, C.A.; Puntarulo, S.; Fernandez Lorenzo de Mele, M.A. Cytotoxicity of corrosion products of degradable Fe-based stents: Relevance of pH and insoluble products. *Colloids Surf. B Biointerfaces* **2015**, *128*, 480–488. [CrossRef] [PubMed]

49. Moravej, M.; Purnama, A.; Fiset, M.; Couet, J.; Mantovani, D. Electroformed pure iron as a new biomaterial for degradable stents: In vitro degradation and preliminary cell viability studies. *Acta Biomater.* **2010**, *6*, 1843–1851. [CrossRef] [PubMed]

50. Drynda, A.; Hoehn, R.; Peuster, M. Influence of Fe(II) and Fe(III) on the expression of genes related to cholesterol- and fatty acid metabolism in human vascular smooth muscle cells. *J. Mater. Sci. Mater. Med.* **2010**, *21*, 1655–1663. [CrossRef] [PubMed]

51. Zhu, S.; Huang, N.; Xu, L.; Zhang, Y.; Liu, H.; Sun, H.; Leng, Y. Biocompatibility of pure iron: In vitro assessment of degradation kinetics and cytotoxicity on endothelial cells. *Mater. Sci. Eng. C* **2009**, *29*, 1589–1592. [CrossRef]

52. Mueller, P.P.; May, T.; Perz, A.; Hauser, H.; Peuster, M. Control of smooth muscle cell proliferation by ferrous iron. *Biomaterials* **2006**, *27*, 2193–2200. [CrossRef] [PubMed]

53. Kraus, T.; Moszner, F.; Fischerauer, S.; Fiedler, M.; Martinelli, E.; Eichler, J.; Witte, F.; Willbold, E.; Schinhammer, M.; Meischel, M.; et al. Biodegradable Fe-based alloys for use in osteosynthesis: Outcome of an in vivo study after 52 weeks. *Acta Biomater.* **2014**, *10*, 3346–3353. [CrossRef] [PubMed]

54. Mueller, P.P.; Arnold, S.; Badar, M.; Bormann, D.; Bach, F.W.; Drynda, A.; Meyer-Lindenberg, A.; Hauser, H.; Peuster, M. Histological and molecular evaluation of iron as degradable medical implant material in a murine animal model. *J. Biomed. Mater. Res. A* **2012**, *100*, 2881–2889. [CrossRef] [PubMed]

55. Voigtlaender, H. [Wandering of foreign bodies. (Grenate fragment in the common bile duct causing jaundice)]. *Chirurg* **1975**, *46*, 467–469. [PubMed]

56. Ghislain, P.D. Spontaneous extrusion of hand grenade fragments from the face 60 years after injury. *JAMA* **2003**, *290*, 1317–1318. [CrossRef] [PubMed]

57. Gaber, Y. [Grenade splinter injury simulating thrombophlebitis]. *Vasa* **2003**, *32*, 40–42. [CrossRef] [PubMed]

58. Mostaed, E.; Sikora-Jasinska, M.; Drelich, J.W.; Vedani, M. Zinc-based alloys for degradable vascular stent applications. *Acta Biomater.* **2018**, *71*, 1–23. [CrossRef] [PubMed]

59. Jin, H.; Zhao, S.; Guillory, R.; Bowen, P.K.; Yin, Z.; Griebel, A.; Schaffer, J.; Earley, E.J.; Goldman, J.; Drelich, J.W. Novel high-strength, low-alloys Zn-Mg (<0.1 wt% Mg) and their arterial biodegradation. *Mater. Sci. Eng. C* **2018**, *84*, 67–79. [CrossRef] [PubMed]

60. Bowen, P.K.; Shearier, E.R.; Zhao, S.; Guillory, R.J., 2nd; Zhao, F.; Goldman, J.; Drelich, J.W. Biodegradable Metals for Cardiovascular Stents: From Clinical Concerns to Recent Zn-Alloys. *Adv. Healthc. Mater.* **2016**, *5*, 1121–1140. [CrossRef] [PubMed]

61. Bowen, P.K.; Drelich, J.; Goldman, J. Zinc Exhibits Ideal Physiological Corrosion Behavior for Bioabsorbable Stents. *Adv. Mater.* **2013**, *25*, 2577–2582. [CrossRef] [PubMed]

62. Bowen, P.K.; Guillory, R.J., 2nd; Shearier, E.R.; Seitz, J.M.; Drelich, J.; Bocks, M.; Zhao, F.; Goldman, J. Metallic zinc exhibits optimal biocompatibility for bioabsorbable endovascular stents. *Mater. Sci. Eng. C Mater. Biol. Appl.* **2015**, *56*, 467–472. [CrossRef] [PubMed]

63. Huse, E.C. A New Ligature? *Chic. Med. J. Examin.* **1878**, *37*, 171–172.

64. Agarwal, S.; Curtin, J.; Duffy, B.; Jaiswal, S. Biodegradable magnesium alloys for orthopaedic applications: A review on corrosion, biocompatibility and surface modifications. *Mater. Sci. Eng. C Mater. Biol. Appl.* **2016**, *68*, 948–963. [CrossRef] [PubMed]

65. Waizy, H.; Seitz, J.-M.; Reifenrath, J.; Weizbauer, A.; Bach, F.-W.; Meyer-Lindenberg, A.; Denkena, B.; Windhagen, H. Biodegradable magnesium implants for orthopedic applications. *J. Mater. Sci.* **2013**, *48*, 39–50. [CrossRef]

66. Meagher, P.; O'Cearbhaill, E.D.; Byrne, J.H.; Browne, D.J. Bulk Metallic Glasses for Implantable Medical Devices and Surgical Tools. *Adv. Mater.* **2016**, *28*, 5755–5762. [CrossRef] [PubMed]

67. Zberg, B.; Uggowitzer, P.J.; Loffler, J.F. MgZnCa glasses without clinically observable hydrogen evolution for biodegradable implants. *Nat. Mater.* **2009**, *8*, 887–891. [CrossRef] [PubMed]

68. Frankel, G.S. Magnesium alloys: Ready for the road. *Nat. Mater.* **2015**, *14*, 1189–1190. [CrossRef] [PubMed]

69. Ma, E.; Xu, J. Biodegradable Alloys: The glass window of opportunities. *Nat. Mater.* **2009**, *8*, 855–857. [CrossRef] [PubMed]

70. Li, Y.; Liu, L.; Wan, P.; Zhai, Z.; Mao, Z.; Ouyang, Z.; Yu, D.; Sun, Q.; Tan, L.; Ren, L.; et al. Biodegradable Mg-Cu alloy implants with antibacterial activity for the treatment of osteomyelitis: In vitro and in vivo evaluations. *Biomaterials* **2016**, *106*, 250–263. [CrossRef] [PubMed]

71. Xu, L.; Pan, F.; Yu, G.; Yang, L.; Zhang, E.; Yang, K. In vitro and in vivo evaluation of the surface bioactivity of a calcium phosphate coated magnesium alloy. *Biomaterials* **2009**, *30*, 1512–1523. [CrossRef] [PubMed]

72. Wong, H.M.; Yeung, K.W.; Lam, K.O.; Tam, V.; Chu, P.K.; Luk, K.D.; Cheung, K.M. A biodegradable polymer-based coating to control the performance of magnesium alloy orthopaedic implants. *Biomaterials* **2010**, *31*, 2084–2096. [CrossRef] [PubMed]

73. Li, X.; Chu, C.L.; Liu, L.; Liu, X.K.; Bai, J.; Guo, C.; Xue, F.; Lin, P.H.; Chu, P.K. Biodegradable poly-lactic acid based-composite reinforced unidirectionally with high-strength magnesium alloy wires. *Biomaterials* **2015**, *49*, 135–144. [CrossRef] [PubMed]

74. Thormann, U.; Alt, V.; Heimann, L.; Gasquere, C.; Heiss, C.; Szalay, G.; Franke, J.; Schnettler, R.; Lips, K.S. The Biocompatibility of Degradable Magnesium Interference Screws: An Experimental Study with Sheep. *BioMed Res. Int.* **2015**, *2015*, 15. [CrossRef] [PubMed]

75. Hort, N.; Huang, Y.; Fechner, D.; Stormer, M.; Blawert, C.; Witte, F.; Vogt, C.; Drucker, H.; Willumeit, R.; Kainer, K.U.; et al. Magnesium alloys as implant materials–principles of property design for Mg-RE alloys. *Acta Biomater.* **2010**, *6*, 1714–1725. [CrossRef] [PubMed]

76. Kirkland, N.T.; Birbilis, N.; Staiger, M.P. Assessing the corrosion of biodegradable magnesium implants: A critical review of current methodologies and their limitations. *Acta Biomater.* **2012**, *8*, 925–936. [CrossRef] [PubMed]

77. Huehnerschulte, T.A.; Angrisani, N.; Rittershaus, D.; Bormann, D.; Windhagen, H.; Meyer-Lindenberg, A. In Vivo Corrosion of Two Novel Magnesium Alloys ZEK100 and AX30 and Their Mechanical Suitability as Biodegradable Implants. *Materials (Basel)* **2011**, *4*, 1144–1167. [CrossRef] [PubMed]

78. Krause, A.; von der Höh, N.; Bormann, D.; Krause, C.; Bach, F.-W.; Windhagen, H.; Meyer-Lindenberg, A. Degradation behaviour and mechanical properties of magnesium implants in rabbit tibiae. *J. Mater. Sci.* **2009**, *45*, 624. [CrossRef]

79. Zeng, R.; Dietzel, W.; Witte, F.; Hort, N.; Blawert, C. Progress and Challenge for Magnesium Alloys as Biomaterials. *Adv. Eng. Mater.* **2008**, *10*, B3–B14. [CrossRef]

80. Kirkland, N.T.; Lespagnol, J.; Birbilis, N.; Staiger, M.P. A survey of bio-corrosion rates of magnesium alloys. *Corros. Sci.* **2010**, *52*, 287–291. [CrossRef]

81. Jahnen-Dechent, W.; Ketteler, M. Magnesium basics. *Clin. Kidney J.* **2012**, *5* (Suppl. S1), i3–i14. [CrossRef] [PubMed]

82. De Baaij, J.H.; Hoenderop, J.G.; Bindels, R.J. Magnesium in man: Implications for health and disease. *Physiol. Rev.* **2015**, *95*, 1–46. [CrossRef] [PubMed]

83. Han, P.; Cheng, P.; Zhang, S.; Zhao, C.; Ni, J.; Zhang, Y.; Zhong, W.; Hou, P.; Zhang, X.; Zheng, Y.; et al. In vitro and in vivo studies on the degradation of high-purity Mg (99.99 wt.%) screw with femoral intracondylar fractured rabbit model. *Biomaterials* **2015**, *64*, 57–69. [CrossRef] [PubMed]

84. Badar, M.; Lunsdorf, H.; Evertz, F.; Rahim, M.I.; Glasmacher, B.; Hauser, H.; Mueller, P.P. The formation of an organic coat and the release of corrosion microparticles from metallic magnesium implants. *Acta Biomater.* **2013**, *9*, 7580–7589. [CrossRef] [PubMed]

85. Rahim, M.I.; Weizbauer, A.; Evertz, F.; Hoffmann, A.; Rohde, M.; Glasmacher, B.; Windhagen, H.; Gross, G.; Seitz, J.M.; Mueller, P.P. Differential magnesium implant corrosion coat formation and contribution to bone bonding. *J. Biomed. Mater. Res. A* **2017**, *105*, 697–709. [CrossRef] [PubMed]

86. Zhang, Y.; Xu, J.; Ruan, Y.C.; Yu, M.K.; O'Laughlin, M.; Wise, H.; Chen, D.; Tian, L.; Shi, D.; Wang, J.; et al. Implant-derived magnesium induces local neuronal production of CGRP to improve bone-fracture healing in rats. *Nat. Med.* **2016**, *22*, 1160–1169. [CrossRef] [PubMed]

87. Witte, F.; Kaese, V.; Haferkamp, H.; Switzer, E.; Meyer-Lindenberg, A.; Wirth, C.J.; Windhagen, H. In vivo corrosion of four magnesium alloys and the associated bone response. *Biomaterials* **2005**, *26*, 3557–3563. [CrossRef] [PubMed]

88. Kuhlmann, J.; Bartsch, I.; Willbold, E.; Schuchardt, S.; Holz, O.; Hort, N.; Hoche, D.; Heineman, W.R.; Witte, F. Fast escape of hydrogen from gas cavities around corroding magnesium implants. *Acta Biomater.* **2013**, *9*, 8714–8721. [CrossRef] [PubMed]

89. Noviana, D.; Paramitha, D.; Ulum, M.F.; Hermawan, H. The effect of hydrogen gas evolution of magnesium implant on the postimplantation mortality of rats. *J. Orthop. Transl.* **2016**, *5*, 9–15. [CrossRef]

90. Chen, Y.; Xu, Z.; Smith, C.; Sankar, J. Recent advances on the development of magnesium alloys for biodegradable implants. *Acta Biomater.* **2014**, *10*, 4561–4573. [CrossRef] [PubMed]

91. Grünewald, T.A.; Rennhofer, H.; Hesse, B.; Burghammer, M.; Stanzl-Tschegg, S.E.; Cotte, M.; Löffler, J.F.; Weinberg, A.M.; Lichtenegger, H.C. Magnesium from bioresorbable implants: Distribution and impact on the nano- and mineral structure of bone. *Biomaterials* **2016**, *76*, 250–260. [CrossRef] [PubMed]

92. Seitz, J.-M.; Lucas, A.; Kirschner, M. Magnesium-Based Compression Screws: A Novelty in the Clinical Use of Implants. *JOM* **2016**, *68*, 1177–1182. [CrossRef]

93. Plaass, C.; von Falck, C.; Ettinger, S.; Sonnow, L.; Calderone, F.; Weizbauer, A.; Reifenrath, J.; Claassen, L.; Waizy, H.; Daniilidis, K.; et al. Bioabsorbable magnesium versus standard titanium compression screws for fixation of distal metatarsal osteotomies—3 year results of a randomized clinical trial. *J. Orthop. Sci.* **2018**, *23*, 321–327. [CrossRef] [PubMed]

94. Windhagen, H.; Radtke, K.; Weizbauer, A.; Diekmann, J.; Noll, Y.; Kreimeyer, U.; Schavan, R.; Stukenborg-Colsman, C.; Waizy, H. Biodegradable magnesium-based screw clinically equivalent to titanium screw in hallux valgus surgery: Short term results of the first prospective, randomized, controlled clinical pilot study. *Biomed. Eng. Online* **2013**, *12*, 62. [CrossRef] [PubMed]

95. Ezechieli, M.; Meyer, H.; Lucas, A.; Helmecke, P.; Becher, C.; Calliess, T.; Windhagen, H.; Ettinger, M. Biomechanical Properties of a Novel Biodegradable Magnesium-Based Interference Screw. *Orthop. Rev. (Pavia)* **2016**, *8*, 6445. [CrossRef] [PubMed]

96. Biber, R.; Pauser, J.; Brem, M.; Bail, H.J. Bioabsorbable metal screws in traumatology: A promising innovation. *Trauma Case Rep.* **2017**, *8*, 11–15. [CrossRef] [PubMed]

97. Ang, H.Y.; Huang, Y.Y.; Lim, S.T.; Wong, P.; Joner, M.; Foin, N. Mechanical behavior of polymer-based vs. metallic-based bioresorbable stents. *J. Thorac. Dis.* **2017**, *9* (Suppl. S9), S923–S934. [CrossRef] [PubMed]

98. Zhao, N.; Watson, N.; Xu, Z.G.; Chen, Y.J.; Waterman, J.; Sankar, J.; Zhu, D.H. In Vitro Biocompatibility and Endothelialization of Novel Magnesium-Rare Earth Alloys for Improved Stent Applications. *PLoS ONE* **2014**, *9*, e98674. [CrossRef] [PubMed]

99. Zhu, S.J.; Liu, Q.; Qian, Y.F.; Sun, B.; Wang, L.G.; Wu, J.M.; Guan, S.K. Effect of different processings on mechanical property and corrosion behavior in simulated body fluid of Mg-Zn-Y-Nd alloy for cardiovascular stent application. *Front. Mater. Sci.* **2014**, *8*, 256–263. [CrossRef]

100. Mao, L.; Shen, L.; Chen, J.; Zhang, X.; Kwak, M.; Wu, Y.; Fan, R.; Zhang, L.; Pei, J.; Yuan, G.; et al. A promising biodegradable magnesium alloy suitable for clinical vascular stent application. *Sci. Rep.* **2017**, *7*, 46343. [CrossRef] [PubMed]

101. Mao, L.; Shen, L.; Chen, J.; Wu, Y.; Kwak, M.; Lu, Y.; Xue, Q.; Pei, J.; Zhang, L.; Yuan, G.; et al. Enhanced bioactivity of Mg-Nd-Zn-Zr alloy achieved with nanoscale MgF2 surface for vascular stent application. *ACS Appl. Mater. Interfaces* **2015**, *7*, 5320–5330. [CrossRef] [PubMed]

102. Waksman, R.; Erbel, R.; Di Mario, C.; Bartunek, J.; de Bruyne, B.; Eberli, F.R.; Erne, P.; Haude, M.; Horrigan, M.; Ilsley, C.; et al. Early- and long-term intravascular ultrasound and angiographic findings after bioabsorbable magnesium stent implantation in human coronary arteries. *JACC Cardiovasc. Interv.* **2009**, *2*, 312–320. [CrossRef] [PubMed]

103. Erbel, R.; Di Mario, C.; Bartunek, J.; Bonnier, J.; de Bruyne, B.; Eberli, F.R.; Erne, P.; Haude, M.; Heublein, B.; Horrigan, M.; et al. Temporary scaffolding of coronary arteries with bioabsorbable magnesium stents: A prospective, non-randomised multicentre trial. *Lancet* **2007**, *369*, 1869–1875. [CrossRef]

104. Rapetto, C.; Leoncini, M. Magmaris: A new generation metallic sirolimus-eluting fully bioresorbable scaffold: Present status and future perspectives. *J. Thorac. Dis.* **2017**, *9* (Suppl. S9), S903–S913. [CrossRef] [PubMed]

105. Haude, M.; Ince, H.; Kische, S.; Abizaid, A.; Tolg, R.; Alves Lemos, P.; Van Mieghem, N.M.; Verheye, S.; von Birgelen, C.; Christiansen, E.H.; et al. Sustained safety and clinical performance of a drug-eluting absorbable metal scaffold up to 24 months: Pooled outcomes of BIOSOLVE-II and BIOSOLVE-III. *EuroIntervention* **2017**, *13*, 432–439. [PubMed]

106. Haude, M.; Erbel, R.; Erne, P.; Verheye, S.; Degen, H.; Bose, D.; Vermeersch, P.; Wijnbergen, I.; Weissman, N.; Prati, F.; et al. Safety and performance of the drug-eluting absorbable metal scaffold (DREAMS) in patients with de-novo coronary lesions: 12 month results of the prospective, multicentre, first-in-man BIOSOLVE-I trial. *Lancet* **2013**, *381*, 836–844. [CrossRef]

107. Willbold, E.; Weizbauer, A.; Loos, A.; Seitz, J.M.; Angrisani, N.; Windhagen, H.; Reifenrath, J. Magnesium alloys: A stony pathway from intensive research to clinical reality. Different test methods and approval-related considerations. *J. Biomed. Mater. Res. A* **2017**, *105*, 329–347. [CrossRef] [PubMed]

108. Arciola, C.R.; Campoccia, D.; Speziale, P.; Montanaro, L.; Costerton, J.W. Biofilm formation in Staphylococcus implant infections. A review of molecular mechanisms and implications for biofilm-resistant materials. *Biomaterials* **2012**, *33*, 5967–5982. [CrossRef] [PubMed]

109. Rahim, M.I.; Eifler, R.; Rais, B.; Mueller, P.P. Alkalization is responsible for antibacterial effects of corroding magnesium. *J. Biomed. Mater. Res. A* **2015**, *103*, 3526–3532. [CrossRef] [PubMed]

110. Feng, H.; Wang, G.; Jin, W.; Zhang, X.; Huang, Y.; Gao, A.; Wu, H.; Wu, G.; Chu, P.K. Systematic Study of Inherent Antibacterial Properties of Magnesium-based Biomaterials. *ACS Appl. Mater. Interfaces* **2016**, *8*, 9662–9673. [CrossRef] [PubMed]

111. Lock, J.Y.; Wyatt, E.; Upadhyayula, S.; Whall, A.; Nunez, V.; Vullev, V.I.; Liu, H. Degradation and antibacterial properties of magnesium alloys in artificial urine for potential resorbable ureteral stent applications. *J. Biomed. Mater. Res. A* **2014**, *102*, 781–792. [CrossRef] [PubMed]

112. Robinson, D.A.; Griffith, R.W.; Shechtman, D.; Evans, R.B.; Conzemius, M.G. In vitro antibacterial properties of magnesium metal against *Escherichia coli*, Pseudomonas aeruginosa and Staphylococcus aureus. *Acta Biomater.* **2010**, *6*, 1869–1877. [CrossRef] [PubMed]

113. Rahim, M.I.; Rohde, M.; Rais, B.; Seitz, J.-M.; Mueller, P.P. Susceptibility of metallic magnesium implants to bacterial biofilm infections. *J. Biomed. Mater. Res. A* **2016**, *104*, 1489–1499. [CrossRef] [PubMed]
114. Hou, P.; Zhao, C.; Cheng, P.; Wu, H.; Ni, J.; Zhang, S.; Lou, T.; Wang, C.; Han, P.; Zhang, X.; et al. Reduced antibacterial property of metallic magnesium in vivo. *Biomed. Mater.* **2016**, *12*, 015010. [CrossRef] [PubMed]
115. Gristina, A.G.; Naylor, P.; Myrvik, Q. Infections from biomaterials and implants: A race for the surface. *Med. Prog. Technol.* **1988**, *14*, 205–224. [PubMed]
116. Church, D.; Elsayed, S.; Reid, O.; Winston, B.; Lindsay, R. Burn wound infections. *Clin. Microbiol. Rev.* **2006**, *19*, 403–434. [CrossRef] [PubMed]
117. Stern, G.A. Pseudomonas keratitis and contact lens wear: The lens/eye is at fault. *Cornea* **1990**, *9* (Suppl. S1), S39–S40. [CrossRef] [PubMed]
118. Qin, S.; Xu, K.; Nie, B.; Ji, F.; Zhang, H. Approaches based on passive and active antibacterial coating on titanium to achieve antibacterial activity. *J. Biomed. Mater. Res. A* **2018**. [CrossRef] [PubMed]
119. Liu, Z.D.; Schade, R.; Luthringer, B.; Hort, N.; Rothe, H.; Muller, S.; Liefeith, K.; Willumeit-Romer, R.; Feyerabend, F. Influence of the Microstructure and Silver Content on Degradation, Cytocompatibility, and Antibacterial Properties of Magnesium-Silver Alloys In Vitro. *Oxid. Med. Cell. Longev.* **2017**, *2017*, 8091265. [CrossRef] [PubMed]
120. Zhang, X.D.; Yi, J.H.; Zhao, G.W.; Huang, L.L.; Yan, G.J.; Chen, Y.S.; Liu, P. Layer-by-layer assembly of silver nanoparticles embedded polyelectrolyte multilayer on magnesium alloy with enhanced antibacterial property. *Surf. Coat. Technol.* **2016**, *286*, 103–112. [CrossRef]
121. Yu, W.L.; Chen, D.Y.; Ding, Z.Y.; Qiu, M.L.; Zhang, Z.W.; Shen, J.; Zhang, X.N.; Zhang, S.X.; He, Y.H.; Shi, Z.M. Synergistic effect of a biodegradable Mg-Zn alloy on osteogenic activity and anti-biofilm ability: An in vitro and in vivo study. *RSC Adv.* **2016**, *6*, 45219–45230. [CrossRef]
122. Zhao, C.; Hou, P.; Ni, J.; Han, P.; Chai, Y.; Zhang, X. Ag-Incorporated FHA Coating on Pure Mg: Degradation and in Vitro Antibacterial Properties. *ACS Appl. Mater. Interfaces* **2016**, *8*, 5093–5103. [CrossRef] [PubMed]
123. Qin, H.; Zhao, Y.; An, Z.; Cheng, M.; Wang, Q.; Cheng, T.; Wang, Q.; Wang, J.; Jiang, Y.; Zhang, X.; et al. Enhanced antibacterial properties, biocompatibility, and corrosion resistance of degradable Mg-Nd-Zn-Zr alloy. *Biomaterials* **2015**, *53*, 211–220. [CrossRef] [PubMed]
124. Tie, D.; Feyerabend, F.; Hort, N.; Hoeche, D.; Kainer, K.U.; Willumeit, R.; Mueller, W.D. In vitro mechanical and corrosion properties of biodegradable Mg-Ag alloys. *Mater. Corros.* **2014**, *65*, 569–576. [CrossRef]
125. Zhang, X.B.; Ba, Z.X.; Wang, Z.Z.; He, X.C.; Shen, C.; Wang, Q. Influence of silver addition on microstructure and corrosion behavior of Mg-Nd-Zn-Zr alloys for biomedical application. *Mater. Lett.* **2013**, *100*, 188–191. [CrossRef]

metals

MDPI

Article

Influence of the Composition of the Hank's Balanced Salt Solution on the Corrosion Behavior of AZ31 and AZ61 Magnesium Alloys

Jakub Tkacz [1,2,*], Karolína Slouková [1], Jozef Minda [1], Juliána Drábiková [1], Stanislava Fintová [1,3], Pavel Doležal [1,4] and Jaromír Wasserbauer [1]

[1] Materials Research Centre, Faculty of Chemistry, Brno University of Technology, 612 00 Brno, Czech Republic; xcsloukova@fch.vut.cz (K.S.); xcminda@fch.vut.cz (J.M.); xcdrabikovaj@fch.vut.cz (J.D.); fintova@ipm.cz (S.F.); dolezal@fme.vutbr.cz (P.D.); wasserbauer@fch.vut.cz (J.W.)
[2] Research Centre, University of Žilina, 010 08 Žilina, Slovakia
[3] Institute of Physics of Materials AS CR v. v. i., Žižkova 22, 616 62 Brno, Czech Republic
[4] Institute of Materials Science and Engineering, Faculty of Mechanical Engineering, Brno University of Technology, Technická 2896/2, 616 69 Brno, Czech Republic
* Correspondence: tkacz@fch.vut.cz; Tel.: +420-54-114-9469

Received: 31 August 2017; Accepted: 25 October 2017; Published: 1 November 2017

Abstract: The electrochemical corrosion characteristics of AZ31 and AZ61 magnesium alloys were analyzed in terms of potentiodynamic tests and electrochemical impedance spectroscopy. The influence of the solution composition and material surface finish was examined also through the analysis of corrosion products created on the samples' surface after electrochemical measurements in terms of scanning electron microscopy using energy-dispersive spectroscopy. Obtained data revealed the differences in the response of the magnesium alloys to enriched Hank's Balanced Salt Solution—HBSS+ (with Mg^{2+} and Ca^{2+} ions) and Hank's Balanced Salt Solution—HBSS (without Mg^{2+} and Ca^{2+} ions). Both examined alloys exhibited better corrosion resistance from the thermodynamic and kinetic point of view in the enriched HBSS+. AZ61 magnesium alloy reached higher values of polarization resistance than AZ31 magnesium alloy in both the used corrosion solutions. Phosphate-based corrosion products were characteristic for the AZ31 and AZ61 alloys tested in the HBSS (without Mg^{2+} and Ca^{2+} ions). The combination of phosphate-based corrosion products and clusters of MgO and $Mg(OH)_2$ was typical for the surface of samples tested in the enriched HBSS+ (with Mg^{2+} and Ca^{2+} ions). Pitting corrosion attack was observed only in the case of enriched HBSS+.

Keywords: magnesium alloy; AZ31; AZ61; HBSS; HBSS+; EIS; potentiodynamic test

1. Introduction

Magnesium is an essential element for living organisms, however, for technical purposes is the magnesium used mainly in the form of alloys. Alloying elements improve magnesium mechanical properties and it can be used to control its reactivity. Due to the suitable combination of physico-mechanical properties, biocompatibility and non-toxicity specific magnesium alloys are investigated for medical applications. In the case of orthopedic implants physical and mechanical properties of magnesium alloys are also important. These properties are similar to the properties of a human bone (e.g., density, compressive yield strength, ultimate tensile strength). Magnesium alloy implants are moreover biocompatible and biodegradable [1–11]. As a result of chemical reactions with the biological environment non-toxic corrosion products are created on the surface of the implants. In the human body the magnesium alloy implants dissolve and are absorbed, which prevents surgical removal of the implants after tissue healing [5,6]. The disadvantage of magnesium and magnesium alloys is their high reactivity at the physiological pH (7.4–7.6) as well as in physiological media containing high concentrations of chloride ions, which could cause rapid disintegration of the implant in the biological environment [7,8]. Furthermore, during the corrosion process of magnesium and its alloys, the release of hydrogen gas may be too fast to be endured by the host tissues [9].

One way to influence corrosion resistance and mechanical properties of magnesium alloys is by using high purity alloys that maintain metal impurities such as iron, nickel and copper below limits. Examples of alloying elements of magnesium alloys for biodegradable implants for improvement of the corrosion resistance of the alloys are calcium, zinc, etc. [10–15]. On the other hand, even the magnesium alloys for medical applications have to have good mechanical properties and they have to also contain other alloying elements. One of the basic alloying elements for magnesium mechanical and corrosion properties improvement is aluminum. Even though Al has a positive effect on magnesium alloys properties, the amount of Al added must be controlled in the case of alloys for medical applications. A high concentration of Al was considered to possibly cause neurotoxic illnesses such as dementia or Alzheimer's disease. However, when Al is introduced into the human body in small concentrations, for instance during dietary ingestion or consumption from natural or urban water supplies, then it is naturally excreted through urine or in the form of bile [10,16–21].

Hank's balanced salt solutions (HBSSs), which are one of the options for simulating the corrosive environment of the body of living organisms are often used for analysis of the corrosion behavior of magnesium alloys that are expected to be used in medicine applications. Mainly due to higher chloride concentrations, HBSSs are more aggressive medium compared to artificial plasma. Sulfate ions contained in HBSS can also result in higher corrosion rate of magnesium and its alloys compared to other corrosion media used for material corrosion properties characterization [22–31].

The reactivity of material in the corrosion environment is, besides many aspects, influenced with the chemical and phase composition of the material and chemical composition of the testing solution.

Electrochemical corrosion behavior of wrought AZ31 and AZ61 alloys in HBSS characterized by Tkacz et al. in [32] revealed different response of the materials due to their chemical and phase composition and surface finish. Based on the EIS (electrochemical impedance spectroscopy) measurement results, AZ61 magnesium alloy was considered as more corrosion resistant when compared to AZ31 magnesium alloy, while opposite conclusion can be considered based on the potentiodynamic test results. Potentiodynamic tests revealed minor influence of the surface finish in the case of AZ31 magnesium alloy represented for example by corrosion potential (E_{corr}) values. The samples with polished surface (E_{corr} = −1.676 ± 0.003 V) were characteristic with more positive value of E_{corr} comparing to the ground samples (E_{corr} = −1.701 ± 0.003 V). On the other hand, no influence of surface finish was observed in the case of AZ61 magnesium alloy (E_{corr} = −1.708 ± 0.004 V for ground sample and E_{corr} = −1.708 ± 0.003 V for polished sample). No significant influence of surface finish was observed based on the EIS test results. The increase of polarization resistance up to 24 h of immersion of the samples in the HBSS to the values of approximately 4000 $\Omega \cdot cm^2$ was characteristic for both the surface states of AZ31 magnesium alloy, while increasing exposure time to the corrosion environment did not have any significant influence on the polarization resistance and the value remained stable. In the case of AZ61 magnesium alloy a significant increase of the values of polarization resistance up to the 48 h of exposure of the samples to the corrosion environment was observed for both the material states, reaching the value of polarization resistance of approximately 9000 $\Omega \cdot cm^2$. Following an increase of immersion time resulted in an additional increase of the R_p to the value of approximately 21,000 $\Omega \cdot cm^2$ for ground and 17,000 $\Omega \cdot cm^2$ for polished samples.

The influence of the chloride ions on corrosion behavior of Mg-Al-Zn based alloys was reported in several studies. Ambat et al. studied in [33] the influence of chloride ion concentration and pH on the corrosion and electrochemical behavior of die-cast and ingot-cast AZ91D alloy. The effect of chloride ion concentration was studied in NaCl (0–10%) solutions at pH 7.25. The effect of pH was analyzed in solutions with the chloride ion concentration kept constant at 3.5% while the pH was varied from 1.0–12.0. Material behavior was analyzed with immersion and potentiodynamic testing. Increase in chloride ion concentration increased the corrosion rate of AZ91D magnesium alloy at pH 7.25 and 12.0 for both the material states, however, at pH 2.0, the effect of chloride ion was found to be negligible. High corrosion rate was observed for both the material states in highly acidic solutions, while the corrosion rate was found to be low in neutral pH and alkaline conditions. The corrosion rate determined from immersion tests was much higher than that obtained from electrochemical measurements. The observation was attributed to the negative difference effect and to the physical removal of β phase during corrosion. The differences between the response of material states to the corrosion environment were discussed in terms of microstructural differences.

The effect of chloride ion concentration and pH on the corrosion (immersion tests) and electrochemical behavior (potentiodynamic tests) of AZ63 alloy were studied in NaCl solutions at different concentrations (0.01, 0.2, 0.6, 1 and 2 M) and pH values (2, 3, 8, 11 and 11.5) were studied by Altun and Sen in [34]. Authors observed that the corrosion rate increased with the increase in concentration of NaCl solution. However, it was observed, that with the increase in chloride ion concentration, the rising rate of corrosion rate decreased. The corrosion rate increase with increasing chloride ion concentration was attributed to the participation of chloride ions in the dissolution reaction. Chloride ions are aggressive for both magnesium and aluminum. The adsorption of chloride ions to oxide covered magnesium surface transforms $Mg(OH)_2$ to easily soluble $MgCl_2$. Authors also observed a shift of the corrosion potential to more negative (more active) values with the increase in chloride ion concentration. The explanation for this behavior was found in adsorption of these ions on the alloy surface at weak parts of the oxide film. Corrosion potential was observed to be shifted to more negative (more active) values with the decrease in pH value of the solution. Higher pH values were discussed to favor the formation of $Mg(OH)_2$ which protects the alloy from corrosion.

Merino et al. studied in [35] the influence of chloride ion concentration and temperature on the corrosion of Mg-Al alloys in a salt fog with the focus on the effect of Al content in the alloy. The results of their investigation showed that the corrosion attack of Mg, AZ31, AZ80 and AZ91D materials under the salt fog test increases with increasing temperature and chloride anion concentration, while the effect of temperature was considered to be more noticeable than that of chloride concentration. Authors also analyzed the influence of the Al content and resulting presence of intermetallic phases on material corrosion behavior. In the case of the wrought AZ31 only negligible influence of the present AlMn based phase was observed, while in the case of cast AZ80 and AZ91 alloys the creation of galvanic couples between Al-Mn and $β-Mg_{17}Al_{12}$ phases with the Mg matrix resulted in more pronounced corrosion attack. Authors also observed the influence of the distribution, size and morphology of the β phase and resulting Al-rich corrosion products layer created on the material surface during the corrosion.

Influence of sulfate anion concentration and pH on the corrosion of Mg-Al-Zn-Mn (GA9) magnesium alloy was investigated by Shetty et al. [36]. The studies were carried out in sodium sulfate solutions with concentrations range of 0.1–2 M; and at different temperatures of 30–50 °C and pH of 3.0–12.0. According to the experimental data, the corrosion rate of the alloy increased with the increase in temperature, and also with the increase in the concentration of sodium sulfate in the medium. It was observed that the rate of corrosion decreased with the increase in pH.

Even thought, magnesium alloys corrosion properties are widely investigated, most of the studies are performed in NaCl and Na_2SO_4 solutions simulating corrosion environment in engineering applications [33–36]. Corrosion behavior of magnesium alloys was studied in several types of HBSS [22–32], however, the influence of the chemical composition of HBSS on corrosion processes is not available in the literature according to the authors' knowledge.

Corrosion characteristics of metallic materials can be analyzed in different ways. In this work, electrochemical methods have been used to investigate the corrosion behavior of magnesium alloys. Potentiodynamic tests and electrochemical impedance spectroscopy were used for the description of the material response to the HBSS and enriched HBSS+. Obtained data were discussed with the aim to identify the influence of material chemical and phase composition, surface roughness and composition of the corrosion solution on corrosion behavior. Values of corrosion potential (E_{corr}) expressing thermodynamics of the corrosion process and corrosion current density values (i_{corr}) expressing the kinetic of the corrosion process were obtained by potentiodynamic measurements. Corrosion potential characteristic for the material expresses the thermodynamic stability of the system and the conditions for material corrosion and its resistance against corrosion process. Kinetic of the corrosion process can be shown by the evolution of the corrosion rate (v_{corr}) which can be calculated from the i_{corr}. Polarization resistance (R_p) values were obtained by electrochemical impedance spectroscopy (EIS). Evolution of the corrosion products on the specimen surface was analyzed in terms of scanning electron microscopy and correlated with the composition of the used corrosion solution.

The presented results show differences in electrochemical corrosion behavior of AZ31 and AZ61 alloys in HBSS and enriched HBSS+ with the aim to characterize material behavior and different response of ground and polished materials on the different chemical composition of the corrosion solution.

2. Materials and Methods

2.1. Material

Wrought AZ31 and AZ61 magnesium alloys plates were used for the experiments. Metallographic analysis and verification of chemical composition of the magnesium alloys were performed by scanning electron microscope (SEM) (ZEISS EVO LS 10, Cambridge, UK) with energy dispersive spectrometer (EDS) (OXFORD Instruments X-MAX 80 mm², Abingdon, UK). Metallographic samples for microstructural analysis were prepared in terms of a conventional procedure consisting of grinding, polishing (diamond paste 1 μm) and etching (picral solution [37]).

2.2. Electrochemical Measurements

Wrought AZ31 and AZ61 alloys plates were cut to samples with dimensions of 20 × 20 × 2 mm³. One batch of the samples of each magnesium alloy was ground with 1200 grit SiC paper with a particle size of ~15 μm (marked #1200) and the second batch was additionally polished with diamond pastes up to 0.25 μm (marked 0.25 μm). Wetting agent during polishing was isopropyl alcohol.

Electrochemical characteristics of the ground and polished samples were measured by potentiostat/galvanostat BioLogic VSP-300 (BioLogic, Seyssinet-Pariset, France). Three-electrode system was used for electrochemical measurements, where the magnesium alloy sample served as a working electrode (WE), saturated calomel electrode (SCE) was used as a reference electrode (RE) and platinum gauze was used as a counter electrode (CE). Used corrosion environment was enriched HBSS+ (with the addition of Mg^{2+} and Ca^{2+} ions) and HBSS (without Mg^{2+} and Ca^{2+} ions) from [32] was used to show the dependence of the response of the magnesium alloys on corrosion solution composition. Experiments were performed at the temperature of 37 ± 1 °C. The composition of the used HBSS and enriched HBSS+ is given in Table 1. An area of 1.0 cm² of the sample was exposed to the corrosion environment during the electrochemical testing. The stabilization time of the sample exposed to the corrosion environment before the measurement was 5 min. Each electrochemical measurement was performed on three specimens with adequate surface finish.

Table 1. Chemical composition of HBSS and enriched HBSS+ used in [32].

Solutions	Composition (mg·dm^{-3})							
	NaCl	KCl	KH$_2$PO$_4$	Glucose	Na$_2$HPO$_4$	MgSO$_4$	CaCl$_2$	Na$_2$CO$_3$
HBSS	8000	400	60	1000	48	-	-	350
Enriched HBSS+	8000	400	60	1000	48	98	140	350

Potentiodynamic measurements were performed by polarizing of the sample surface in the range from -100 mV to $+200$ mV vs. open circuit potential (E_{OCP}). Scan rate was 1 mV·s^{-1}.

Electrochemical impedance spectroscopy (EIS) measurements were performed after exposure of the sample surface to the enriched HBSS+ with Mg^{2+} and Ca^{2+} ions for 5 min, 1, 2, 4, 8, 12, 24, 48, 72, 96 and 168 h. EIS scan frequency ranged from 100 kHz to 100 mHz, and the perturbation amplitude was 5 mV.

With the aim to determine the influence of the enriched HBSS+ with Mg^{2+} and Ca^{2+} ions on the experimental material electrochemical corrosion behavior were the EIS measurements performed also in the solution not containing addition Mg^{2+} and Ca^{2+} ions (Table 1).

The chemical composition of the corrosion products created on the surface of the AZ31 and AZ61 alloys during EIS measurements was analyzed with SEM ZEISS EVO LS 10 with EDS OXFORD Instruments X-MAX 80 mm^2. The samples were rinsed with isopropyl alcohol and dried with air before the analysis.

3. Results

3.1. Microstructural Analysis

The microstructure of the AZ31 magnesium alloy is consisted of polyhedral grains of the substitutional solid solution α-Mg in which AlMn intermetallic phase particles were observed (Figure 1) [38–40]. The distribution of the basic alloying elements revealed with the mapping mode by EDS is shown in Figure 1. The chemical composition of the AZ31 magnesium alloy verified by EDS is provided in Table 2. The content of the main alloying elements (Al, Zn and Mn) agrees with the standard ASTM B90M [41].

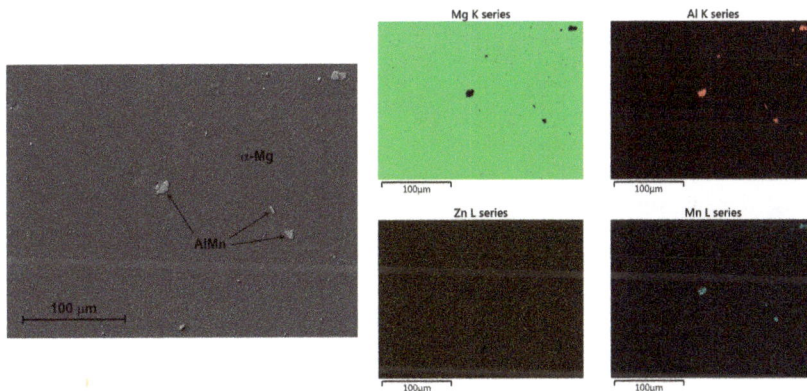

Figure 1. Microstructure of AZ31 magnesium alloy SEM (Scanning Electron Microscope) with EDS (Energy Dispersive Spectrometer) maps of the elements distribution: magnesium (Mg), aluminum (Al), zinc (Zn) and manganese (Mn).

Table 2. Chemical composition of AZ31 magnesium alloy (EDS: Energy Dispersive Spectrometer).

Element	Al	Zn	Mn	Mg
Chemical composition (wt %)	3.2	0.9	0.4	balance

The microstructure of the AZ61 magnesium alloy is consists of polyhedral grains of the substitution solid solution (α-Mg), β-phase (particles of Mg$_{17}$Al$_{12}$ intermetallic phase) and AlMn intermetallic phase particles (Figure 2). The EDS mapping mode was used for the investigation of the distribution of the basic alloying elements in the alloy (Figure 2). The chemical composition of the

AZ61 magnesium alloy was verified by EDS (Table 3). The content of the main alloying elements (Al, Zn and Mn) agrees with the standard ASTM B107M [42].

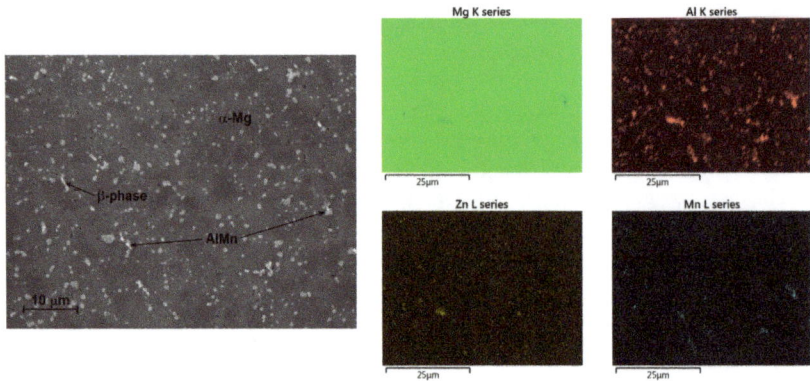

Figure 2. Microstructure of AZ61 magnesium alloy (SEM) with EDS maps of the elements distribution: magnesium (Mg), aluminum (Al), zinc (Zn) and manganese (Mn).

Table 3. Chemical composition of AZ61 magnesium alloy (EDS).

Element	Al	Zn	Mn	Mg
Chemical composition (wt %)	6.0	0.7	0.3	balance

3.2. Potentiodynamic Measurements

Figure 3 shows typical potentiodynamic curves obtained with linear polarization of ground and polished samples of AZ31 and AZ61 alloys. The pitting corrosion attack of all the tested samples was revealed by an increase of current density at the anodic branch of the curves which belongs to the value of pitting potential (E_{pit}) [38]. The values of the corrosion current density (i_{corr}) for the AZ31 magnesium alloy were determined only from the cathodic branch of the polarization curves applying the Tafel extrapolation. The Tafel region for the anodic branches of the potentiodynamic curves characterizing AZ31 magnesium alloy was insufficient to use the Tafel extrapolation to obtain relevant values of i_{corr}, following the rule that the plane region has to be at least 50 mV [38] from the E_{corr}. On the other hand, i_{corr} values for the AZ61 magnesium alloy were determined from both the branches of the obtained polarization curves applying Tafel extrapolation.

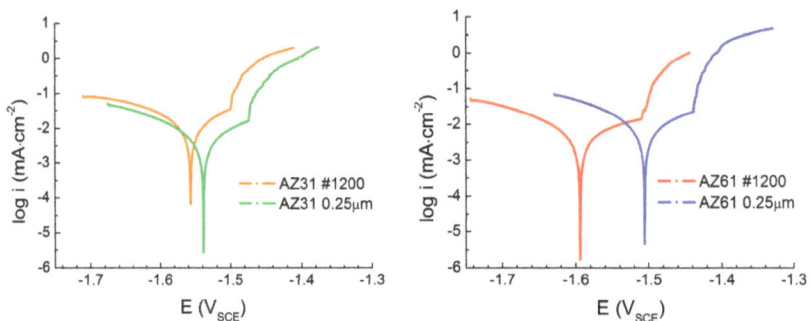

Figure 3. Potentiodynamic curves of ground (#1200) and polished (0.25 μm) surface of AZ31 and AZ61 magnesium alloys in enriched HBSS+ (with Mg^{2+} and Ca^{2+} ions).

For both the tested materials the finer surface (polishing with 0.25 µm diamond paste) resulted in the potentiodynamic curve shift to more positive values of potential comparing to the curves characterizing the ground samples.

The electrochemical characteristics estimated by potentiodynamic measurements: E_{OCP}, E_{corr}, E_{pit}, i_{corr} and corrosion rate v_{corr} are given in Table 4. From the obtained data, a larger influence of the surface finish can be observed in the case of AZ61 magnesium alloy when compared to the AZ31 magnesium alloy.

Table 4. Results of potentiodynamic tests performed in enriched HBSS+ (with Mg^{2+} and Ca^{2+} ions).

Alloy Finish	E_{OCP} (V)	E_{corr} (V)	E_{pit} (V)	i_{corr} ($\mu A \cdot cm^{-2}$)	v_{corr} ($\mu m \cdot year^{-1}$)
AZ31—#1200	-1.721 ± 0.015	-1.556 ± 0.012	-1.500 ± 0.014	10.5 ± 0.8	234.6 ± 56.6
AZ61—#1200	-1.717 ± 0.014	-1.594 ± 0.014	-1.509 ± 0.009	5.5 ± 0.6	123.9 ± 40.6
AZ31—0.25 µm	-1.674 ± 0.005	-1.538 ± 0.008	-1.474 ± 0.011	4.9 ± 0.5	112.5 ± 26.8
AZ61—0.25 µm	-1.643 ± 0.013	-1.505 ± 0.011	-1.431 ± 0.011	7.4 ± 0.6	171.7 ± 43.4

3.3. Electrochemical Impedance Spectroscopy

Nyquist plots representing the data obtained with EIS for ground and polished samples of AZ31 and AZ61 alloys were analyzed applying equivalent circuits shown in Figure 4. These equivalent circuits consist of resistance of the corrosion environment (solution) R_s, the resistance of layer of corrosion products R_1 and resistance of the magnesium alloy base material R_2. In some cases, the inductance L was present in the equivalent circuit, according to the character of the obtained Nyquist plot. Constant phase element (CPE) represents capacity formed between the corrosion environment and the corrosion products created on materials surface or the corrosion products and the magnesium alloy, respectively. Polarization resistance R_p values were calculated according to the equations presented under the equivalent circuits in Figure 4.

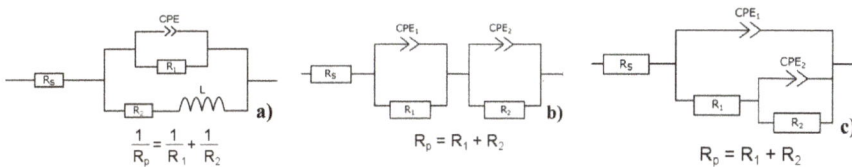

Figure 4. Equivalent circuits used to the evaluation of the obtained Nyquist plots: (**a**) with an inductive loop in the equivalent circuit, (**b**) serial connection in the equivalent circuit and (**c**) parallel connection in the equivalent circuit (CPE: Constant phase element).

Figure 5 shows Nyquist plots characterizing electrochemical corrosion behavior of ground and polished samples of AZ31 and AZ61 alloys. Immersion times of the alloys in the corrosion environment were from 5 min to 168 h. In all the plots at different immersion times, semicircles at high and low frequencies are present. Obtained impedance data were analyzed using EC-Lab software and best-fitted using the appropriate equivalent circuit model (Figure 4). The determined resulting R_p values are given in Table 5.

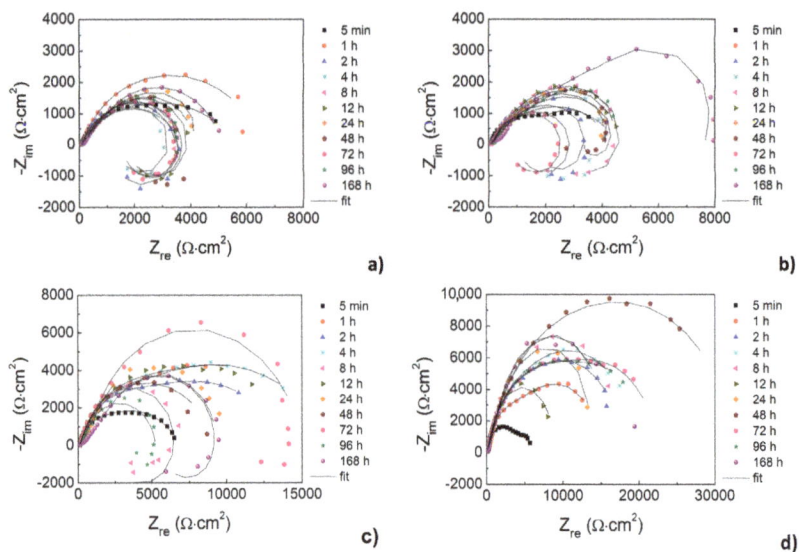

Figure 5. Nyquist plots of ground (#1200) and polished (0.25 μm) samples of AZ31 and AZ61 magnesium alloys obtained in enriched HBSS+ (with Mg^{2+} and Ca^{2+} ions): (**a**) AZ31—0.25 μm, (**b**) AZ31—#1200, (**c**) AZ61—0.25 μm and (**d**) AZ61—#1200.

Table 5. Polarization resistance values obtained from EIS (electrochemical impedance spectroscopy) measurements performed in enriched HBSS+ (with Mg^{2+} and Ca^{2+} ions).

Samples	R_p ($\Omega \cdot cm^2$)					
	5 min	1 h	2 h	4 h	8 h	12 h
AZ31—#1200	5209 ± 531	3626 ± 773	3495 ± 40	3489 ± 101	3924 ± 36	5331 ± 217
AZ61—#1200	6256 ± 19	15,458 ± 31	16,501 ± 58	18,402 ± 68	15,509 ± 333	8800 ± 564
AZ31—0.25 μm	4805 ± 142	6604 ± 695	3544 ± 53	3711 ± 115	3962 ± 445	5102 ± 964
AZ61—0.25 μm	5916 ± 37	14,404 ± 207	12,277 ± 232	14,203 ± 33	9193 ± 280	15,471 ± 575
Samples	24 h	48 h	72 h	96 h	168 h	-
AZ31—#1200	5201 ± 516	4163 ± 356	2667 ± 776	4949 ± 624	7481 ± 115	-
AZ61—#1200	27,005 ± 396	24,392 ± 768	22,903 ± 769	22,326 ± 850	15,691 ± 2500	-
AZ31—0.25 μm	4139 ± 478	4405 ± 706	3455 ± 171	3668 ± 92	4853 ± 532	-
AZ61—0.25 μm	9097 ± 501	9786 ± 1020	14,501 ± 1236	12,693 ± 126	10,759 ± 318	-

3.4. Characterization of Corrosion Products

Figure 6 shows the SEM images of the surface of the AZ31 and AZ61 alloys after EIS measurements (168 h of immersion in enriched HBSS+ with Mg^{2+} and Ca^{2+} ions).

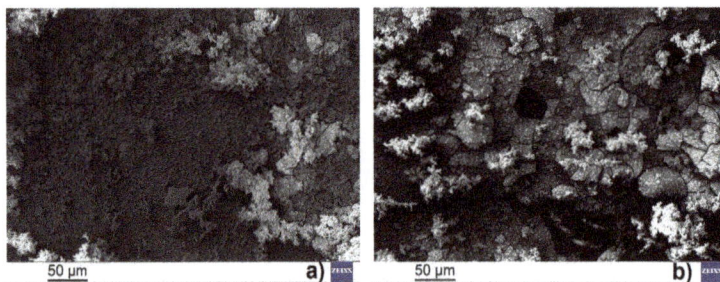

Figure 6. *Cont.*

Figure 6. Morphology of the corrosion products present on the surface of tested samples after EIS measurement (168 h of immersion in enriched HBSS+ with Mg^{2+} and Ca^{2+} ions): (**a**) AZ31—0.25 μm, (**b**) AZ31—#1200, (**c**) AZ61—0.25 μm and (**d**) AZ61—#1200.

Chemical composition (Table 6) of corrosion products present on the surfaces of the tested samples shows the presence of higher amount of oxygen and phosphorus indicating the presence of phosphates, oxides and hydroxides of magnesium (or calcium respectively).

Table 6. Chemical composition of corrosion product on the surface of AZ31 and AZ61 magnesium alloys after EIS measurements in enriched HBSS+ (with Mg^{2+} and Ca^{2+} ions).

Samples	Chemical Composition (wt %)								
	Mg	Al	Cl	P	O	C	Na	Ca	S
AZ31—#1200	7.3	0.3	0.3	15.3	42.8	5.5	0.4	28.0	0.1
AZ61—#1200	14.4	6.2	0.5	10.6	44.2	13.6	0.4	9.8	0.3
AZ31—0.25 μm	5.1	0.1	0.2	15.6	42.4	5.0	0.4	31.1	0.1
AZ61—0.25 μm	14.4	4.6	0.6	11.8	43.5	11.3	0.5	13.1	0.2

Comparing the surface morphology of AZ31 magnesium alloy with polished and ground surface more pronounced cracking of the layer of corrosion products created on the sample surface can be seen for the ground sample (Figure 6b). The surface of the samples was covered by corrosion products based on MgO, $Mg(OH)_2$ [43,44] and due to the used corrosion environment also phosphate-based corrosion products [45]. While the amount of MgO and $Mg(OH)_2$ products is comparable for both the sample surfaces, the phosphate-based layer of corrosion products seems to be thicker for the ground sample (cracked layer presented on the sample surface, below the clusters of MgO and $Mg(OH)_2$ products).

Significantly smaller amount of MgO and $Mg(OH)_2$ products were observed on the surface of AZ61 magnesium alloy when compared to AZ31 magnesium alloy. Similarly, as in the case of AZ31 magnesium alloy also in the case of AZ61 magnesium alloy a layer of phosphate-based corrosion products was observed on samples surface. Thicker layer (larger features and cracks) seems to appear on the surface of ground AZ61 magnesium alloy comparing of the polished AZ61 magnesium alloy (Figure 6c,d).

Figure 7 shows the SEM images of the surface of the AZ31 and AZ61 alloys after EIS measurements (168 h of immersion in HBSS without Mg^{2+} and Ca^{2+} ions).

Figure 7. Morphology of the surface after EIS (electrochemical impedance spectroscopy) measurements (168 h of immersion in HBSS without Mg^{2+} and Ca^{2+} ions): (**a**) AZ31—0.25 μm, (**b**) AZ31—#1200, (**c**) AZ61—0.25 μm and (**d**) AZ61—#1200.

Chemical composition (Table 7) of corrosion products present on the surfaces of the tested samples from Figure 7 shows the presence of oxygen and phosphorus indicating the presence of phosphates, oxides and hydroxides of magnesium.

Table 7. Chemical composition of corrosion product on the surface of AZ31 and AZ61 magnesium alloys after EIS measurements in HBSS (without Mg^{2+} and Ca^{2+} ions).

Samples	Chemical Composition (wt %)						
	Mg	Al	Cl	P	O	C	Na
AZ31—#1200	31.4	4.1	0.8	7.7	43.2	12.1	0.7
AZ61—#1200	36.3	6.1	0.6	6.5	36.3	13.3	0.9
AZ31—0.25 μm	35.8	4.7	0.5	7.1	37.9	13.1	0.9
AZ61—0.25 μm	26.6	7.9	0.5	7.1	45.2	11.8	0.9

The layer of phosphate-based corrosion products was observed on all samples' surfaces. Thicker layer (larger features and cracks) seems to appear on the surface of polished AZ61 magnesium alloy comparing of the ground AZ61 magnesium alloy (Figure 7c,d). In the phosphate-based layer on the AZ31 magnesium alloy surface were regions without any visible corrosion products (Figure 7b). On the polished surface of the AZ31 magnesium alloy (Figure 7a) were moreover present small amount of MgO and $Mg(OH)_2$ products.

4. Discussion

The effect of the surface treatment (ground vs. polished surface) was observed during potentiodynamic measurements on both types of tested magnesium alloys. In both the cases, more positive value of corrosion potential, E_{corr} (Table 4), was determined for the polished samples (0.25 µm) when compared to the ground samples (#1200). The microstructure of the AZ61 magnesium alloy contains a higher number of intermetallic phases ($Mg_{17}Al_{12}$ and AlMn, Figure 2) than the microstructure of the AZ31 magnesium alloy containing only AlMn particles (Figure 1). All of these intermetallic phase particles have more positive potential [46–48] than the substitution solid solution (α-Mg) which caused the formation of microcells which usually result in the acceleration of the corrosion process. The observations are in agreement with [49] where a positive effect of decreasing surface roughness on AZ91 alloy electrochemical corrosion properties was presented. However, authors in [49] performed the EIS measurement only 2 h of exposure of the material in 0.5 wt % NaCl solution and cannot detect the influence of evolution of the layer of corrosion products and its adhesion to the material surface.

The difference in the E_{corr} values determined for ground and polished surface was more significant in the case of AZ61 magnesium alloy (Figure 8). The grinding process results in the higher roughness of the treated surface comparing to the polished surface resulting in a larger real surface area exposed to the corrosion environment [49]. E_{corr} reaches more negative value comparing to the polished surface and therefore the ground surface can be considered as less stable from a thermodynamic point of view than the polished surface. While only a small amount of intermetallic phase particles was present in the case of the AZ31 magnesium alloy, the effect of the chemical heterogeneity and surface roughness was smaller compared to the AZ61 magnesium alloy samples. Corrosion potential E_{corr} of the AZ31 and AZ61 alloys determined from measurements in HBSS without Mg^{2+} and Ca^{2+} ions reported by authors in [32] have more negative values than data obtained in enriched HBSS+ with Mg^{2+} and Ca^{2+} ions (Figure 8). In the [32] was not observed pitting corrosion attack of the samples (no pitting potential E_{pit} was observed on the curves) which indicate higher reactivity of tested magnesium alloys in enriched HBSS+ probably caused by the content of Mg^{2+}, Ca^{2+} and sulphate ions.

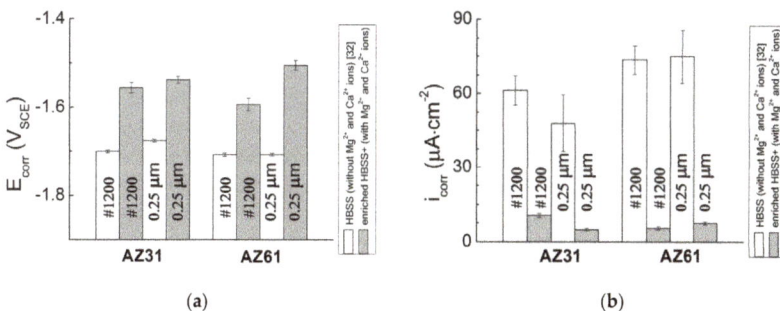

Figure 8. Comparison of (a) E_{corr} and (b) i_{corr} of ground (#1200) and polished (0.25 µm) surface AZ31 and AZ61 magnesium alloys in enriched HBSS+ (with Mg^{2+} and Ca^{2+} ions) and in HBSS (without Mg^{2+} and Ca^{2+} ions [32]).

While in the case of AZ31 magnesium alloy a positive effect of polishing on the material response of the corrosion environment from the kinetic point of view can be considered, opposite behavior was observed in the case of AZ61 magnesium alloy. Polishing of the surface of AZ31 alloy resulted in 50% decrease in the corrosion current density (i_{corr}). However, the same treatment resulted in 50% increase of the i_{corr} in the case of the AZ61 alloy. However, this behavior can be also explained by the presence on a large amount of intermetallic phase particles in the microstructure of AZ61 alloy and higher real surface area after grinding process.

Corrosion current density (i_{corr}) and the resulting corrosion rate (v_{corr}) were higher in the corrosion environment of HBSS without Mg^{2+} and Ca^{2+} ions reported in [32] (Figure 8). Comparing to the presented data the differences in the values were mostly about ten times. This effect could be caused by the presence or absence of especially Mg^{2+} ions in the solution. In the case of HBSS without Mg^{2+} and Ca^{2+} ions could play a role the concentration gradient [50–52]. Magnesium in the alloys reacts with the corrosion environment to produce Mg^{2+} ions. In the beginning of the immersion of the samples no Mg^{2+} ions are present in the solution. Thanks to the concentration gradient the Mg^{2+} ions migrate from the place with high concentration place of the ions (the surface of the samples) to the place with low concentration of the ions (HBSS without Mg^{2+} and Ca^{2+} ions). This migration of the Mg^{2+} ions supports reactions of the corrosion environment with the surface of AZ31 and AZ61 alloys. On the other hand, the concentration gradient of the enriched HBSS+ with Mg^{2+} and Ca^{2+} ions should be lower compared to the solution without the ions because of the primary presence of the Mg^{2+} ions in the corrosion environment. The migration of the Mg^{2+} ions from the surface of the alloys to the corrosion environment is not so fast (like in HBSS without Mg^{2+} and Ca^{2+} ions) which reduces a number of the amount of reactions of the corrosion environment with the surface of the magnesium alloys.

This theory is supported by the results presented in [33,34,36]. The authors describe the effect of the chlorides and sulphate ions concentration on the corrosion resistance of magnesium alloys. With the increasing content of these ions, the i_{corr} values increased and v_{corr}, respectively. However, when comparing i_{corr} values (Figure 8) in HBSS and enriched HBSS+, this phenomenon did not occur. Enriched HBSS+ contains more chloride and sulphate ions (Table 1) but i_{corr} values are lower. Corrosion behavior of magnesium alloys in corrosion environment with sulphate ions is usually measured in sodium sulphate solution [36] where the absence of Mg^{2+} ions could cause higher reactivity of the alloys.

Potentiodynamic measurements are only short-time measurements. All the measurements take approximately 10 min (5 min of stabilization time and 5 min of measurement itself). Therefore, the potentiodynamic measurements are affected by conditions at the beginning of the measurement like the concentration of specific ions in the solution, concentration gradient, etc. Moreover, if the corrosion potential (E_{pit}) appear, no Tafel region could be observed in the anodic branch of the polarization curves [38].

From the long-time point of view the corrosion behavior of AZ31 and AZ61 alloys can be characterized by EIS. Electrochemical corrosion behavior of the magnesium alloys in enriched HBSS+ with Mg^{2+} and Ca^{2+} ions represented by the evolution of polarization resistance is shown in Figure 9a.

While almost no influence of the surface finish was observed in the case of AZ31 magnesium alloy, higher values of polarization resistance were observed for ground AZ61 magnesium alloy comparing to the polished material state.

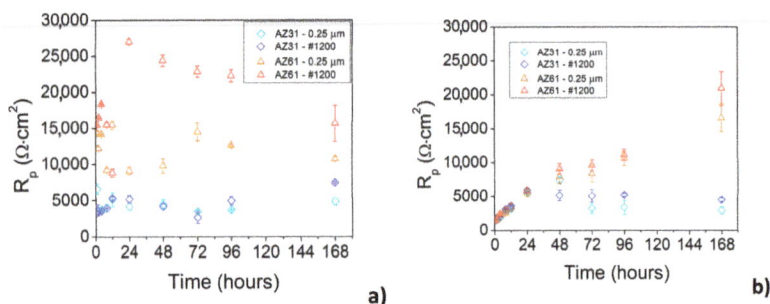

Figure 9. Polarization resistance R_p of ground (#1200) and polished (0.25 μm) surface of AZ31 and AZ61 magnesium alloys: (**a**) in enriched HBSS+ (with Mg^{2+} and Ca^{2+} ions) and (**b**) in HBSS (without Mg^{2+} and Ca^{2+} ions [32]).

In the case of AZ31 magnesium alloy the values of polarization resistance oscillated around $4000 \ \Omega \cdot cm^2$ for all the times of measurement for both the material states. The similar resistivity of both the material states can be connected with similar corrosion products (amount and character) observed on material surfaces (Figure 6a,b). On the surfaces of both the analyzed samples was observed the presence of a layer of magnesium phosphate ($Mg_3(PO_4)_2$) and hydroxiapatite ($Ca_{10}(PO_4)_6(OH)_2$) primarily created on the sample surface and covered by large number of clusters of MgO and $Mg(OH)_2$ products [45,53]. The assumption that the layer of the magnesium phosphate ($Mg_3(PO_4)_2$) and hydroxiapatite ($Ca_{10}(PO_4)_6(OH)_2$) is thicker (Figure 6) in the case of ground sample correlate with slightly higher values of R_p determined for the material state (Table 5). However, material surface finish did not show any significant influence on AZ31 magnesium alloy corrosion resistivity in enriched HBSS+ with Mg^{2+} and Ca^{2+} ions from the long-time point of view.

In the case of ground AZ61 magnesium alloy an increase of R_p up to 24 h of exposure followed by its decrease back to the value characteristic for the beginning of the experiment was observed (Table 5 and Figure 8a). In the case of polished AZ61 magnesium alloy a decrease of R_p up to 24 h of exposure, it follows an increase up to 72 h of exposure followed by an additional decrease to the value lower than the resistance of material on the beginning of the exposure was observed (Table 5 and Figure 8). Higher values of R_p comparing to the AZ31 magnesium alloy can be explained by material higher chemical heterogeneity and thus faster evolution of corrosion products on the material surface. On the other hand, the differences between the polarization resistance of the ground and polished AZ61 magnesium alloy can be corresponding to the higher material surface roughness. Higher real surface area of the rough ground samples can be connected by faster growth of the layer of magnesium phosphate ($Mg_3(PO_4)_2$) and hydroxiapatite ($Ca_{10}(PO_4)_6(OH)_2$) on sample surface comparing to the polished sample surface. This was also observed in Figure 6c,d, where the thicker layer of corrosion products was assumed for the ground sample when compared to the polished sample. The not stable evolution of the R_p in time can correlate with the thicker layer of corrosion products and its cracking, corrosion or remove from the samples surface, which was observed only in a minor form in the case of AZ31 magnesium alloy with the stable evolution of R_p in time. The presence of inductive loop L [54,55] in the equivalent circuit implies the occurrence of pitting corrosion on the magnesium alloy's surface.

The values of polarization resistance R_p are quite high at the beginning of exposition of the samples in the corrosion environment. The values of R_p support the theory of Mg^{2+} ions concentration gradient as well as in the case of explanation for i_{corr} change (v_{corr} respectively). Magnesium reacts more slowly when Mg^{2+} ions are present in the corrosion environments as itself (enriched HBSS+ with Mg^{2+} and Ca^{2+} ions). Helmholtz double layer created on the interface magnesium alloy/corrosion environment is more stable as in the case of HBSS without Mg^{2+} ions and the resistance of the material is high.

The opposite phenomenon can be observed in the case of HBSS without Mg^{2+} and Ca^{2+} ions. At the beginning of the exposition of the samples to this corrosion environment the R_p values are lower (Figure 9b) than in the case of the enriched HBSS+ with Mg^{2+} and Ca^{2+} ions and the value increase with increasing time of exposure to the corrosion environment.

Up to 24 h of exposure comparable values of R_p were determined for both the alloys and for both the material states in HBSS without Mg^{2+} and Ca^{2+} ions (Figure 9b). With increasing exposure time the values of R_p slowly decreased for AZ31 magnesium alloy, while slightly higher values of R_p were determined for the ground samples. This observation corresponds to the smaller number of cracks in the layer of corrosion products present on the surface of ground samples comparing to the polished samples (Figure 7a,b). An increase of R_p above 24 h of exposure to the corrosion environment was observed for AZ61 magnesium alloy, while the highest values were determined for 168 h of exposure to the HBSS without Mg^{2+} and Ca^{2+} ions (Figure 9b). Same as in the case of the AZ31 alloy, also in the case of AZ61 alloy slightly higher values of R_p were determined for the ground samples. Figure 7c,d showing the surface of the samples after the EIS measurements can exhibit the explanation for this material behavior. While the smaller number of finer cracks were observed on the surface of the layer of corrosion products created on the ground sample compared to the deep and large cracks observed on the surface of the polished sample.

In the both corrosion environments, the influence of corrosion products must be taken into account when predicting material corrosion behavior, because magnesium and its alloys are very reactive by themselves. The corrosion layer behaves as a barrier against the further access of corrosion environment to the surface of the magnesium alloys. The layers of the corrosion products are composed mainly of the oxides, hydroxides, phosphates and chlorides [45]. However, the layer is porous and uncompacted (cracks are present in Figures 6 and 7) which leads to other reactions in the corrosion environments of the samples. Even though, different corrosion products are created on the surface of AZ31 and AZ61 alloys due to the chemical composition of the used corrosion solution, more stable material behavior was observed for the AZ31 magnesium alloy samples in both of the solutions. AZ61 magnesium alloy exhibited more stable corrosion response in the HBSS without Mg^{2+} and Ca^{2+} ions [32] comparing to its behavior in enriched HBSS+ with Mg^{2+} and Ca^{2+} ions.

5. Conclusions

This work was focused on the evaluation of corrosion behavior of AZ31 and AZ61 magnesium alloys in enriched HBSS+ (with Mg^{2+} and Ca^{2+} ions) by electrochemical methods and the results were compared with results presented in [32]. The results of the work can be summarized as follows:

(1) Electrochemical corrosion properties of AZ31 and AZ61 alloys are dependent on corrosion environment composition.

(2) AZ31 and AZ61 alloys had more positive corrosion potential E_{corr} and significantly lower values of corrosion current density i_{corr} in the enriched HBSS+ (with Mg^{2+} and Ca^{2+} ions) comparing to the standard HBSS (without Mg^{2+} and Ca^{2+} ions). It was also observed that the pitting corrosion attack with pitting potential E_{pit} in the enriched HBSS+ solution (with Mg^{2+} and Ca^{2+} ions) which resulted in less uniform corrosion attack (less predictable material behavior) of the material compared to material behavior in HBSS.

(3) No significant influence of the surface treatment (ground vs. polished) was observed during EIS measurements of AZ31 magnesium alloy performed in both types of the HBSSs.

(4) Polarization resistance R_p of AZ61 magnesium alloy in enriched HBSS+ (with Mg^{2+} and Ca^{2+} ions) was affected by surface treatment. R_p values of ground samples were higher than R_p of polished samples, after 24 h of exposure.

(5) AZ61 magnesium alloy reached higher R_p values than AZ31 magnesium alloy in both the used HBSSs which indicated better corrosion resistivity.

(6) While only phosphate-based corrosion products layer was characteristic for magnesium alloys EIS tested in HBSS, the combination of phosphate-based corrosion products layer and clusters of MgO and $Mg(OH)_2$ products were presented on the surface of specimens tested in enriched HBSS+ (with Mg^{2+} and Ca^{2+} ions).

Acknowledgments: This work was supported by the project "Materials Research Centre at FCH BUT—Sustainability and Development", REG LO1211, with financial support from National Programme for Sustainability I (Ministry of Education, Youth and Sports), Czech Republic; NETME center plus, projects of Ministry of Education, Youth and Sports of the Czech Republic under the "National sustainability program"; and by the Slovak Research and Development Agency for support in experimental by the projects No. APVV-14-0284, No. APVV-14-0772 and No. APVV-14-0096.

Author Contributions: Jakub Tkacz, Pavel Doležal and Jaromír Wasserbauer conceived and designed the experiments; Karolína Slouková and Jozef Minda performed electrochemical measurements; Jakub Tkacz and Juliána Drábiková performed SEM analysis; Pavel Doležal provided experimental materials; Jakub Tkacz and Stanislava Fintová wrote the paper.

Conflicts of Interest: The authors declare no conflict of interest.

References

1. Jung, O.; Smeets, R.; Porchetta, D.; Kopp, A.; Ptock, C.; Müller, U.; Heiland, M.; Schwade, M.; Behr, B.; Kröger, N.; et al. Optimized in vitro procedure for assessing the cytocompatibility of magnesium-based biomaterials. *Acta Biomater.* **2015**, *23*, 354–363. [CrossRef] [PubMed]

2. Farraro, K.F.; Kim, K.E.; Woo, S.L.; Flowers, J.R.; McCullough, M.B. Revolutionizing orthopedic biomaterials: The potential of biodegradable and bioresorbable magnesium-based materials for functional tissue engineering. *J. Biomech.* **2014**, *47*, 1979–1986. [CrossRef] [PubMed]

3. Mhaede, M.; Pastorek, F.; Hadzima, B. Influence of shot peening on corrosion properties of biocompatible magnesium alloy AZ31 coated by dicalcium phosphate dihydrate (DCPD). *Mater. Sci. Eng. C* **2014**, *39*, 330–335. [CrossRef] [PubMed]

4. Gu, X.N.; Zheng, Y.F. A review on magnesium alloys as biodegradable materials. *Front. Mater. Sci. China* **2010**, *4*, 111–115. [CrossRef]

5. Witte, F.; Kaese, V.; Haferkamp, H.; Switzer, E.; Meyer-Lindenberg, A.; Wirth, C.J.; Windhagen, H. In vivo corrosion of four magnesium alloys and the associated bone response. *Biomaterials* **2005**, *26*, 3557–3563. [CrossRef] [PubMed]

6. Li, W.; Guan, S.; Chen, J.; Hu, J.; Chen, S.; Wang, L.; Zhu, S. Preparation and in vitro degradation of the composite coating with high adhesion strength on biodegradable Mg-Zn-Ca alloy. *Mater. Charact.* **2011**, *62*, 1158–1165. [CrossRef]

7. Homayun, B.; Afshar, A. Microstructure, mechanical properties, corrosion behavior and cytotoxicity of Mg-Zn-Al-Ca alloys as biodegradable materials. *J. Alloys Compd.* **2014**, *607*, 1–10. [CrossRef]

8. Bakhsheshi-Rad, H.R.; Idris, M.H.; Abdul-Kadir, M.R.; Ourdjini, A.; Medraj, M.; Daroonparvar, M.; Hamzah, E. Mechanical and bio-corrosion properties of quaternary Mg-Ca-Mn-Zn alloys compared with binary Mg-Ca alloys. *Mater. Des.* **2014**, *53*, 283–292. [CrossRef]

9. Staiger, M.P.; Pietak, A.M.; Huadmai, J.; Dias, G. Magnesium and its alloys as orthopedic biomaterials: A review. *Biomaterials* **2006**, *27*, 1728–1734. [CrossRef] [PubMed]

10. Song, G. Control of biodegradation of biocompatable magnesium alloys. *Corros. Sci.* **2007**, *49*, 1696–1701. [CrossRef]

11. Salleh, E.M.; Zuhailawati, H.; Ramakrishnan, S.; Gepreel, M.A.-H. A statistical prediction of density and hardness of biodegradable mechanically alloyed Mg-Zn alloy using fractional factorial design. *J. Alloys Compd.* **2015**, *644*, 476–484. [CrossRef]

12. Plum, L.M.; Rink, L.; Haase, H. The essential toxin: Impact of zinc on human health. *Int. J. Environ. Res. Public Health* **2010**, *7*, 1342–1365. [CrossRef] [PubMed]

13. Atkins, P. *Physical Chemistry*, 8th ed.; Freeman and Company: New York, NY, USA, 2006; pp. 1–1053.

14. Gray, J.E.; Luan, B. Protective coatings on magnesium and its alloys—A critical review. *J. Alloys Compd.* **2002**, *336*, 88–113. [CrossRef]

15. Song, G.L.; Atrens, A. Corrosion mechanism of magnesium alloys. *Adv. Eng. Mater.* **1999**, *1*, 11–13. [CrossRef]

16. Ferero López, A.D.; Lehr, I.L.; Saidman, S.B. Anodisation of AZ91D magnesium alloy in molybdate solution for corrosion protection. *J. Alloys Compd.* **2017**, *702*, 338–345. [CrossRef]

17. Kannan, M.B.; Raman, R.K.S. In vitro degradation and mechanical integrity of calcium-containing magnesium alloys in modified-simulated body fluid. *Biomaterials* **2008**, *29*, 2306–2314. [CrossRef] [PubMed]

18. Xin, Y.; Hu, T.; Chu, P.K. In vitro studies of biomedical magnesium alloys in a simulated physiological environment: A review. *Acta Biomater.* **2011**, *7*, 1452–1459. [CrossRef] [PubMed]

19. Proudfoot, A.T. Aluminium and zinc phosphide poisoning. *Clin. Toxicol.* **2009**, *47*, 89–100. [CrossRef] [PubMed]

20. Seitz, J.M.; Eifler, R.; Bach, F.W.; Maier, H.J. Magnesium degradation products: Effects on tissue and human metabolism. *J. Biomed. Mater. Res. A* **2014**, *102*, 3744–3753. [CrossRef] [PubMed]

21. Adeknmbi, I.; Mosher, C.Z.; Lu, H.H.; Reihle, M.; Kubba, H.; Tanner, K.E. Mechanical behaviour of biodegradable AZ31 magnesium alloy after long term in vitro degradation. *Mater. Sci. Eng. C* **2017**, *77*, 1135–1144. [CrossRef] [PubMed]

22. Li, Q.; Jiang, G.; Wang, C.; Dong, J.; He, G. Mechanical degradation of porous titanium with entangled structure filled with biodegradable magnesium in Hanks' solution. *Mater. Sci. Eng. C* **2015**, *57*, 349–354. [CrossRef] [PubMed]

23. Brar, H.S.; Wong, J.; Manuel, M.V. Investigation of the mechanical and degradation properties of Mg-Sr and Mg-Zn-Sr alloys for use as potential biodegradable implant materials. *J. Mech. Behav. Biomed. Mater.* **2012**, *7*, 87–95. [CrossRef] [PubMed]

24. Berglund, I.S.; Brar, H.S.; Dolgova, N.; Acharya, A.P.; Keselowsky, B.G.; Sarntinoranont, M.; Manuel, M.V. Synthesis and characterization of Mg-Ca-Sr alloys for biodegradable orthopedic implant applications. *J. Biomed. Mater. Res. Part B Appl. Biomater.* **2012**, *100*, 1524–1534. [CrossRef] [PubMed]

25. Zhang, F.; Ma, A.; Song, D.; Jiang, J.; Lu, F.; Zhang, L.; Yang, D.; Chen, J. Improving in vitro biocorrosion resistance of Mg-Zn-Mn-Ca alloy in Hank's solution through addition of cerium. *J. Rare Earths* **2015**, *33*, 93–101. [CrossRef]

26. Johnston, S.; Shi, Z.; Atrens, A. The influence of pH on the corrosion rate of high-purity Mg, AZ91 and ZE41 in bicarbonate buffered Hanks' solution. *Corros. Sci.* **2015**, *101*, 182–192. [CrossRef]

27. Zainal Abidin, N.I.; Rolfe, B.; Owen, H.; Malisano, J.; Martin, D.; Hofstetter, J.; Uggowitzer, P.J.; Atrens, A. The in vivo and in vitro corrosion of high-purity magnesium and magnesium alloys WZ21 and AZ91. *Corros. Sci.* **2013**, *75*, 354–366. [CrossRef]

28. Han, G.; Lee, J.-Y.; Kim, Y.-C.; Park, J.H.; Kim, D.-I.; Han, H.-S.; Yang, S.-J.; Seok, H.-K. Preferred crystallographic pitting corrosion of pure magnesium in Hanks' solution. *Corros. Sci.* **2012**, *63*, 316–322. [CrossRef]

29. Zeng, R.-C.; Sun, L.; Zheng, Y.-F.; Cui, H.-Z.; Han, E.-H. Corrosion and characterisation of dual phase Mg-Li-Ca alloy in Hank's solution: The influence of microstructural features. *Corros. Sci.* **2014**, *79*, 69–82. [CrossRef]

30. Ng, W.F.; Chiu, K.Y.; Cheng, F.T. Effect of pH on the in vitro corrosion rate of magnesium degradable implant material. *Mater. Sci. Eng. C* **2010**, *30*, 898–903. [CrossRef]

31. Bukovinová, L.; Hadzima, B. Electrochemical characteristics of magnesium alloy AZ31 in Hank's solution. *Corros. Eng. Sci. Technol.* **2012**, *47*, 352–357. [CrossRef]

32. Tkacz, J.; Slouková, K.; Minda, J.; Drábiková, J.; Fintová, S.; Doležal, P.; Wasserbauer, J. Corrosion behavior of wrought magnesium alloys AZ31 and AZ61 in Hank's solution. *Koroze a Ochrana Materiálu* **2016**, *60*, 101–106. [CrossRef]

33. Ambat, R.; Aung, N.N.; Zhou, W. Studies on the influence of chloride ion and pH on the corrosion and electrochemical behaviour of AZ91D magnesium alloy. *J. Appl. Electrochem.* **2000**, *30*, 865–874. [CrossRef]

34. Altun, H.; Sen, S. Studies on the influence of chloride ion concentration and pH on the corrosion and electrochemical behaviour of AZ63 magnesium alloy. *Mater. Des.* **2004**, *25*, 637–643. [CrossRef]

35. Merino, M.C.; Pardo, A.; Arrabal, R.; Merino, S.; Casajus, P.; Mohedano, M. Influence of chloride ion concentration and temperature on the corrosion of Mg-Al alloys in salt fog. *Corros. Sci.* **2010**, *52*, 1696–1704. [CrossRef]

36. Shetty, S.; Nayak, J.; Shetty, N. Influence of sulfate ion concentration and pH on the corrosion of Mg-Al-Zn-Mn (GA9) magnesium alloy. *J. Magnes. Alloys* **2015**, *3*, 258–270. [CrossRef]

37. Vander Voort, G.F. *Metallography, Principles and Practice*, 1st ed.; ASM International: Materials Park, OH, USA, 1999; pp. 1–752.

38. Tkacz, J.; Minda, J.; Fintová, S.; Wasserbauer, J. Comparison of electrochemical methods for the evaluation of cast AZ91 magnesium alloy. *Materials* **2016**, *9*, 925. [CrossRef] [PubMed]

39. Liu, F.; Song, Y.; Shan, D.; Han, E. Corrosion behavior of AZ31 magnesium alloy in simulated acid rain solution. *Trans. Nonferr. Met. Soc. China* **2010**, *20*, 638–642. [CrossRef]

40. Deng, J.; Huang, G.; Zhao, Y.; Wang, B. Electrochemical performance of AZ31 magnesium alloy under different processing conditions. *Rare Met. Mater. Eng.* **2014**, *43*, 316–321.

41. *Standard Specification for Magnesium-Alloy Sheet and Plate*; ASTM B90/B90M-15; ASTM International: West Conshohocken, PA, USA, 2015. Available online: www.astm.org (accessed on 20 October 2017).

42. *Standard Specification for Magnesium-Alloy Extruded Bars, Rods, Profiles, Tubes, and Wire*; ASTM B107/B107M-13; ASTM International: West Conshohocken, PA, USA, 2013. Available online: www.astm.org (accessed on 20 October 2017).

43. Harandi, S.E.; Banerjee, P.C.; Easton, C.D.; Raman, R.K.S. Influence of bovine serum albumin in Hanks' solution on the corrosion and stress corrosion cracking of a magnesium alloy. *Mater. Sci. Eng. C* **2017**, *80*, 335–345. [CrossRef] [PubMed]

44. Zhu, Y.; Wu, G.; Zhang, Y.H.; Zhao, Q. Growth and characterization of Mg(OH)$_2$ film on magnesium alloy AZ31. *Appl. Surf. Sci.* **2011**, *257*, 6129–6137. [CrossRef]
45. Song, Y.; Shan, D.; Chen, R.; Zhang, F.; Han, E.-H. Biodegradable behaviors of AZ31 magnesium alloy in simulated body fluid. *Mater. Sci. Eng. C* **2009**, *29*, 1039–1045. [CrossRef]
46. Mathieu, S.; Rapin, C.; Steinmetz, J.; Steinmetz, P. A corrosion study of the main constituent phases of AZ91 magnesium alloys. *Corros. Sci.* **2003**, *45*, 2741–2755. [CrossRef]
47. Pardo, A.; Merino, M.C.; Coy, A.E.; Arrabal, R.; Viejo, F.; Matykina, E. Corrosion behaviour of magnesium/aluminium alloys in 3.5 wt % NaCl. *Corros. Sci.* **2008**, *50*, 823–834. [CrossRef]
48. Feliu, S.; Llorente, I. Corrosion product layers on magnesium alloys AZ31 and AZ61: Surface chemistry and protective ability. *Appl. Surf. Sci.* **2015**, *347*, 736–746. [CrossRef]
49. Walter, R.; Kannan, M.B. Influence of surface roughness on the corrosion behaviour of magnesium alloy. *Mater. Des.* **2011**, *32*, 2350–2354. [CrossRef]
50. Esmaily, M.; Svensson, J.E.; Fajardo, S.; Birbilis, N.; Frankel, G.S.; Virtanen, S.; Arrable, R.; Thomas, S.; Johansson, L.G. Fundamentals and advances in magnesium alloy corrosion. *Prog. Mater. Sci.* **2017**, *89*, 92–193. [CrossRef]
51. Zhou, M.; Yan, L.; Ling, H.; Diao, Y.; Pang, X.; Wang, Y. Design and fabrication of enhanced corrosion resistance Zn-Al layered double hydroxides films based anion-exchange mechanism on magnesium alloys. *Appl. Surf. Sci.* **2017**, *404*, 246–253. [CrossRef]
52. Tian, J.; Huang, H.L.; Pan, Z.Q.; Zhou, H. Effect of flow velocity on corrosion behavior of AZ91D magnesium alloy at elbow of loop system. *Trans. Nonferr. Met. Soc. China* **2016**, *26*, 2857–2867. [CrossRef]
53. Zhu, B.; Wang, S.; Wang, L.; Yang, Y.; Liang, J.; Cao, B. Preparation of hydroxyapatite/tannic acid coating to enhance the corrosion resistance and cytocompatibility of AZ31 magnesium alloys. *Coatings* **2017**, *7*, 105. [CrossRef]
54. Jayaraj, J.; Raj, S.A.; Srinivasan, A.; Ananthakumar, S.; Pillay, U.T.S.; Dhaipule, N.G.K.; Mudali, U.K. Composite magnesium phosphate coatings for improved corrosion resistance of magnesium AZ31 alloy. *Corros. Sci.* **2016**, *113*, 104–115. [CrossRef]
55. Yagi, S.; Sengoku, A.; Kubota, K.; Matsubara, E. Surface modification of ACM522 magnesium alloy by plasma electrolytic oxidation in phosphate electrolyte. *Corros. Sci.* **2012**, *57*, 74–80. [CrossRef]

Article

Characterization of Powder Metallurgy Processed Pure Magnesium Materials for Biomedical Applications

Matěj Březina [1,*], Jozef Minda [1], Pavel Doležal [1,2], Michaela Krystýnová [1], Stanislava Fintová [1,3], Josef Zapletal [2], Jaromír Wasserbauer [1] and Petr Ptáček [1]

[1] Materials Research Centre, Faculty of Chemistry, Brno University of Technology, Purkyňova 464/118, 61200 Brno, Czech Republic; xcminda@fch.vut.cz (J.M.); dolezal@fme.vutbr.cz (P.D.); xckrystynovam@fch.vut.cz (M.K.); fintova@ipm.cz (S.F.); wasserbauer@fch.vut.cz (J.W.); ptacek@fch.vut.cz (P.P.)

[2] Institute of Materials Science and Engineering, Faculty of Mechanical Engineering, Brno University of Technology, Technická 2896/2, 61669 Brno, Czech Republic; zapletal@fme.vutbr.cz

[3] Institute of Physics of Materials, Academy of Sciences of the Czech Republic, Žižkova 22, 61662 Brno, Czech Republic

* Correspondence: xcbrezinam@fch.vutbr.cz; Tel.: +420-54-114-9469

Received: 14 September 2017; Accepted: 24 October 2017; Published: 31 October 2017

Abstract: Magnesium with its mechanical properties and nontoxicity is predetermined as a material for biomedical applications; however, its high reactivity is a limiting factor for its usage. Powder metallurgy is one of the promising methods for the enhancement of material mechanical properties and, due to the introduced plastic deformation, can also have a positive influence on corrosion resistance. Pure magnesium samples were prepared via powder metallurgy. Compacting pressures from 100 MPa to 500 MPa were used for samples' preparation at room temperature and elevated temperatures. The microstructure of the obtained compacts was analyzed in terms of microscopy. The three-point bending test and microhardness testing were adopted to define the compacts' mechanical properties, discussing the results with respect to fractographic analysis. Electrochemical corrosion properties analyzed with electrochemical impedance spectroscopy carried out in HBSS (Hank's Balanced Salt Solution) and enriched HBSS were correlated with the metallographic analysis of the corrosion process. Cold compacted materials were very brittle with low strength (up to 50 MPa) and microhardness (up to 50 HV (load: 0.025 kg)) and degraded rapidly in both solutions. Hot pressed materials yielded much higher strength (up to 250 MPa) and microhardness (up to 65 HV (load: 0.025 kg)), and the electrochemical characteristics were significantly better when compared to the cold compacted samples. Temperatures of 300 °C and 400 °C and high compacting pressures from 300 MPa to 500 MPa had a positive influence on material bonding, mechanical and electrochemical properties. A compacting temperature of 500 °C had a detrimental effect on material compaction when using pressure above 200 MPa.

Keywords: magnesium; powder metallurgy; cold pressing; hot pressing; EIS (Electrochemical impedance spectroscopy); three-point bending test; corrosion

1. Introduction

Magnesium and its alloys are modern lightweight materials applicable in a wide range of industrial fields from aerospace and automotive to biomedical applications. Its main advantages are a good strength to weight ratio and biocompatibility in combination with biodegradability. However, due to the high reactivity of pure Mg and the mechanical properties, not really sufficient for engineering applications, mainly magnesium alloys are used [1–4].

Good mechanical properties of magnesium and its alloys can be furthermore significantly upgraded by decreasing the grain size, nowadays performed mainly via severe plastic deformation (SPD) techniques [3,5], powder metallurgy (PM) processing [1,2,6,7] or by a combination of both methods. Decreasing the metallic grain size within the material volume, either by alloying elements, SPD methods, such as extrusion, equal channel angular pressing (ECAP) and high-pressure torsion (HPT) or PM, leads to increase of hardness, tensile and yield strength, but the plasticity of the material was shown to be decreased [6–10].

PM processing of magnesium is influenced by its high affinity to oxygen, which results in the formation of a thermodynamically-stable layer of corrosion products on the magnesium powder particles' surface. The created layer dramatically inhibits the diffusion processes required for material densification during PM processing [4]. Because of the high affinity of magnesium to oxygen, a protective atmosphere (usually argon or nitrogen) has to be used for handling of magnesium powders and specimens, as well as for subsequent sintering [4].

Applying high pressures at elevated temperatures (hot pressing) to process magnesium results in high plastic deformation of powder particles. Powder particles' deformation leads to cracking of the layer of corrosion products normally present on the powder particles' surface [11–13]. The applied plastic deformation results in an increase of the contact area of powder particles and, in combination with the applied temperature, enhances the diffusion processes. Increasing compacting pressures usually also lead to a decrease in the porosity of the processed bulk material [11–13]. The porosity of the processed bulk material is usually considered as a disadvantage of the PM processing techniques. However, the porous biocompatible material can be incorporated well into the tissue and can degrade at a specific rate, which provides a tool for the tailoring properties of PM processed materials for biomedical applications [14,15]. The functional porosity of a PM processed magnesium-based implant would support the primary fixation and the degradation of the implant by enabling the ingrowth of bone cells (osteointegration) into the degrading implant. Furthermore, the corrosion products of magnesium, created during implant biodegradation, support osteoconductivity of the bone [4].

Changes in porosity also have a significant effect on corrosion resistance and corrosion attack progression within the material volume. Highly porous materials corrode very rapidly, as a larger area of the material surface is exposed to the corrosion environment. Corrosion resistance of either pure magnesium or magnesium alloys is seldom suitable for technical applications or even biomedical applications [3,9,16–20]. Magnesium corrosion resistance can be improved by alloying the metal with aluminum, zinc or rare earth metal elements; however, for significantly better corrosion resistance, another way of reducing the degradation rate must be considered. Conversion coatings are widely studied as corrosion protection for magnesium and its alloys. Fluoride and calcium phosphate-based conversion coatings have a great potential of reducing the corrosion rate of biomedical magnesium implants [1–4,12–24].

Electrochemical characterization of PM magnesium materials is commonly carried out either in NaCl solution or in Hank's Balanced Salt Solution (HBSS) or enriched HBSS [21,25–29]. Electrochemical impedance spectroscopy (EIS) measurements provide complex information of the material degradation characteristics during exposure in corrosion medium in time. From the chemical point of view, several specific reactions can occur in HBSS and enriched HBSS medium due to the additional content of several ions when compared to NaCl solution [30].

The properties of pure magnesium materials prepared via PM methods are seldom studied [15,27,31–33] or the studies refer to wrought pure magnesium materials [5]. Most of the available studies characterize the PM processed pure magnesium and magnesium-based materials in terms of mechanical and corrosion properties; however, the analyses are focused on the influence of the material porosity due to its possible use for biomedical applications in the form of the porous implants.

Porous magnesium samples with different porosity (due to the addition of an ammonium bicarbonate powder) structures and mechanical properties were analyzed in [31]. Magnesium powder samples with different porosities, due to the spacer addition, were prepared by uniaxial cold pressing

(265 MPa) with additional heat treatment (for removal of hexane and sintering at 550 °C for 6 h). In the case of pure magnesium compacts, the porosity of 12 vol % corresponded to a flexural strength of 38 MPa. The increase of the compacts' porosity was observed to be accompanied by the mechanical properties' decrease (the porosity of 38 vol % corresponded to the flexural strength of 4.4 MPa). The negative influence of the porosity was observed also in the case of material immersion into 9 g/L NaCl solution.

The same authors studied also the influence of the purity of the argon atmosphere on final PM processed magnesium-based material properties in [32], varying the sintering times as 0 h, 3 h, 6 h, 12 h and 24 h. While the samples prepared using technical Ar reached similar porosities in the sintering time range, the porosity of the samples prepared using gettered Ar was observed to decrease with increasing sintering time. Under a gettered Ar atmosphere, a prolonged sintering time enhanced diffusion connections between magnesium particles and improved the mechanical properties of the samples (flexural strength of 9 MPa for 3 h and 115.4 MPa for 24 h), whereas under a technical Ar atmosphere (flexural strength of 5 MPa for 3 h and 3.4 MPa for 24 h), oxidation at the particle surfaces caused deterioration in the mechanical properties of the samples.

Mechanical properties and corrosion resistance of cold pressed magnesium (310 MPa) and extruded at 420 °C employing an extrusion ratio of 16:1 were studied in [24]. Processed material reached the ultimate tensile strength of 320 MPa, a yield stress of 280 MPa and 2% ductility. The reached values were higher when compared to a coarse-grained cast AZ31 magnesium alloy, except material ductility. PM processed pure magnesium mechanical properties were superior to those of cast pure magnesium as provided by the authors. Silane film and anticorrosive paint were shown to enhance the corrosion behavior of PM magnesium during the first hours of immersion, but their protection effectiveness completely disappears after two days. For longer immersion times, the fluoride conversion coating prepared in HF solution was shown to be an effective barrier to protect PM magnesium from degradation.

Pre-rolling of magnesium powder prior to spark plasma sintering and hot extrusion was used for material grain refinement in [6]. The grain size obtained for the material extruded from the powders after 5 and 10 rolling pre-treatments decreased from 9.2 μm (only extruded Mg) to 2.9 μm and 2.1 μm, respectively. The grain refinement resulted in an ultimate tensile strength increase from 242 MPa for extruded pure magnesium to a value of 270.6 MPa for material extruded after sintering of 10 times pre-rolled powder. Yield stress was improved from 170 MPa to 206.4 MPa by the pre-rolling of powder.

The aim of this study is to evaluate the microstructural, mechanical and electrochemical properties of pure magnesium-based PM processed materials when varying the processing parameters. Materials were prepared via hot pressing at temperatures of 300 °C, 400 °C and 500 °C applying pressures of 100 MPa, 200 MPa, 300 MPa, 400 MPa and 500 MPa. For comparison, cold compacted materials prepared under the same compaction pressures were analyzed in the same manner. The mechanical properties of compacts were analyzed in terms of the three-point bending test containing fractographic analysis and with microhardness testing. Electrochemical corrosion characteristics of the processed materials in HBSS and enriched HBSS were analyzed using electrochemical impedance spectroscopy, and the data were extended by metallographic analysis of the corrosion attack within the material volume.

2. Materials and Methods

Magnesium powder used in this study (Figure 1) was irregularly shaped with an average particle size of approximately 30 μm. The purity of the base material was 99.8% as declared by supplier Goodfellow (Huntingdon, UK); however, an oxide layer was found on the surface of powder particles using the scanning electron microscope (SEM, ZEISS EVO LS 10, Carl Zeiss Ltd., Cambridge, UK) using energy dispersive spectroscopy (EDS, Oxford Instruments plc, Abingdon, UK). This layer is to be expected on the surface of magnesium, and it was probably presented on the particles from the powder preparation.

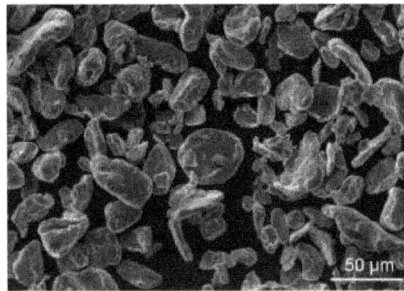

Figure 1. Magnesium powder SEM (scanning electron microscope): powder particles' morphology and size distribution.

For metallographic and electrochemical analysis, cylindrical compacts of 5 mm in height and 20 mm in diameter were prepared using a steel die. The preparation of the magnesium powder into the steel die for compaction was carried out under a nitrogen atmosphere to avoid further oxygen contamination. Magnesium powder (2.7 g) inserted into the steel die was compacted into compacts applying different uniaxial pressures; 100 MPa, 200 MPa, 300 MPa, 400 MPa and 500 MPa. Compaction was carried out using the Zwick Z250 Allround-Line machine (Zwick GmbH&Co. KG in Germany, Ulm, Germany) at room temperature and at elevated temperatures of 300 °C, 400 °C and 500 °C for one hour.

The microstructure of prepared samples was studied on cross-sectional cuts using SEM (ZEISS EVO LS 10) and a light optical microscope (LM, Zeiss Axio Observer Z1m, Carl Zeiss AG, Oberkochen, Germany). The internal structure of the compacts and the created grain size were analyzed with the electron backscattered diffraction (EBSD) technique (SEM, Tescan LYRA 3 XMU FEG/SEMxFIB, TESCAN Brno, s.r.o., Brno, Czech Republic). The step size of the method was 0.3 μm. Samples' cross-sections intended for metallographic analysis were mold into a polymeric resin at room temperature, ground and polished according to the standard procedures of metallographic samples' preparation. To reveal the microstructure, a 5% Nital etchant (5% nitric acid in ethanol) was used for several seconds. Compacts' porosity was estimated from calculated density (calculated from sample dimensions and weight) considering a density of cast pure magnesium as the 0 porosity standard.

Microhardness testing was conducted on cross-sections of prepared compacts using the LECO AMH LM 248 microhardness tester (LECO Corporation, Saint Joseph, MI, USA) according to the ISO 6507-1 standard. For all the measurement, 25 g of load and a 10-s dwell time were used. The three-point bending test was carried out on samples with dimensions of 4 mm in height, 4 mm in width and 18 mm in length cut from the compacts, according to the ISO 7438 standard. The test machine used for the three-point bending test was the Zwick Z020 (Zwick GmbH&Co. KG in Germany, Ulm, Germany); the radius of the supports and the point was 2.5 mm; the distance between supports was 16 mm; and the velocity of applied loading was 1 mm/min. The three-point bending test was performed on one sample for each material preparation method.

Potentiostatic electrochemical impedance spectroscopy (EIS) was used as a method for the characterization of the prepared magnesium samples' electrochemical corrosion characteristics. The three electrode cell system was used for the measurement with the Pt electrode as the counter electrode and the calomel electrode as the reference electrode, and the sample (1 cm^2 exposed area) served as the working electrode. With the aim to remove a layer of corrosion products created on the compacts' surface, each sample was ground using 4000 SiC paper and rinsed with isopropanol just before the measurement. Measurements were carried out in HBSS and enriched HBSS (containing Mg^{2+} and Ca^{2+} ions). The composition of the used solutions is given in Table 1. The frequency used for the measurements was in the range from 100 kHz to 10 MHz, with the signal amplitude of 10 mV. All the measurements were carried out at laboratory temperature. EIS data were obtained after 5 min, 1 h, 2 h, 4 h, 8 h, 12 h, 24 h, 48 h, 72 h and 96 h of immersion. Each measurement time was 25.8 min. Due to the open porosity of the cold pressed samples, only cold pressed samples prepared under 500 MPa compacting pressure were used for the electrochemical corrosion characterization. In the case of hot pressing, samples prepared under 100 MPa and 500 MPa prepared at 400 °C were used for EIS measurements with the aim to reveal differences in material electrochemical corrosion behavior according to the processing conditions. Each material type was characterized by measurements performed on three samples.

Table 1. Composition of the HBSS (Hank's Balanced Salt Solution) and enriched HBSS (supplied by GE (General Electric) Healthcare) used.

Component	Concentration (mg·dm^{-3})	
	HBSS	**Enriched HBSS**
NaCl	8000	8000
KCl	400	400
KH_2PO_4	60	60
Glucose	1000	1000
Na_2HPO_4	48	48
$MgSO_4$	-	98
$CaCl_2$	-	140
Na_2CO_3	350	350

3. Results

3.1. Microstructure Characterization

The microstructure of the selected samples of pure magnesium compacts and details showing characteristic features of the material are presented in Figure 2. Microstructural analysis revealed that with increasing compaction pressure, the deformation of the powder particle increased and the porosity decreased. This trend was apparent in all the prepared samples compacted at elevated and room temperature.

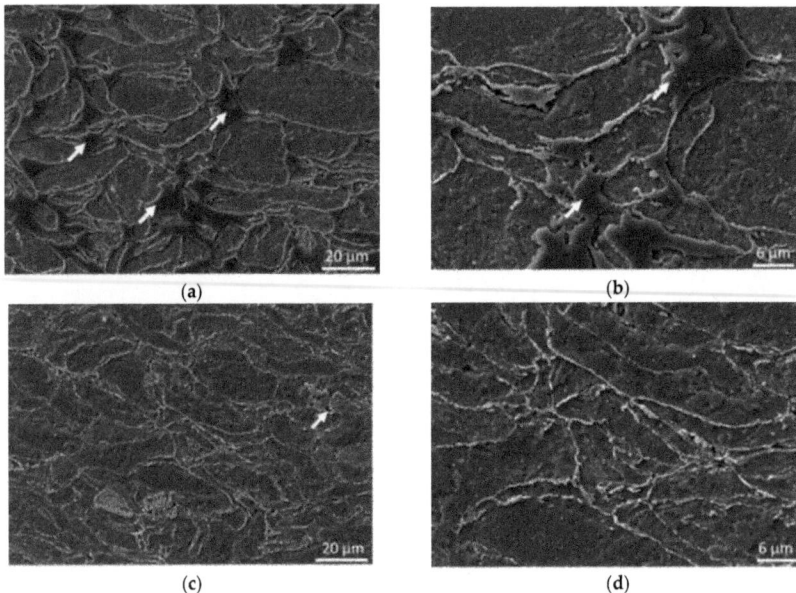

(a)

(b)

(c)

(d)

Figure 2. *Cont.*

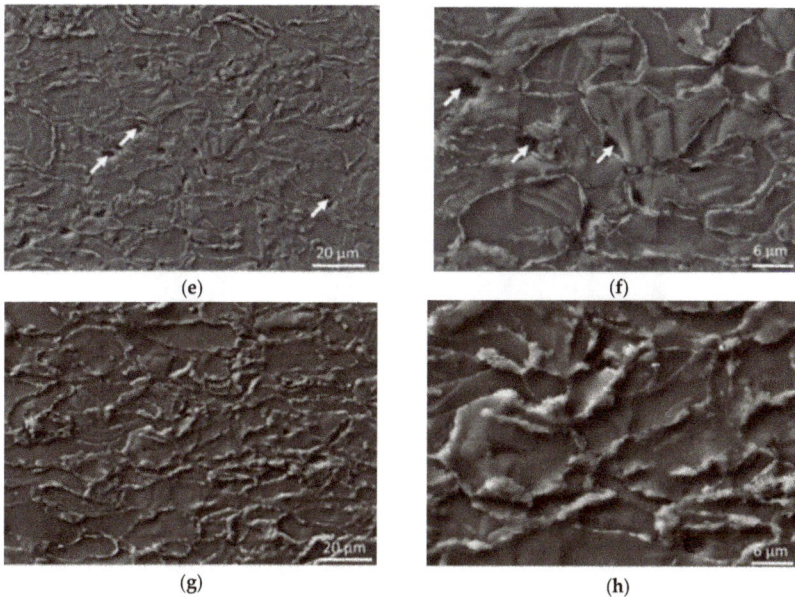

Figure 2. Microstructure of selected samples: (**a,b**) cold compacted at 100 MPa; (**c,d**) cold compacted at 500 MPa; (**e,f**) hot pressed at 100 MPa, 400 °C; (**g,h**) hot pressed at 500 MPa, 400 °C; pores are indicated with arrows.

A significant change in porosity is visible when cold pressed (Figure 2a,b) and hot pressed materials (Figure 2c,d) prepared under compacting pressure of 100 MPa are compared. In cold compacted material, the porosity is open, and metallographic resin used for the sample molding is present in the microstructure (arrows in Figure 2a,b). The hot pressed sample contained also some pores in the microstructure (Figure 2e); however, they were much smaller when compared to the cold pressed sample (Figure 2a). Only the oxide layer present on the powder particles' surface can be seen in Figure 2f; no metallographic resin is observed in the pores. Hot pressing under 500 MPa at 400 °C led to a further decrease in porosity (Figure 2g,h).

Applying higher pressures during sample preparation led to a significant decrease of porosity in the cold compacted samples, as well as in the hot pressed samples. The level of deformation of the powder particles increased with increasing the compacting pressure. However, the change was observed to be increasing from 100 MPa up to 300 MPa compacting pressures, while the shape and the level of deformation of the powder particles in the case of cold and hot pressed samples did not change for compacting pressures from 300 MPa to 500 MPa. The size of metallic grains within the powder particles was impossible to calculate using light microscopy (Figure 2). Electron backscattered diffraction was therefore applied to reveal the basic metallic structure of the samples (Figure 3).

The main difference in the microstructure between cold and hot pressed samples applying higher pressures than 300 MPa was revealed by EBSD. In Figure 3, EBSD maps of cold compacted and hot pressed (400 °C, 1 h) samples prepared under 400 MPa are shown. In the case of the cold compacted sample, very fine grains were created in the powder particle during the material compaction (Figure 3a,b). However, the material on the powder particles' interface was not identified by EBSD, which indicates that the powder particles are not diffusion bonded. Larger grains were created in the magnesium powder particles compacted at 400 °C (Figure 3c,d). The observed metallic grains are connected to the metallic grains created in the neighboring powder particles, which indicates the good connection of the material structure and promotes powder particles' bonding.

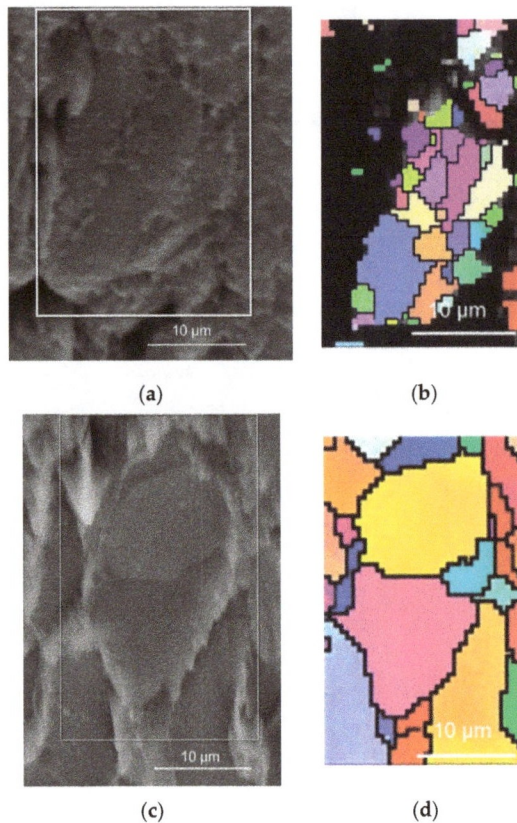

Figure 3. Electron backscattered diffraction (EBSD) analysis of cold compacted and hot pressed material: (**a**) SEM image and (**b**) EBSD map of the cold compacted sample prepared under 400 MPa; (**c**) SEM image and (**d**) EBSD map of the hot pressed sample prepared under 400 MPa at 400 °C. Colored-metallic grains of magnesium: dark—grain boundary (lines), pores or oxides (wide gaps).

3.2. Mechanical Characterization

3.2.1. Three-Point Bending Test and Microhardness Test Results

Three-point bending test revealed an increasing trend of flexural strength of prepared materials corresponding to the increasing samples' compacting pressure (Figure 4a). The highest flexural strength was obtained for hot pressed samples compacted at 400 °C. Samples prepared at 300 °C followed the same trend as the samples prepared at 400 °C; however, the flexural strength was lower (Figure 4a). Hot pressing magnesium powder at 500 °C resulted in lower values of flexural strength compared to lower compaction temperatures (except RT (room temperature)). Furthermore, during the preparation of the samples pressed under 400 MPa at 500 °C, deformation of the steel mold (used for sample preparation) occurred; therefore, the samples' preparation was not possible. The cold pressed samples' test results show significantly lower values of flexural strength compared to the hot pressed samples. Except the flexural strength measured on hot pressed samples prepared under 500 MPa, the value of flexural strength was observed to increase with increasing compaction pressure.

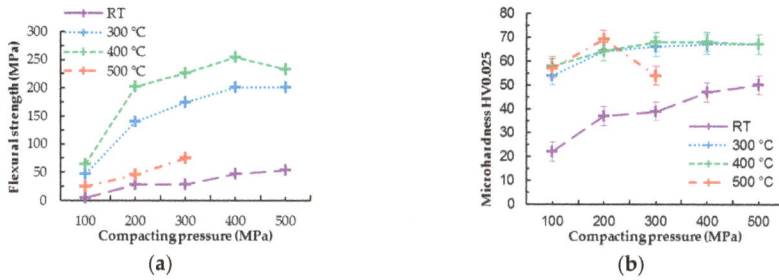

Figure 4. Three-point bending tests and microhardness measurement results; (**a**) flexural strength, (**b**) microhardness reuslts (RT: room temperature).

Microhardness test results revealed a significant difference between cold and hot pressed samples (Figure 4b). The increase of the microhardness corresponding to the increasing compaction pressure was observed for cold compacted and hot pressed samples prepared at 300 °C and 400 °C (Figure 4b), while the identical trend of microhardness evolution was observed for both hot pressed materials. The influence of compaction pressure on the microhardness was more pronounced for cold compacted samples, however, higher values of microhardness were characteristic for hot pressed samples. A specific situation was observed in the case of hot pressed samples prepared at 500 °C, where the primary microhardness increase observed when increasing compaction pressure from 100 MPa to 200 MPa was followed by a decrease of the microhardness for samples prepared under 300 MPa (Figure 4b).

3.2.2. Fractographic Analysis

Fracture surfaces of samples broken during the three-point bending test were analyzed in the region where tensile loading was applied. The fractographic analysis revealed differences between cold compacted and hot pressed materials (Figure 5).

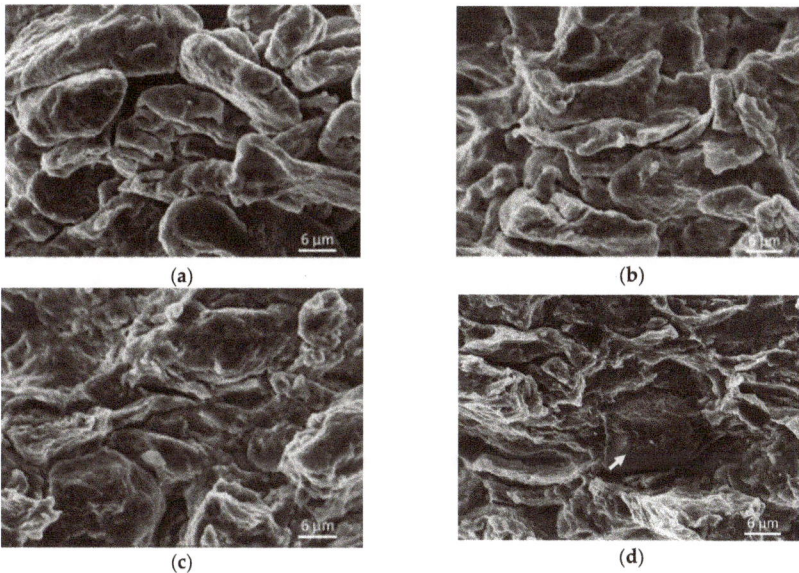

Figure 5. Selected fracture surfaces of samples' failure during three-point bending test: (**a**) cold compacted 100 MPa; (**b**) cold compacted 500 MPa; (**c**) hot pressed 100 MPa at 400 °C; (**d**) hot pressed 500 MPa at 400 °C.

Cold compacted materials' fracture surfaces exhibited only trans-granular failure regardless of the applied compaction pressure. Different applied compacting pressure resulted only in differences of deformation of the powder particles present on the fracture surface (Figure 5a,b). Samples compacted under 100 MPa revealed only minimal plastic deformation of the powder particles, and the fracture surface exhibited a quite large amount of secondary cracks following magnesium powder particles' boundaries (Figure 5a). Samples compacted under higher pressures (300 MPa and higher) exhibited significant plastic deformation of powder particles and a smaller amount of secondary cracks compared to the low pressure compacted samples, as was observed on the samples' fracture surfaces (Figure 5b).

Hot pressed samples contained significantly deformed powder particles, and only trans-granular failure was observed for samples compacted under lower pressures (100 MPa and 200 MPa) (Figure 5c). With the increase of the compaction pressure in the case of hot pressed samples, also the inter-granular failure of the powder particles was observed on the samples' fracture surfaces (Figure 5d, marked with an arrow).

3.3. Electrochemical Characterization

The corrosion resistance of the studied materials was evaluated by EIS. As the corrosion medium, HBSS and enriched HBSS (containing Mg^{2+} and Ca^{2+} ions) were used (the composition is given in Table 1). Based on the obtained data represented by Nyquist plots, three equivalent circuits (EC) shown in Figure 6 were used for the data analysis and determination of materials' electrochemical corrosion characteristics.

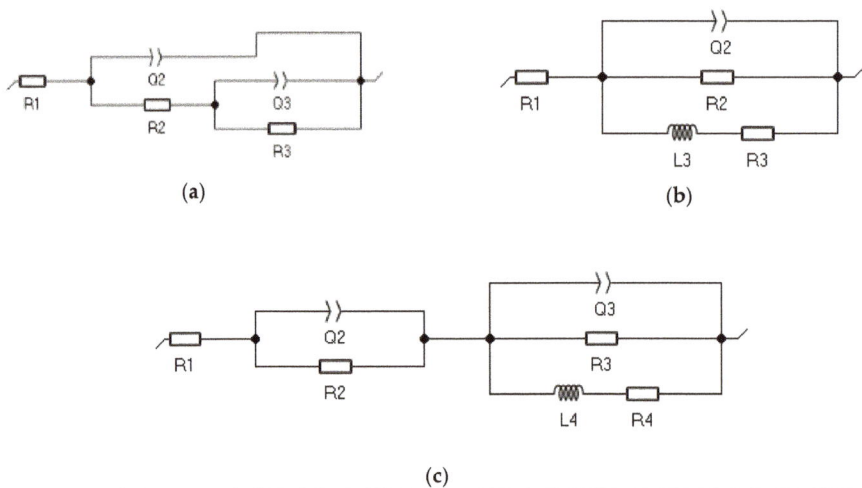

Figure 6. (a–c) Equivalent circuits used for evaluation of the Nyquist plots.

The EC given in Figure 6a was used for the description of the behavior of materials exhibiting two capacitive loops on the Nyquist plot. One capacitance loop was obtained for high frequencies and the second capacitance loop for low frequencies. Two capacitance loops characterize the response of the material by the creation of a layer of corrosion products on its surface. The EC in the Figure 6a can be from the electrochemical point of view explained as the simple electrical response of the damaged (porous and cracked) layer of corrosion products created on the metal surface to the corrosion environment (partially blocked electrode [33]). Damage of the layer can be understood for example as cracks in the layer created due to its porous nature and thickness increasing due to the material response to the corrosion environment, while some parts of the damaged layer can be removed from the material surface, and the metallic surface can be revealed to the corrosion environment again. EC presented in Figure 6a consists of elements representing solution resistance R_1; Q_2 is a constant phase element (CPE), which is a component modeling the behavior of the present corrosion products (taking into account its porous origin and its damage); R_2 represents the resistance of the corrosion products' layer against the solution including the resistance of the corrosion products and resistance of the material against the solution penetrating through the corrosion products' layer damaged areas and pores; Q_3 is a constant phase element (CPE), which is a component modeling the behavior of the interface between the solution and metal surface including the influence of the damaged layer of corrosion products, and R_3 is the resistance of the material against the solution (charge transfer in the double layer).

The resulting polarization resistance R_p of this model can be calculated according to the Equation (1):

$$R_p = R_2 + R_3 \tag{1}$$

EC including an inductive element is shown in Figure 6b. The model was used for the evaluation of the Nyquist plots containing one high-frequency capacitance loop and one low-frequency inductive loop. The element R_1 represents solution resistance; Q_2 represents double layer capacitance expressed by the CPE element; R_2 is the resistance against the charge transfer of the solution and the created layer of corrosion products; L_3 is inductance connected with the negative difference effect (NDE) caused in the case of magnesium by adsorption of H^+, Mg^+ [30]; and R_3 is the resistance of this inductive element and also the resistance of charge transfer on the sample surface. The resulting polarization resistance R_p value of this model is given by Equation (2):

$$\frac{1}{R_p} = \frac{1}{R_2} + \frac{1}{R_3} \tag{2}$$

In the case of Nyquist plots consisting of the high-frequency capacitance loop, medium frequency capacitance loop and low-frequency inductive loop, an EC shown in the Figure 6c was used for the data analysis. In this case, the creation of the layer of corrosion products (porous in its nature) on the material surface accompanied with additional creation of the secondary layer (outer) of corrosion products on its surface closing the pores in the primary (inner) layer is assumed. The element R_1 represents the solution resistance; Q_2 can be connected with the outer layer of corrosion products; R_2 is the resistance against charge transfer of the outer layer; Q_3 represents the capacitance between the outer and inner porous layer of corrosion products; R_3 is the resistance against charge transfer between the outer and inner porous layers of corrosion products; L_4 is inductance equivalent to the response of the adsorbate species on the sample surface due to the NDE; and R_4 is resistance against the charge transfer of interfacial reaction on the sample surface (resistance caused by adsorbed species on the metal surface). The final R_p in the model is given by Equation (3):

$$R_p = R_2 + \frac{R_3 \cdot R_4}{R_3 + R_4} \tag{3}$$

3.3.1. EIS Measured in HBSS

The EIS spectra for cold compacted samples measured in HBSS medium in the form of Nyquist plots are shown in Figure 7a, and data obtained by fitting the plots are given in Table 2. The plots contain three loops including an inductive one. Based on the obtained plots' character assuming the same material response to HBSS in time, the EC model shown in Figure 6c was used for the electrochemical corrosion characteristics' determination. Measurement of the cold compacted sample was stopped after 12 h due to the fatal corrosion damage of the samples. The very low capacitance response between the layer of corrosion products and compacts was observed at the beginning of the exposure of the samples to the corrosion environment. This could be explained by difficulties in the creation of a coherent layer of corrosion products on the sample's surface due to the high porosity of the cold compacted material. The values of R_p were only slowly increasing up to the end of the EIS measurements. The initial R_p value was determined as 109 $\Omega \cdot cm^2$. An increase of R_p was observed with increasing the immersion time to 2 h; however, this increase was followed by a drop of R_p to a value comparable with the initial polarization resistance. This drop was followed by a slow increase of the polarization resistance of cold compacted samples up to 252 $\Omega \cdot cm^2$ obtained after 12 h. Changing values of n_2 and n_3, characterizing the stability of the created outer and inner layer of corrosion products on the sample surface indicate the non-uniformity of the layer. At the beginning of the measurement, the inner layer of corrosion products exhibited higher stability, and its higher contribution to the protectivity of the samples was changed in immersion times of 2 h and 4 h when the outer layer with higher stability could protect the material against the following corrosion attack. However, exceeding 4 h of exposure of the material to HBSS, the created layers exhibited only very low stability (expressed by coefficients n_2 and n_3) (Table 2).

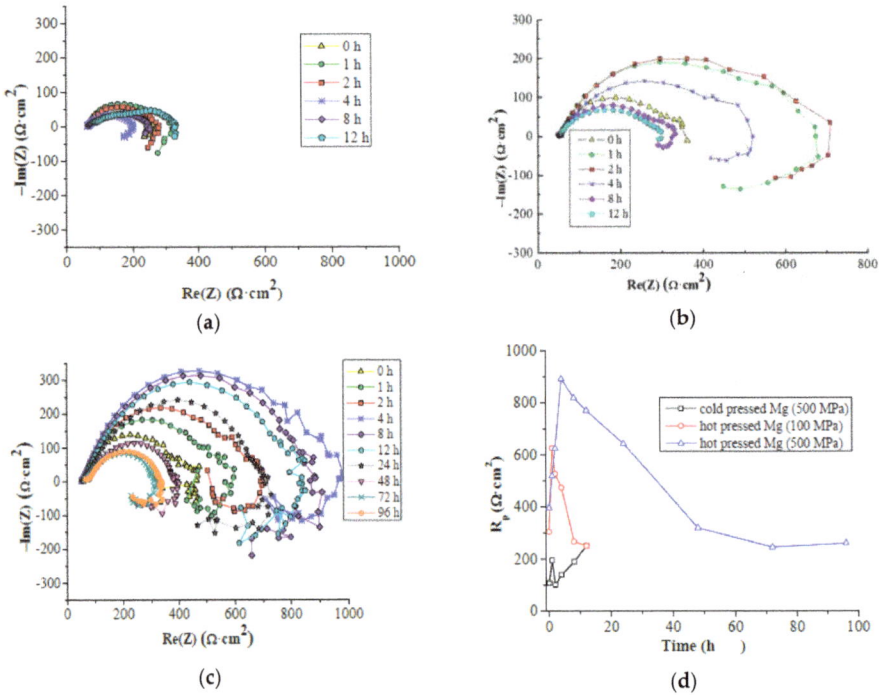

Figure 7. Electrochemical behavior of powder metallurgy (PM) magnesium materials in HBSS: (**a**) cold pressed 500 MPa; (**b**) hot pressed 100 MPa at 400 °C; (**c**) hot pressed 500 MPa at 400 °C; (**d**) comparison of R_p values in time.

Table 2. Fitted EIS (electrochemical impedance spectroscopy) data for the cold compacted sample prepared under 500 MPa, HBSS.

Time (h)	R_1 ($\Omega \cdot cm^2$)	R_2 ($\Omega \cdot cm^2$)	R_3 ($\Omega \cdot cm^2$)	R_4 ($\Omega \cdot cm^2$)	R_p ($\Omega \cdot cm^2$)	Q_2 ($F \cdot s^{n-1} \cdot 10^{-6}$)	Q_3 ($F \cdot s^{n-1} \cdot 10^{-6}$)	n_2	n_3	L_4 (H)
0	65	130	64	−16	109	0.50	72.50	0.61	0.97	390
1	63	244	0	0	196	279.00	715.00	0.67	0.60	0
2	66	51	161	74	101	284	0.17	0.91	0.74	2
4	63	54	145	219	141	22.16	0.00	1.00	0.54	1
8	60	190	−1	−8	190	520.00	1.58	0.47	0.01	42
12	77	189	95	189	252	221	5.72	0.67	0.56	1

60

Metals **2017**, *7*, 461

The characteristic Nyquist plots for the hot pressed sample prepared under 100 MPa obtained with EIS in HBSS are shown in Figure 7b, and the data obtained after fitting of these curves applying adequate EC are shown in the Table 3. The initial and the last measured electrochemical response of the sample tested in HBSS (measurements at 0 and 12 h) correspond to behavior that can be described by the EC model given in Figure 6a. In other cases, three loops were characteristic for Nyquist plots, and the compacts' behavior can be described with EC in Figure 6c. This type of electrical response includes also an inductive loop. The hot pressed 100 MPa powder compacts were measurable only within 12 h due to the high degradation of the material in HBSS. The R_p values rose from the beginning of the experiment (305 $\Omega \cdot cm^2$) to the maximum value (627 $\Omega \cdot cm^2$) obtained after 1 h of exposure of the material to HBSS. After 1 h of exposure, the R_p continuously decreased until the end of the experiment (251 $\Omega \cdot cm^2$). The value of n_2 characterizing the stability of the outer layer of corrosion products decreased with increasing the time of exposure of the compacts to the corrosion environment. Nevertheless, the value of n_3 characterizing the inner porous layer of corrosion products (created in the areas of contact of the corrosion solution with the damaged primary created layer of corrosion products) was changing in the whole experiment time, indicating its reaction with the compacted material. However, the primary created layer exhibited good stability and could exhibit the protection of the material in the corrosion in times of 0 h, 4 h and 12 h of exposure (Table 3). However, the decreasing n_3 values indicate the primary layer damage. Changing values of n_2 and n_3 characterize the stability of layers of corrosion products created on the material surface.

Table 3. Fitted EIS data for the hot pressed sample prepared under 100 MPa at 400 °C, HBSS.

Time (h)	R_1 ($\Omega \cdot cm^2$)	R_2 ($\Omega \cdot cm^2$)	R_3 ($\Omega \cdot cm^2$)	R_4 ($\Omega \cdot cm^2$)	R_p ($\Omega \cdot cm^2$)	Q_2 (F·s^{n-1}·10^{-6})	Q_3 (F·s^{n-1}·10^{-6})	n_2	n_3	L_4 (H)
0	54	263	43	-	305	73.9	2676.0	0.83	1.00	-
1	50	627	-1	-1	627	132.1	8887.0	0.74	0.69	0
2	49	655	11	-10	527	136.4	7759.0	0.74	0.39	4
4	51	474	3969	0	474	176.8	6082.0	0.70	0.99	136
8	48	270	162	-2	268	206.9	5617.0	0.68	0.03	2
12	53	-3	254	-	251	262.6	26.6	0.55	0.86	-

The Nyquist curves for hot pressed samples prepared under 500 MPa are shown in Figure 7c, and the data obtained after fitting of the curves are shown in Table 4. Only one type of electrochemical response was identified for all the measurements of the samples. The suitable EC model for material behavior analysis is shown in Figure 6b. The EC in Figure 6b is composed of two time-independent loops and contains one inductive loop. The inductive electrical response of the material to the HBSS was clear for the whole experiment time. The material polarization resistance R_p increased from the initial (396 $\Omega \cdot cm^2$) measurement and reached the maximum value (892 $\Omega \cdot cm^2$) at the measurement at 4 h of exposure. This initial increase of R_p was followed by a gradual decrease until 72 h of exposure (246 $\Omega \cdot cm^2$). The last R_p value after 96 h of exposure was 261 $\Omega \cdot cm^2$. The value of n_2, in Table 4, indicates similar stability of the created layer of corrosion products on the material surface in the whole exposure time.

The comparison of the evolution of R_p of the studied samples in HBSS in time is shown in Figure 7d. In all the cases, an initial increase of R_p was observed; however, in the case of cold compacted and hot pressed material prepared under 100 MPa, this increase was observed only until 1 h of exposure. Cold pressed samples reached lower values of polarization resistance when compared to the hot pressed samples. The primary increase of R_p was followed by a decrease for 2 h of exposure, and the following increase was observed for cold compacted samples. The value of polarization resistance for hot pressed samples prepared under 100 MPa was after the maximum reached after 1 h of exposure only decreasing. A similar trend was characteristic also for the hot pressed samples prepared under 500 MPa; however, in this case, the maximal value of R_p was reached after 4 h of exposure. All the samples reached similar values of polarization resistance (before samples' fatal degradation) at the maximal time of exposure.

Table 4. Fitted EIS data for the hot pressed sample prepared under 500 MPa at 400 °C, HBSS.

Time (h)	R_1 ($\Omega\cdot cm^2$)	R_2 ($\Omega\cdot cm^2$)	R_3 ($\Omega\cdot cm^2$)	R_4 ($\Omega\cdot cm^2$)	R_p ($\Omega\cdot cm^2$)	Q_2 ($F\cdot s^{n-1}\cdot 10^{-6}$)	Q_3 ($F\cdot s^{n-1}\cdot 10^{-6}$)	n_2	n_3	L_3 (H)
0	54	289	107	-	396	62.1	-	0.79	-	1005
1	55	354.5	164	-	519	62.2	-	0.79	-	762
2	56	456.5	169	-	626	58.0	-	0.79	-	540
4	61	647	245	-	892	40.4	-	0.81	-	733
8	65	512	307	-	819	31.6	-	0.82	-	2067
12	66	487	283.5	-	771	30.3	-	0.81	-	2081
24	69	368	276	-	644	31.6	-	0.79	-	1725
48	74	164	156	-	320	48.9	-	0.76	-	796
72	72	114	132	-	246	69.0	-	0.74	-	703
96	78	130	131	-	261	69.0	-	0.75	-	744

3.3.2. EIS Measured in Enriched HBSS

The electrochemical corrosion tests were also performed on the prepared samples in the enriched HBSS with Mg^{2+} and Ca^{2+}.

The EC in Figure 6b with three loops was selected for analysis of the Nyquist plots obtained for cold pressed samples prepared under 500 MPa up to 12 h of exposure to enriched HBSS. The Nyquist plot characterizing material behavior in 24 h of exposure to the corrosion environment had a different character, and the EC for the partially blocked electrode shown in Figure 6a was used for the data determination. After 24 h of exposure of the samples to the corrosion environment, substantial corrosion damage occurred. The determined data are shown in the Table 5, and the obtained Nyquist plots are shown in Figure 8a. The R_p obtained at the beginning of the measurement (473 $\Omega\cdot cm^2$) was followed by its decrease to a value of 104 $\Omega\cdot cm^2$ obtained after 8 h of exposure. The value of polarization resistance started to increase again, and the value of 530 $\Omega\cdot cm^2$ was determined at the end of the experiment (24 h). The stability of the layer of corrosion products created on the samples' surface is characterized by the values of n_2 and n_3. Based on the Nyquist plots' character (EC in Figure 6b and EC in Figure 6a), the stability of both layers is changing in time, which indicates their changing reactivity and degradation in enriched HBSS in time.

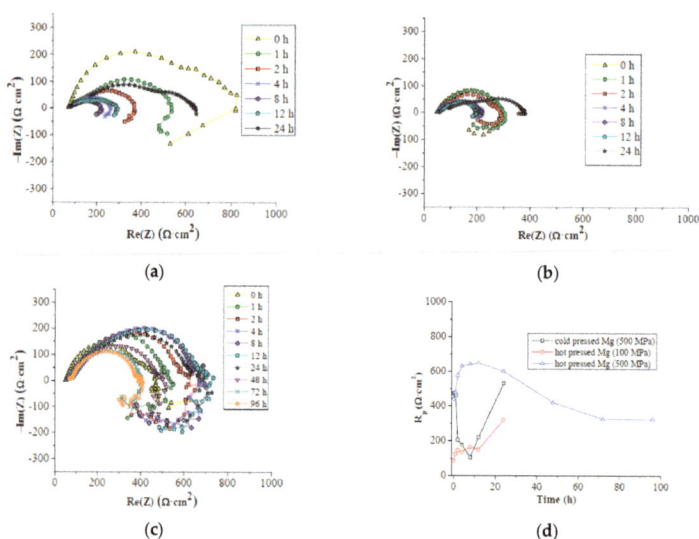

Figure 8. Electrochemical behavior of PM magnesium materials in enriched HBSS: (**a**) cold pressed 500 MPa; (**b**) hot pressed 100 MPa 400 °C; (**c**) hot pressed 500 MPa 400 °C; (**d**) comparison of R_p values in time.

Table 5. Fitted EIS data for the cold compacted sample prepared under 500 MPa, enriched HBSS.

Time (h)	R_1 ($\Omega \cdot cm^2$)	R_2 ($\Omega \cdot cm^2$)	R_3 ($\Omega \cdot cm^2$)	R_4 ($\Omega \cdot cm^2$)	R_p ($\Omega \cdot cm^2$)	Q_2 ($F \cdot s^{n-1} \cdot 10^{-6}$)	Q_3 ($F \cdot s^{n-1} \cdot 10^{-6}$)	n_2	n_3	L (H)
0	56	703	55	−45	473	63.6	126.3	0.72	0.10	71
1	67	327	376	204	459	105.3	95.9	0.71	0.47	0
2	60	208	129	−3	205	689.8	151.1	0.28	0.80	996
4	77	28	156	2866	176	24.3	285.4	0.65	0.63	0
8	76	147	6	−5	104	644.1	276.0	0.44	0.95	2
12	71	137	115	291	219	135.4	915.5	0.43	0.60	0
24	80	423	107	-	530	28.6	652.6	0.49	0.68	-

The Nyquist plots characterizing the behavior of hot pressed samples prepared under 100 MPa at 400 °C are presented in Figure 8b. Based on the character of the obtained plots, the EC given in Figure 6b (up to 4 h of exposure) and Figure 6a (8 to 24 h of exposure) were used for the data determination. The measurement was performed for 24 h because of the high level of sample corrosion degradation after this time. The initial R_p value was determined as 83 $\Omega \cdot cm^2$ by fitting analysis. The initial state was followed by a slight increase and decrease of R_p values until the end of the measurement when the maximal value of R_p was determined as 319 $\Omega \cdot cm^2$. The stability of the layer of corrosion products (expressed with the n_2 value given in Table 6) created on the sample's surface was decreasing with increasing exposure time; however, after 8 h of exposure, the layer became porous, and some reactions took place also in the pores where the corrosion environment reached the surface of the samples.

Table 6. Fitted EIS data for the hot pressed sample prepared under 100 MPa at 400 °C, enriched HBSS.

Time (h)	R_1 ($\Omega \cdot cm^2$)	R_2 ($\Omega \cdot cm^2$)	R_3 ($\Omega \cdot cm^2$)	R_4 ($\Omega \cdot cm^2$)	R_p ($\Omega \cdot cm^2$)	Q_2 ($F \cdot s^{n-1} \cdot 10^{-6}$)	Q_3 ($F \cdot s^{n-1} \cdot 10^{-6}$)	n_2	n_3	L (H)
0	50	243	125	-	83	63.9	-	0.75	-	2799
1	52	255	231	-	121	109.8	-	0.69	-	2368
2	52	240	360	-	144	220.2	-	0.60	-	2693
4	51	176	539	-	133	366.5	-	0.53	-	3512
8	56	122	41	-	163	191.6	91.4	0.58	0.92	-
12	54	25	123	-	147	72.4	190.1	0.61	0.65	-
24	70	219	100	-	319	63.9	171.1	0.41	0.77	-

The EC given in Figure 6a was used for the fitting of the Nyquist plots characterizing the behavior of hot pressed samples prepared under 500 MPa at 400 °C at the beginning of the exposure (0 h) (Figure 8c). Increasing the exposure time, the character of the Nyquist plots changed, and the EC shown in Figure 6b was used for the plots' fitting. Except the first measurement, there is a clear inductive response in all the Nyquist plots. The obtained electrochemical corrosion characteristics are given in Table 7. The R_p increases from the initial (436 $\Omega \cdot cm^2$) measurement and reached the maximum value (647 $\Omega \cdot cm^2$) at 12 h of measurement. This increase of R_p was followed by a continual decrease until the end of the experiment at 96 h of exposure (321 $\Omega \cdot cm^2$). The stability of the layer of corrosion products created on the samples' surface characterized with the n_2 value slightly decreased after 2 h of exposure to the enriched HBSS, however, remained stable, and its stability started to slightly increase after 48 h of exposure (Table 7).

R_p evolution in enriched HBSS in time determined for magnesium PM processed samples is shown in Figure 8d. The value of R_p determined for the cold compacts at the beginning of the exposure decreased with time of exposure up to 8 h, while the following exposure resulted in the R_p increase to a value slightly higher than the primary value. In the case of hot pressed samples prepared under 100 MPa at 400 °C, a slight increase of the R_p from the beginning of the exposure up to 12 h was observed, while additional exposure to the enriched HBSS was connected with the following increase of R_p. In the case of hot pressed samples prepared under 500 MPa at 400 °C, an increase of R_p up to 12 h of exposure was observed, while additional exposure resulted in a polarization resistance decrease to values below the R_p measured at the beginning of the exposure. However, the values of R_p determined for hot pressed samples prepared under 500 MPa were higher compared to other material states.

Table 7. Fitted EIS data for the hot pressed sample prepared under 500 MPa at 400 °C, enriched HBSS.

Time (h)	R_1 ($\Omega \cdot cm^2$)	R_2 ($\Omega \cdot cm^2$)	R_3 ($\Omega \cdot cm^2$)	R_4 ($\Omega \cdot cm^2$)	R_p ($\Omega \cdot cm^2$)	Q_2 ($F \cdot s^{n-1} \cdot 10^{-6}$)	Q_3 ($F \cdot s^{n-1} \cdot 10^{-6}$)	n_2	n_3	L (H)
0	53	332	104	-	436	32.3	2832.0	0.85	0.49	-
1	64	262	215	-	476	45.2	-	0.74	-	1526
2	65	261	314	-	575	54.6	-	0.68	-	2367
4	67	302	326	-	629	53.8	-	0.66	-	2520
8	72	288	352	-	640	53.6	-	0.65	-	2268
12	72	260	387	-	647	53.5	-	0.65	-	3215
24	84	300	300	-	600	52.8	-	0.65	-	2246
48	83	194	226	-	419	57.0	-	0.71	-	2139
72	84	168	157	-	325	37.3	-	0.77	-	1410
96	77	156	165	-	321	42.4	-	0.76	-	1565

3.4. Corrosion Mechanism Analysis

The surface of the compacts tested by the EIS method was documented with the aim to compare the mechanism of corrosion attack and its dependence on material structure and the corrosion environment used (Figures 9 and 10). Metallographic cross-sections of the tested compacts were prepared with the aim to identify the mechanism of the corrosion process within the material volume. Samples' cross-sections were molded into the resin and prepared by the standard metallographic procedures. Due to the molding into the resin and the porosity and incompactness of the corroded compacts and corrosion products on their surface, gas bubbles were present on the mold specimens (above the surface of the sample).

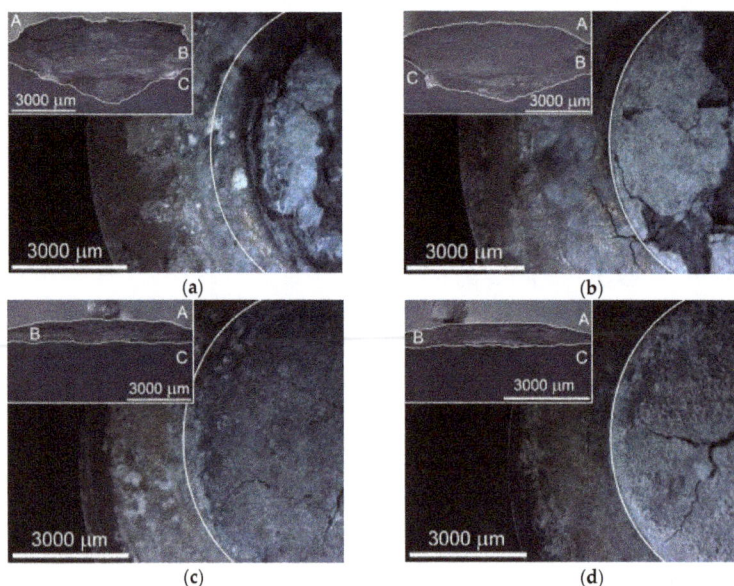

Figure 9. *Cont.*

(e)

(f)

Figure 9. Corroded surface of PM magnesium compacts (the area exposed during EIS measurement is outlined) and specimens cross-sections (A, resin; B, layer of corrosion products; C, compact); (**a**) cold compacted under 500 MPa, 12 h in HBSS; (**b**) cold compacted under 500 MPa, 24 h in enriched HBSS; (**c**) hot pressed under 100 MPa, 400 °C, 12 h in HBSS; (**d**) hot pressed under 100 MPa, 400 °C, 24 h in enriched HBSS; (**e**) hot pressed under 500 MPa, 400 °C, 96 h in HBSS; (**f**) hot pressed under 500 MPa, 400 °C, 96 h in enriched HBSS.

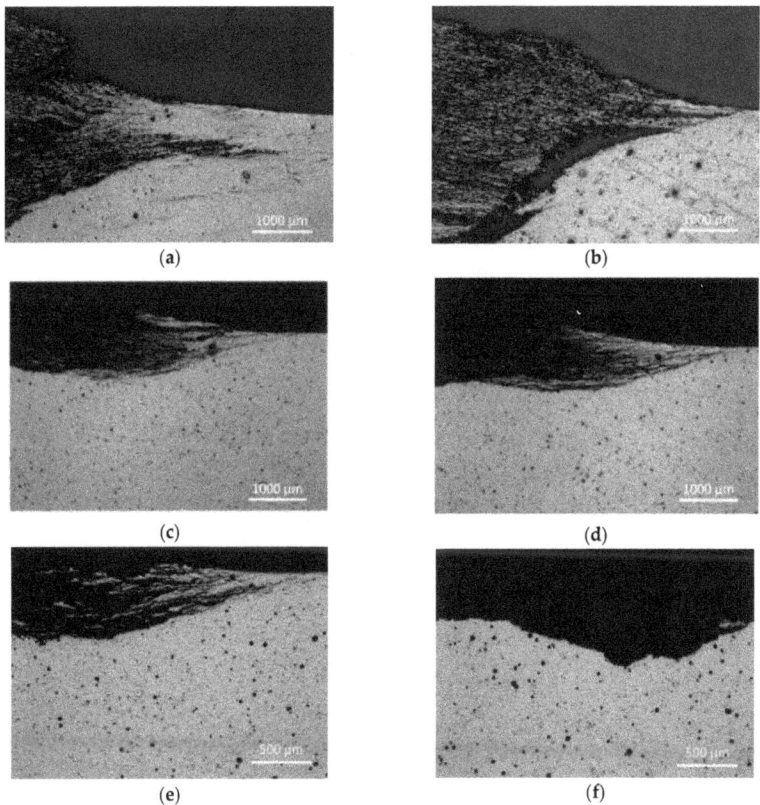

(a)

(b)

(c)

(d)

(e)

(f)

Figure 10. Corroded cross-section of pure magnesium compacts: (**a**) cold compacted 500 MPa, 12 h in HBSS; (**b**) cold compacted 500 MPa, 24 h in enriched HBSS; (**c**) hot pressed 100 MPa, 400 °C, 12 h in HBSS; (**d**) hot pressed 100 MPa, 400 °C, 24 h in enriched HBSS; (**e**) hot pressed 500 MPa, 400 °C, 96 h in HBSS; (**f**) hot pressed 500 MPa, 400 °C, 96 h in enriched HBSS.

In all cases, the area exposed to the corrosion environment during the EIS measurement was covered with a layer of corrosion products. The created layer was not compact and contained a large number of cracks, which were observed also in the layer of corrosion products.

The surface of the area of the compacts exposed to the HBSS and enriched HBSS was fully covered with grey-/white-colored corrosion products during the EIS measurement (outlined areas). Based on the surface observation, the corrosion mechanism can be considered as the same in the HBSS and enriched HBSS, as the corrosion products' structure is similar and no specific material behavior was observed. The corrosion mechanism, however, was dependent on the sample preparation.

In the case of cold pressed samples, the created corrosion products covered the whole area exposed to the corrosion environments (Figure 9a,b). On the details of the metallographic cross-section given in Figure 9a,b, quite pronounced corrosion attack of the material can be seen. The layer of corrosion products grows to a depth of approximately 4000 µm during the exposure to both corrosion environments. Due to the large thickness of the created layer of corrosion products, its cracking and decohesion with the cold compacted materials can be seen on samples' cross-sections in Figure 9a,b. In the details of Figure 10a,b, layer-like corrosion progression into the material volume can be recognized. The material behavior was the same for HBSS and enriched HBSS. However, from the comparison of Figure 10a,b, eventual removal of the layer of corrosion products from the surface of the compact tested in enriched HBSS can be observed as the time of exposure increased. Therefore, the more degradable influence of enriched HBSS on compacts compared to HBSS can be concluded for cold compacted pure magnesium.

The hot pressed samples (Figure 9c–f) corroded uniformly throughout the whole tested area. Samples prepared under 100 MPa compacting pressure at 400 °C exhibited a uniform layer of corrosion product after exposure in HBSS, with a slightly layer-like structure, which is more apparent from the detail in Figure 10c. As the corrosion products' layer grew into the material volume, the uniform layer cracked, which in long-term exposure leads to very severe corrosion attack. Enriched HBSS seems to have a similar influence on the hot pressed sample. The corrosion products' layer observable on the sample cross-section shown in Figure 9d was comparable to the one created in HBSS (Figure 9c). Both layers had a comparable thickness (less than 1500 µm), and both of them exhibited cracks. Furthermore, the mechanism of the corrosion attack progress through the material shown in Figure 10c,d was similar. The corrosion seems to progress to the material volume layer by layer, following layers of compacted powders (layer-like corrosion progression). The corrosion attack was less pronounced when compared to the cold compacted materials.

The hot pressed samples prepared under 500 MPa at 400 °C exhibit different behaviors in terms of corrosion attack in HBSS and enriched HBSS compared to the cold pressed and hot pressed samples prepared under 100 MPa. The surface of the hot pressed samples prepared under 500 MPa was completely covered with corrosion products; however, the depth to which the samples degraded is much smaller compared to the samples prepared under 100 MPa at 400 °C and cold pressed samples (Figure 9e,f). This shows much better coherence of the powder particles forming the samples. The depth of corrosion attack is small (approximately 500 µm) comparing to the other material states. However, in this case, different responses of compacts on HBSS and enriched HBSS were observed. Even from the cross-sections in Figure 10e,f, the layer-like corrosion progression is visible only for the sample exposed to HBSS. In the case of the exposure of the compacts to the enriched HBSS, the layer of corrosion products was removed from the material surface during the measurement (Figure 10f), which indicates low layer cohesion to the material and/or high volume expansion of the created corrosion products.

4. Discussion

Microstructural, mechanical and electrochemical corrosion characteristics of pure magnesium materials prepared by powder metallurgy were analyzed with the aim to identify the influence of processing parameters. Sets of cold compacted and hot pressed materials (300 °C, 400 °C and 500 °C) were prepared from magnesium powder under 100 MPa, 200 MPa, 300 MPa, 400 MPa and 500 MPa compacting pressures. While the material has potential in biomedical applications, the electrochemical corrosion characteristics of the processed materials were analyzed in HBSS and enriched HBSS using electrochemical impedance spectroscopy, and the data were extended with the analysis of the corrosion attack within the material volume.

Low compacting pressures led only to minimal plastic deformation of magnesium powder particles during cold compaction or hot pressing (Figure 2). Increasing the compaction pressure (above 300 MPa) at RT had an influence on the deformation of the powder particles while increased deformability of particles corresponded with increasing compaction pressure. The influence was also sufficient for changing the character of the porosity from open porosity to closed porosity and decreased the number of pores on the samples' cross-sections, which is in agreement with [11–13]. Hot pressing led to a significant decrease of the porosity of the compacts and also to a change in the character of the porosity (Figure 2) [4,32]. The open porosity characteristic for cold pressed samples was replaced with closed localized pores observed in hot pressed samples' microstructure. This change can be attributed to better plastic deformation of magnesium at a higher temperature, resulting in higher plastic deformation of powder particles and closing the open porosity present on particles' boundaries due to the shape factor (spherical particles cannot fill the space 100%). The change of the magnesium deformability results from the hexagonal close-packed (HCP) crystallographic lattice structure of the material. While only one slip system is active at RT, the increase in temperature led to activation of more slip systems. Furthermore, the twinning mechanism is dominant for HCP lattice structures instead of the slip mechanism below 200 °C [34,35].

Besides the improvement of magnesium deformability, the elevated temperature also influenced the evolution of grains in the compacted powder particles, which was revealed with EBSD analysis shown in Figure 3. The cold compacted sample revealed a very fine microstructure within the powder particle, but wide gaps between powder particles were observed. EBSD analysis of the hot pressed sample revealed coarser metallic grains in powder particles; however, the gaps between powder particles were not visible as diffusion bonding at higher temperatures took place between powder particles.

A coarser metallic microstructure generally leads to a decrease in hardness and tensile strength compared to the fine-grained material [4,5]. However, the observed fine microstructure corresponds only to the powder particles, while the particles' bonding was observed to be weak (also, an open porosity was observed). Therefore, this fine microstructure obtained by cold compacting of magnesium did not affect the mechanical properties of the samples in a positive way due to the low compactness of particle boundaries.

Improvement of mechanical properties by hot pressing of pure magnesium is visible in Figures 3 and 4. Hot pressed samples prepared at 300 °C revealed the same trend in increasing flexural strength with increasing compaction pressure as samples prepared at 400 °C; however, the values of flexural strength are slightly higher for the series of samples prepared at 400 °C. This fact can be attributed to a higher diffusion rate of magnesium at 400 °C, which leads to better diffusion bonding between powder particles and possibly more pronounced grain coarsening at higher temperature [11]. Samples prepared at 500 °C reached much lower values of flexural strength compared to other hot pressed samples (comparing samples prepared under the same pressures) and only slightly higher than cold compacted samples (Figure 4a). At the temperature of 500 °C, the diffusion of magnesium is even higher than at 400 °C, which should lead to better powder particle diffusion bonding; however, the higher temperature (500 °C) can lead to intensive coarsening of metallic grains created in the powder particles. As a result, the grain coarsening could be a reason for the flexural strength decrease (Figure 4a), considering good material compaction based on EBSD analysis.

Bonding of powder particles into the coherent sample is also apparent from the microhardness test results. Cold compacted samples follow the same trend of increasing microhardness with increasing compaction pressure as samples prepared via hot pressing; however, the obtained values are significantly lower. This can be attributed to the good cohesion of samples prepared via hot pressing. Cold compacted samples hold together only due to the powder particles' interlocking as a result of plastic deformation, while the hot pressed samples are interlocked and diffusion bonded. The good cohesion of powder particles in the case of hot pressed samples is also visible from the fractographic analysis provided in Figure 5, as inter-granular failure was visible only on fracture surfaces of samples prepared by hot pressing. Fractographic analysis from cold compacted samples revealed only the trans-granular failure of samples, and no inter-granular failure was observed. While the character of fracture surfaces differs, only brittle fracture was reported in all the measured samples. The character of fracture surfaces furthermore confirms the assumption that the samples prepared by cold pressing are not diffusion bonded.

Differences between cold compacted and hot pressed samples were also observed during EIS measurement. Low compactness of cold compacted samples compared to the hot pressed samples resulted in lower corrosion resistance of cold compacted samples. This was proven by the corrosion rate and the depth to which corrosion progressed. It is clearly evident that by dissolving of magnesium under H_2 evolution, the corrosion environment is alkalized (increase pH value), and the corrosion rate of magnesium slows down [30]. Corrosion products created on magnesium are usually based on MgO and $Mg(OH)_2$. The layer of corrosion products, especially the $Mg(OH)_2$ compound, is easily disturbed and dissolved in environments containing Cl^-. The layer offers only very low protection to magnesium against corrosion especially in environments containing aggressive ions [36]. HBSS is a more complex environment containing phosphates and hydrogen phosphates. More compounds in the corrosion environment enable more chemical reactions between the metallic surface, corrosion environment and corrosion products [30]. Products of these reactions can create a layer of corrosion products (outer layer) over the created MgO and $Mg(OH)_2$ inner primary layer created on the sample surface. Both layers develop during the corrosion process and are porous by their nature or the damage of the compact areas can occur. All the types of corrosion products are created during the material exposure to the environment and grow on the material surface at the same time. These layers cannot be exactly distinguished.

According to the Le Chatelier theory, HBSS should be more aggressive for magnesium than enriched HBSS. Comparing the corrosion process character of examined magnesium materials (Figures 9 and 10), this theory was not clearly proven. The similar thickness of the layers of corrosion products (Figure 9), even the progress of corrosion attack within the material volume (Figure 10), was observed considering the same processing parameters. Corrosion attack progress was the only observed difference between the cold compacted and hot pressed samples; whereas the cold compacted samples that were only mechanically-bonded exhibited a more pronounced degradation process when compared to hot pressed samples. The higher compacting pressure used (500 MPa), resulting in even better material compaction, had a positive influence also on material corrosion resistance.

The highest corrosion resistance was reached for hot pressed magnesium materials prepared under 500 MPa in both corrosion mediums. This was proven by higher polarization resistance and duration of the corrosion experiment compared to other material states. The time of experiments indicates more pronounced corrosion attack of samples in HBSS compared to enriched HBSS. This difference was not observed for hot pressed samples prepared under 500 MPa. However, the values of R_p determined for the materials tested in both solution types did not exhibit significant differences, and the higher aggressiveness of HBSS compared to enriched HBSS was not proven with the results. Slightly higher values of R_p reached for different material states in enriched HBSS when compared to HBSS did not show a significant influence of corrosion environment on material properties. However, similar values of R_p were reached at the end of EIS test for all the materials in each corrosion environment. Different material types exhibit similar corrosion behavior regardless of the corrosion environment used.

5. Conclusions

Powder metallurgy processed magnesium compacts prepared at room temperature and at 300 °C, 400 °C and 500 °C under pressures in the range from 100 MPa to 500 MPa were analyzed in terms of the evolution of the microstructure, mechanical properties and electrochemical corrosion characteristics.

Microstructural analysis revealed a positive influence of increasing pressure and temperature on material compaction and porosity. The increase of the compacting pressure used led to increasing of the powder particles' deformation, which corresponds to the decrease of compacts' porosity. Elevating pressing temperature resulted in significant enhancement of powder particles' deformability and a decrease of compacts' porosity. EBSD analysis showed fine-grained structure within cold compacted powder particles; however, it also revealed that the powder particles were not bonded. Elevated temperature resulted in grain coarsening; however, the powder particles were bonded.

Flexural strength was observed to be enhanced up to five times for the hot pressed samples compared to the cold compacted samples. The samples prepared at 500 °C reached similar values of flexural strength as cold compacted samples. While only the inter-granular fracture mechanism was observed for cold compacted and hot pressed samples prepared under lower pressures (below 300 MPa), samples prepared at 400 °C using higher compacting pressures exhibited also a trans-granular fracture. Vickers microhardness of hot pressed samples was comparable for all the preparation temperatures, and the values were higher than for cold compacted samples (up to 40%).

Electrochemical impedance spectroscopy measurements revealed a positive influence of the hot pressing on the electrochemical corrosion behavior of PM magnesium. For both corrosion environments used, the highest values of polarization resistance in time were determined for the hot pressed samples prepared under 500 MPa. The corrosion mechanism was similar to the cold pressed and hot pressed samples prepared under low pressures, where a layer of corrosion products (thicker for cold compacted samples) was observed on the tested samples' surface. A similar response of both sample types was observed using HBSS and enriched HBSS (considering the sample preparation conditions). Less pronounced corrosion attack was observed for hot pressed samples prepared under 500 MPa, and also, a similar material response for both the corrosion environment types was observed.

Applying temperatures of 300 °C and 400 °C and high pressures (300 to 500 MPa) for magnesium powder compaction seems to have a significantly positive influence on material bonding, mechanical and electrochemical corrosion properties. A higher compaction temperature was shown to have a detrimental effect on material compaction when using compacting pressure above 200 MPa.

Acknowledgments: This work was supported by Project No. LO1211, Materials Research Centre and by Project No. LO1202, NETME (New Technologies for Mechanical Engineering) center plus, projects of Ministry of Education, Youth and Sports of the Czech Republic under the "National sustainability program".

Author Contributions: Pavel Doležal and Matěj Březina conceived and designed the experiments. Matěj Březina, Michaela Krystýnová and Jozef Minda performed the experiments. Matěj Březina and Jozef Minda analyzed the data. Stanislava Fintová, Petr Ptáček, Josef Zapletal and Jaromír Wasserbauer contributed reagents/materials/analysis tools. Matěj Březina and Stanislava Fintová wrote the paper.

References

1. Kubásek, J.; Dvorský, D.; Čavojský, M.; Vojtěch, D.; Beronská, N.; Fousová, M. Superior properties of Mg-4Y-3RE-Zr alloy prepared by powder metallurgy. *J. Mater. Sci. Technol.* **2017**, *33*, 652–660. [CrossRef]
2. Zhou, T.; Yang, M.; Zhou, Z.; Hu, J.; Chen, Z. Microstructure and mechanical properties of rapidly solidified/powder metallurgy Mg-6Zn and Mg-6Zn-5Ca at room and elevated temperatures. *J. Alloys Compd.* **2013**, *560*, 161–166. [CrossRef]

3. Yan, Y.; Cao, H.; Kang, Y.; Yu, K.; Xiao, T.; Luo, J.; Deng, Y.; Fang, H.; Xiong, H.; Dai, Y. Effects of Zn concentration and heat treatment on the microstructure, mechanical properties and corrosion behavior of as-extruded Mg-Zn alloys produced by powder metallurgy. *J. Alloys Compd.* **2017**, *693*, 1277–1289. [CrossRef]

4. Chang, I.; Zhao, Y. *Advances in Powder Metallurgy: Properties, Processing and Applications*; Woodhead Publishing: Cambridge, UK, 2013; p. 90.

5. Suwas, S.; Gottstein, G.; Kumar, R. Evolution of crystallographic texture during equal channel angular extrusion (ECAE) and its effects on secondary processing of magnesium. *Mater. Sci. Eng. A* **2007**, *471*. [CrossRef]

6. Shen, J.; Imai, H.; Chen, B.; Ye, X.; Umeda, J.; Kondoh, K. Deformation mechanisms of pure Mg materials fabricated by using pre-rolled powders. *Mater. Sci. Eng. A* **2016**, *658*, 309–320. [CrossRef]

7. Stulikova, I.; Smola, B.; Vlach, M.; Kudrnova, H.; Piesova, J. Influence of powder metallurgy route on precipitation processes in MgTbNd alloy. *Mater. Charact.* **2016**, *112*, 149–154. [CrossRef]

8. Manne, B.; Bontha, S.; Ramesh, M.R.; Krishna, M.; Balla, V.K. Solid state amorphization of Mg-Zn-Ca system via mechanical alloying and characterization. *Adv. Powder Technol.* **2017**, *28*, 223–229. [CrossRef]

9. Bornapour, M.; Celikin, M.; Cerruti, M.; Pekguleryuz, M. Magnesium implant alloy with low levels of strontium and calcium: The third element effect and phase selection improve bio-corrosion resistance and mechanical performance. *Mater. Sci. Eng. C* **2014**, *35*, 267–282. [CrossRef] [PubMed]

10. Nayak, S.; Bhushan, B.; Jayaganthan, R.; Gopinath, P.; Agarwal, R.D.; Lahiri, D. Strengthening of Mg based alloy through grain refinement for orthopaedic application. *J. Mech. Behav. Biomed. Mater.* **2016**, *59*, 57–70. [CrossRef] [PubMed]

11. Wang, X.; Hu, L.; Liu, K.; Zhang, Y. Grain growth kinetics of bulk AZ31 magnesium alloy by hot pressing. *J. Alloys Compd.* **2012**, *527*, 193–196. [CrossRef]

12. Kayhan, S.M.; Tahmasebifar, A.; Koç, M.; Usta, Y.; Tezcaner, A.; Evis, Z. Experimental and numerical investigations for mechanical and microstructural characterization of micro-manufactured AZ91D magnesium alloy disks for biomedical applications. *Mater. Des.* **2016**, *93*, 397–408. [CrossRef]

13. Meng, F.; Rosalie, J.M.; Singh, A.; Tsuchiya, K. Precipitation behavior of an ultra-fine grained Mg-Zn alloy processed by high-pressure torsion. *Mater. Sci. Eng. A* **2015**, *644*, 386–391. [CrossRef]

14. Reddy, T.H.; Pal, S.; Kumar, K.C.; Mohan, M.K.; Kokol, V. Finite element analysis for mechanical response of magnesium foams with regular structure obtained by powder metallurgy method. *Procedia Eng.* **2016**, *149*, 425–430. [CrossRef]

15. Vahid, A.; Hodgson, P.; Li, Y. New porous Mg composites for bone implants. *J. Alloys Compd.* **2017**, *724*, 176–186. [CrossRef]

16. Zhang, S.; Zheng, Y.; Zhang, L.; Bi, Y.; Li, J.; Liu, J.; Yu, Y.; Guo, H.; Li, Y. In vitro and in vivo corrosion and histocompatibility of pure Mg and a Mg-6Zn alloy as urinary implants in rat model. *Mater. Sci. Eng. C* **2016**, *68*, 414–422. [CrossRef] [PubMed]

17. Khalajabadi, S.Z.; Abdul Kadir, M.R.; Izman, S.; Marvibaigi, M. The effect of MgO on the biodegradation, physical properties and biocompatibility of a Mg/HA/MgO nanocomposite manufactured by powder metallurgy method. *J. Alloys Compd.* **2016**, *655*, 266–280. [CrossRef]

18. Bornapour, M.; Muja, N.; Shum-Tim, D.; Cerruti, M.; Pekguleryuz, M. Biocompatibility and biodegradability of Mg-Sr alloys: The formation of Sr-substituted hydroxyapatite. *Acta Biomater.* **2013**, *9*, 5319–5330. [CrossRef] [PubMed]

19. Singh Raman, R.K.; Jafari, S.; Harandi, S.E. Corrosion fatigue fracture of magnesium alloys in bioimplant applications: A review. *Eng. Fract. Mech.* **2015**, *137*, 97–108. [CrossRef]

20. Yazdimamaghani, M.; Razavi, M.; Vashaee, D.; Moharamzadeh, K.; Boccaccini, A.R.; Tayebi, L. Porous magnesium-based scaffolds for tissue engineering. *Mater. Sci. Eng. C* **2017**, *71*, 1253–1266. [CrossRef] [PubMed]

21. Drábiková, J.; Pastorek, F.; Fintová, S.; Doležal, P.; Wasserbauer, J. Improvement of bio-compatible AZ61 magnesium alloy corrosion resistance by fluoride conversion coating. *Koroze Ochrana Mater.* **2016**, *60*. [CrossRef]

22. Shadanbaz, S.; Dias, G.J. Calcium phosphate coatings on magnesium alloys for biomedical applications: A review. *Acta Biomater.* **2012**, *8*, 20–30. [CrossRef] [PubMed]

23. Chiu, K.Y.; Wong, M.H.; Cheng, F.T.; Man, H.C. Characterization and corrosion studies of fluoride conversion coating on degradable Mg implants. *Surf. Coat. Technol.* **2007**, *202*, 590–598. [CrossRef]

24. Carboneras, M.; Hernández, L.S.; Del Valle, J.A.; García-Alonso, M.C.; Escudero, M.L. Corrosion protection of different environmentally friendly coatings on powder metallurgy magnesium. *J. Alloys Compd.* **2010**, *496*, 442–448. [CrossRef]

25. Jafari, S.; Singh Raman, R.K.; Davies, C.H.J. Corrosion fatigue of a magnesium alloy in modified simulated body fluid. *Eng. Fract. Mech.* **2015**, *137*, 2–11. [CrossRef]

26. Tahmasebifar, A.; Kayhan, S.M.; Evis, Z.; Tezcaner, A.; Çinici, H.; Koç, M. Mechanical, electrochemical and biocompatibility evaluation of AZ91D magnesium alloy as a biomaterial. *J. Alloys Compd.* **2016**, *687*, 906–919. [CrossRef]

27. King, A.D.; Birbilis, N.; Scully, J.R. Accurate electrochemical measurement of magnesium corrosion rates; a combined impedance, mass-loss and hydrogen collection study. *Electrochim. Acta* **2014**, *121*, 394–406. [CrossRef]

28. Bender, S.; Goellner, J.; Heyn, A.; Schmigalla, S. A new theory for the negative difference effect in magnesium corrosion. *Mater. Corros.* **2011**. [CrossRef]

29. Liao, J.; Hotta, M.; Mori, Y. Improved corrosion resistance of a high-strength Mg-Al-Mn-Ca magnesium alloy made by rapid solidification powder metallurgy. *Mater. Sci. Eng. A* **2012**, *544*, 10–20. [CrossRef]

30. Song, G.-L. Corrosion of magnesium alloys. In *Woodhead Publishing in Materials*; Woodhead Publishing: Philadelphia, PA, USA, 2011; p. 640.

31. Čapek, J.; Vojtěch, D. Properties of porous magnesium prepared by powder metallurgy. *Mater. Sci. Eng. C* **2013**, *33*, 564–569. [CrossRef] [PubMed]

32. Čapek, J.; Vojtěch, D. Effect of sintering conditions on the microstructural and mechanical characteristics of porous magnesium materials prepared by powder metallurgy. *Mater. Sci. Eng. C* **2014**, *35*, 21–28. [CrossRef] [PubMed]

33. Orazem, M.E.; Tribollet, B. *Electrochemical Impedance Spectroscopy*; Wiley: Hoboken, NJ, USA, 2008; p. 560.

34. Drápala, J. *Hořčík, Jeho Slitiny a Binární Systémy Hořčík—Přímĕs: Magnesium, Its Alloya and Mg-Admixture Binary Systems*; Vysoká škola báňská—Technická Univerzita: Ostrava, Czech Republic, 2004; p. 172.

35. Kang, F.; Li, Z.; Wang, J.T.; Cheng, P.; Wu, H.Y. The activation of $\langle c + a \rangle$ non-basal slip in magnesium alloys. *J. Mater. Sci.* **2012**, *47*, 7854–7859. [CrossRef]

36. Xin, Y.; Huo, K.; Tao, H.; Tang, G.; Chu, P.K. Influence of aggressive ions on the degradation behavior of biomedical magnesium alloy in physiological environment. *Acta Biomater.* **2008**, *4*, 2008–2015. [CrossRef] [PubMed]

![metals logo] *metals*

MDPI

Article

Development of a Novel Degradation-Controlled Magnesium-Based Regeneration Membrane for Future Guided Bone Regeneration (GBR) Therapy

Da-Jun Lin [1], Fei-Yi Hung [1,*], Hung-Pang Lee [2] and Ming-Long Yeh [2,*]

[1] Department of Materials Science and Engineering, National Cheng Kung University, Tainan 701, Taiwan;
     larrylin111@hotmail.com
[2] Department of Biomedical Engineering, National Cheng Kung University, Tainan 701, Taiwan;
     qer6322129@gmail.com
*   Correspondence: fyhung@mail.ncku.edu.tw (F.-Y.H.); mlyeh@mail.ncku.edu.tw (M.-L.Y.);
     Tel.: +886-06-2757575 (ext. 62950) (F.-Y.H.); +886-06-2757575 (ext. 63429) (M.-L.Y.)

Received: 21 September 2017; Accepted: 3 November 2017; Published: 6 November 2017

Abstract: This study aimed to develop and evaluate the ECO-friendly Mg-5Zn-0.5Zr (ECO505) alloy for application in dental-guided bone regeneration (GBR). The microstructure and surface properties of biomedical Mg materials greatly influence anti-corrosion performance and biocompatibility. Accordingly, for the purpose of microstructure and surface modification, heat treatments and surface coatings were chosen to provide varied functional characteristics. We developed and integrated both an optimized solution heat-treatment condition and surface fluoride coating technique to fabricate a Mg-based regeneration membrane. The heat-treated Mg regeneration membrane (ARRm-H380) and duplex-treated regeneration membrane group (ARRm-H380-F24 h) were thoroughly investigated to characterize the mechanical properties, as well as the in vitro corrosion and in vivo degradation behaviors. Significant enhancement in ductility and corrosion resistance for the ARRm-H380 was obtained through the optimized solid-solution heat treatment; meanwhile, the corrosion resistance of ARRm-H380-F24 h showed further improvement, resulting in superior substrate integrity. In addition, the ARRm-H380 provided the proper amount of Mg-ion concentration to accelerate bone growth in the early stage (more than 80% new bone formation). From a specific biomedical application point of view, these research results point out a successful manufacturing route and suggest that the heat treatment and duplex treatment could be employed to offer custom functional regeneration membranes for different clinical patients.

Keywords: Mg alloy; regeneration membrane; guided bone regeneration; heat treatment; fluoride coating; biocompatibility

---

1. Introduction

Periodontitis is a bacterial-mediated inflammatory disease that can lead to damage of the periodontal ligament and gingival tissue, and may also cause alveolar bone resorption [1]. The prevalence of continuous periodontitis growth has been well documented in modern society [2], and in many clinical reports, delaying therapy can eventually result in tooth loss and alveolar atrophy (especially for older patients) [3]. In such cases, the defect area must first be reconstructed before artificial tooth root implantation. To this end, guided bone regeneration (GBR) procedures have been noted as a reliable periodontal regeneration and alveolar augmentation therapy, and have registered high success rates in recent years [4]. Currently, there are two material systems for GBR procedures, namely degradable and non-degradable membrane materials [5]. With degradable regeneration membranes (usually made by poly-lactic acid (PLA) or collagen sheets), secondary surgery for implant removal is not required; however, the low mechanical strength and stiffness do not offer sufficient structural strength. In contrast, non-degradable regeneration membranes (e.g., Ti mesh or Teflon-mesh) are the most commonly used materials; however, their use requires secondary surgery for membrane removal, which increases the risk of bacterial infection. In response to these clinical considerations, this study targeted the development of a new type of regeneration membrane that features both adequate strength and biodegradability.

Magnesium (Mg) is a promising metallic material for biomedical applications due to its unique biodegradability, satisfactory biocompatibility, and excellent biomechanical properties [6]. Moreover, Mg materials possess satisfactory biocompatibility and biofunctionality that can accelerate cell proliferation and wound healing [7,8]. Therefore, Mg materials can be seen as potential candidates for new types of regeneration membranes for dental GBR procedures. However, the poor anti-corrosion behaviors and rapid mechanical fading of Mg materials in a physiological electrolyte environment currently limit its clinical applicability [9,10]. Nevertheless, Cai et al. recently revealed that Mg-5Zn alloy is a good candidate for orthopedic implants with an optimal Zn alloying amount [11]. Song et al. also demonstrated that Mg-5Zn alloy possesses a uniform corrosion behavior and less localized corrosion incidence [12]. Moreover, several related studies have shown that properly modified (including microstructure modification and surface treatment) Mg implants offer stable mechanical retention with reasonable degradation rates both in vitro and in vivo [13–15]. Results from our previous works showed that heat treatment and plastic deformation procedures greatly influence the degradation behavior and biocompatibility of Mg-Zn-Zr series alloys via microstructure transformation [7,16]. Li et al. investigated in vitro and in vivo corrosion, as well as the mechanical properties and biocompatibility of Mg-Zn-Zr alloy; in addition, their research also indicated that the degradation behavior and bone healing behavior could be effectively improved after surface fluoride coating [17].

At present, the global demand of Mg materials is increasing year by year. It therefore follows that the demand will further increase once Mg-based medical devices are widely accepted in clinical practice. However, the casting procedure of Mg alloys always needs to use a great amount of SF_6 (a potent greenhouse gas), which can accelerate global warming [18]. To mitigate this shortcoming, the SF_6-applied casting procedure should be replaced by novel green ECO-casting techniques (using HFC-134a gas). In this study, to fabricate a highly functional medical device, a novel ECO-casting technique, which integrates a solid-solution heat treatment and surface treatment, was developed to produce an optimized biodegradable regeneration membrane for dental application using ECO-friendly Mg-5Zn-0.5Zr (named ECO505) alloy material.

2. Materials and Methods

2.1. ECO-Casting Process and Shaping Process

Mg-20 wt. % Zr, 4N grade (99.99 wt. %) pure Mg and pure Zn ingots were used for producing a Mg-5Zn-0.45Zr alloy billet. The precisely weighted raw materials were melted in a mild steel crucible using an electrical resistance furnace. During the melting and alloying process, a 20% HFC-R134a-80% Ar gas mixture was applied to the melt as a protection gas to prevent self-ignition [19]. The melt was held at 700 °C and stirred to homogenize the melt composition. After the melting process, the melt was cast into a preheated (350 °C) stainless steel mold to produce the ECO505 alloy billet (as shown in Figure 1).

Before the regeneration membrane shaping process, the cast billets were previously homogenized at 380 °C for 12 h and subsequently water quenched. Afterwards, the homogenized billets were directly extruded at 350 °C using a ram speed of 10 mm/s and an extrusion ratio of 35 to form stripe extrudates. The rolling raw materials (40 mm in length, 15 mm in width and 4 mm in height) were cut from the stripe extrudates with the rolling plane parallel to the extrusion direction. During the hot-rolling process, the process conditions were fixed at the rolling strain rate of 1.98 s^{-1} and under 350 °C, for which the reduction rate per rolling pass was 20%. Furthermore, the rolled samples were reheated to 350 °C and kept at that temperature for 5 min prior to each rolling pass. The final thickness of the as-rolled ECO505 regeneration membrane (named as ARRm) samples was controlled at 0.4 mm (as shown in Figure 1).

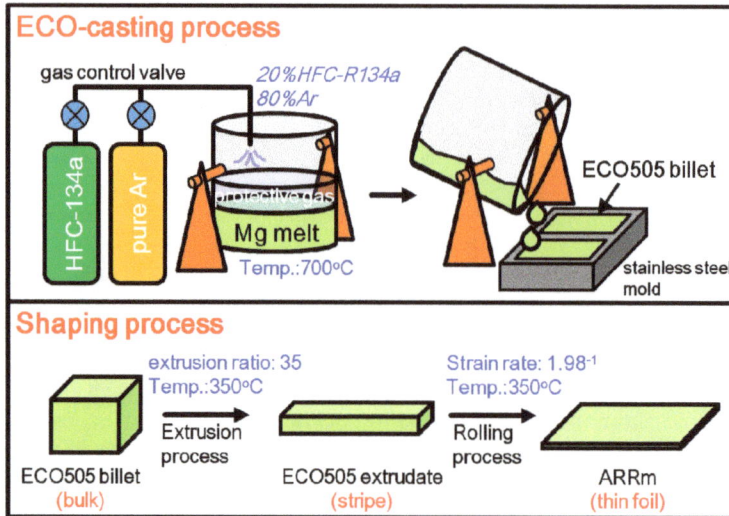

Figure 1. Schematic diagram of ECO-casting process and shaping process.

2.2. Heat Treatment Modification

The ARRm samples were processed into a circular thin foil (12.7 mm in diameter). Solution heat treatment of the ARRm samples was carried out in a tubular vacuum furnace at 380 °C for 1, 2, 4 and 10 h, respectively. After heat treatment, the samples were water quenched at room temperature. The names for the solution heat-treated samples were assigned by heating time as follows: ARRm-380 °C_1 h, ARRm-380 °C_2 h, ARRm-380 °C_4 h and ARRm-380 °C_10 h. The name of the final optimized condition was replaced by ARRm-H380 to show its representative meaning.

2.3. Surface Fluoride Coating

The surfaces of the ARRm and ARRm-H380 samples were coated with a protective MgF_2 layer. Prior to the coating process, all samples were polished successively with 200 to 8000 grit using SiC sandpaper. The polished specimens were rinsed in acetone and ethanol, and subsequently dried in a stream of dry air. The fluoride conversion treatment followed the suggested procedure from a previously published report, which involved the samples being soaked in 42 wt. % HF and placed in an orbital shaker at 90 rpm for 24 h [20]. After the samples were removed from the HF conversion bath, they were rinsed twice with absolute ethanol and dried. Names for the MgF_2-coated samples were assigned according to treatment history, and hereafter referred to as ARRm-F24 h and ARRm-H380-F24 h, respectively.

2.4. Materials Characterization

All samples (as-cast, as-extruded and ARRm) for microstructure characterization were ground and polished to 0.05 μm and finally etched by picric-acetic acid solution (4.2 g picric acid + 20 mL acetic acid + 80 mL ethanol). Optical microscopy (BX41-LED, Olympus, Tokyo, Japan) and scanning electron microscopy (JSM-6510LV, JEOL, Tokyo, Japan) with an energy dispersive spectrometer (EDS, INCA 350, Oxford Instrument, Oxford, UK) were used for microstructure characterization. The average grain size was determined according to the ASTM E112-96 standard. With the aim to characterize the effects of heat treatment on matrix softening, Vicker's hardness measurements (HMV-G21, Shimadzu, Kyoto, Japan) were performed on the polished surface of the heat treated specimens with different treatment conditions, for which the loaded force was 100 g and the holding time was 10 s. Tensile test used a hydraulic-powered mechanical testing system (MTS-810, MTS Systems Corporation, Minneapolis, MN, USA) with a tensile speed of 1 mm/min. A typical dog-bone tensile specimen expressed with a gauge length of 20 mm, a gauge width of 7 mm and a thickness of 0.4 mm was selected for tensile test. At least four samples are tested (Vicker's hardness and tensile test) for each group.

2.5. In Vitro Corrosion Test

The electrochemical corrosion mechanism and performance of the fluoride conversion coating were investigated using an electrochemical station (PARSTAT 2273, Princeton Applied Research, Oak Ridge, TN, USA). Polarization curves were measured using a classic three-electrode cell, where a saturated calomel electrode (SCE, +0.242 V vs. SHE) constituted the reference electrode with a Pt-coated Ti mesh as the counter-electrode. The area of the working electrode exposed to the electrolyte was controlled to within 1 cm^2 by a Teflon holder. The electrolyte used was revised simulated body fluid (r-SBF) solution (which per liter included 5.403 g of NaCl, 0.736 g of NaHCO$_3$, 2.036 g of Na$_2$CO$_3$, 0.225 g of KCl, 0.182 g of K$_2$HPO$_4$, 0.310 g of MgCl$_2$·6H$_2$O, 11.928 g of 4-(2-hydroxyethyl)-1-piperazineethanesulfonic acid (HEPES), 0.293 g of CaCl$_2$, and 0.072 g of Na$_2$SO$_4$ dissolved in deionized water) buffered at pH = 7.4 using HEPES and NaOH, with the environmental temperature controlled at 37 °C. Prior to conducting the experiment, nitrogen gas was bubbled through the r-SBF to remove dissolved oxygen. The polarization curve was acquired with a scanning rate of 1 mV s^{-1} from -1.8 V to -0.8 V.

To characterize the corrosion properties of the Mg-based regeneration membrane, two immersion corrosion examinations were used in this research. Firstly, the normal immersion corrosion (un-bent/r-SBF immersion) test was employed to measure the corrosion rates, behaviors and tendencies of the specimens in r-SBF by following the principle of ASTM G31-72. The r-SBF volume to surface area ratio was fixed at 20 mL/cm^2. Secondly, the actual-simulated immersion corrosion (pre-bent/r-SBF immersion) test was employed to acquire the corrosion rates, behaviors and tendencies of the pre-bent specimens in r-SBF by a custom designed examination. For this examination, several PVC racks (1 cm width and 5 cm long) were used to act as alveolar bone, upon which the pre-bent specimens were fixed by nylon screws, as shown in Figure 2a. Then, the specimens were immersed into r-SBF. The r-SBF volume to surface area ratio of this custom designed examination was fixed at 50 mL/cm^2. The experiments were held for 1 week in an isothermal incubator at 37 °C. The result is the average of five samples ($n = 5$).

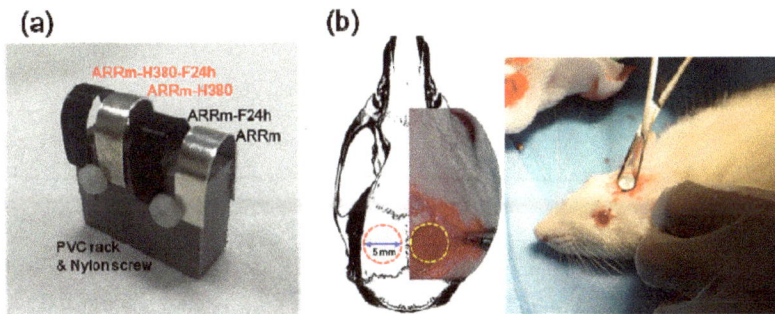

Figure 2. (**a**) Custom designed pre-bent/corrosion device and specimens; (**b**) animal implantation model.

2.6. Animal Model and Cranial Implantation Experiments

In this study, the animal experiments and surgical procedure were approved by the Institutional Animal Care and Use Committee of National Cheng Kung University (approval No. 105258). Male-controlled Sprague Dawley (SD) rats were housed in an environmentally controlled animal feeding room (25 ± 1 °C, 40~70% humidity, with a regular 12 h light cycle per day) to an age of 12~14 weeks. All SD rats were randomly assigned to testing groups prior to surgery. General anesthesia was administered via an intra-abdominal injection of Zoletil 50 (Virbac, Carros, France) and Xylazine (Panion and BF Biotech Inc., Taipei, Taiwan) mixture (0.2 mg/100 g of Zoletil 50 and 0.5 mg/100 g of Xylazine); then, local infiltration anesthesia (xylocaine, 0.2 mL for each surgical incision) was performed at the surgical sites before surgery. An incision along the periphery of the skull was created to peel back to the anterior portion of the skull. Two 5 mm critical size defect (CSD) were drilled in the calvarial bone using a trephine bur and low speed handpiece (as shown in Figure 2b). While drilling the cranial bone, sterilized saline was continuously injected to cool the drilling heat. The defects were covered with the regeneration membrane (a circular foil 7 mm in diameter and 0.4 mm thick) made by ECO505 raw material. Control and experimental groups (ARRm-H380 and ARRm-H380-F24 h) were employed to realize the applicability and performance of the regeneration membranes. After surgery, the wounds were closed by 5–0 non-absorbable silk sutures and antibiotics applied to prevent wound infection. At least four SD rats are examined for each group and time point ($n = 4$).

2.7. Micro-CT Analysis and 3D Image Reconstruction

To obtain both qualitative and quantitative data of the bone regeneration level within the CSD, a μ-CT (Skyscan 1076, Kontich, Belgium) was used to scan the cranial bone. The sham, ARRm-H380 and ARRm-H380-F24 h groups were measured at 4-week and 12-week time points. The voltage and X-ray current were controlled at 90 kV and 110 μA, respectively. The examined cranial bones and ECO505 regeneration membranes were scanned through a 360° rotation angle, with a rotation interval of 1° and pixel size of 18 μm resolution. From the data measurement with SkyScan software (Version 1.4.4, Kontich, Belgium), a cylindrical region of interest (ROI) 5 mm in diameter within the CSD site was selected for analysis. The volume and diameter of the bone growth were measured as the new bone volume fraction (%). The scanned cranial bone data were then reconstructed by Mimics software (Version 4.0, Materialise NV, Leuven, Belgium) to obtain high-quality 3D reconstructions. For clear identification, the original cranial bone and new bone were colored gray and light blue, respectively.

2.8. Statistical Analysis

The results of mechanical and in vivo implantation experiments are given as the mean viability value ± standard deviation. Analysis of one-way variance (ANOVA) was conducted to evaluate the statistical significance of differences. Differences at $p \leq 0.05$ were considered to be statistically significant.

3. Results and Discussion

3.1. Metallographic and Microstructure Observation

Figure 3 presents the optical micrographs of the as-cast, as-homogenized, as-extruded and as-rolled specimens. For the ECO505 billet, abundant $MgZn_2$ particles (the main secondary phase in Mg-Zn-Zr alloy) distributed along the grain boundary can be seen. These brittle secondary phases might affect the hot-working behavior of the Mg alloy. Prasad et al. suggested that applying a homogenization heat treatment can dissolve the intermetallic particles and improve the hot workability [21]. In the present study, after a homogenization treatment at 380 °C for 12 h, most of the $MgZn_2$ phases were dissolved into the Mg matrix. The microstructure of the homogenized specimen was characterized as equiaxed grains with an average grain size of about 90 μm. Figure 3d shows a typical extrusion feature, in which dynamic recrystallization (DRX) grains, with an average grain-size of about 15 μm, can be observed in the matrix. Figure 3e,f show macrograph and optical images of the ARRm. With respect to the macrograph, the ARRm showed no cracks, voids or severe-edge cracking. Its microstructure was fully evolved to a fine DRX microstructure without twins or other harmful defects, the average grain size of which was around 4.8 μm. The alloying elemental compositions of the ARRm are presented in Table 1. As seen, ARRm contains low amounts of noble impurities (such as Fe, Cr, Ni, Cu) and shows an acceptable level, indicating that the metallurgy and plastic deformation process in this work are feasible.

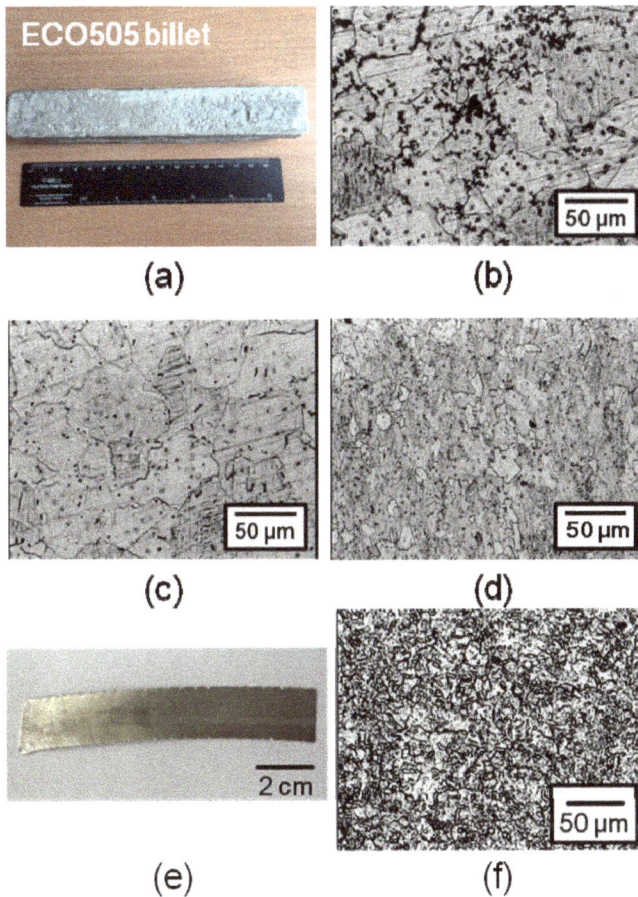

Figure 3. Typical macrograph and/or microstructures of each group: (**a**) macrograph of ECO505 billet; (**b**) microstructure of as-cast ECO505; (**c**) microstructure of as-homogenized ECO505 billet; (**d**) microstructure of as-extruded ECO505; (**e**,**f**) macrograph and microstructure of as-rolled ECO505 (named as ARRm).

Table 1. Elemental composition of ECO505 made regeneration membrane (unit: wt. %).

Group	Zn	Zr	Fe	Cr	Ni	Mn	Cu	Mg
ECO505	5.02	0.48	0.0012	0.0010	0.0023	0.0102	N.D.	Bal.

3.2. Solid-Solution Heat Treatment Modification and Mechanical Properties

Figure 4 shows the hardness variation curves of the ARRm samples isothermally heat treated at 340, 360, 380, 400 and 420 °C for 1 h. For the specimens heat treated at 360 °C, the hardness value slightly decreased to 85 HV, which indicates insufficient solid solution efficiency. The hardness value decreased with increasing heat treatment temperature, particularly for specimens heat treated at temperatures higher than 380 °C. For the treatment temperature of 380 °C, the matrix was significantly softened (65 HV) than 340 °C and 360 °C treated groups. This phenomenon also means that the brittle secondary phases can be easily dissolved at 380 °C; therefore, we selected the temperature control of 380 °C for the ARRm heat-treatment process.

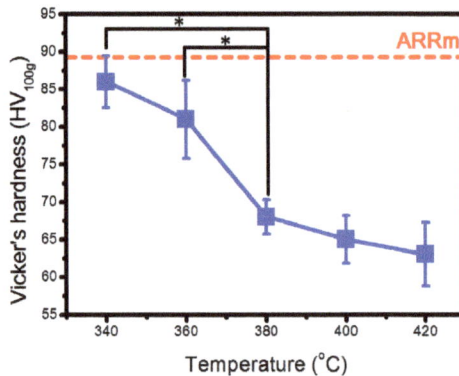

Figure 4. Vicker's hardness variation curves of different heat treatment temperature. (Data presented as mean $\pm$ SD, $n = 4$ and analyzed using a one-way ANOVA, * $p < 0.05$) (red dash line: the Vicker's hardness of un-solution heat treated specimen).

Figure 5 shows the microstructure evolution of ARRm isothermally heat treated at 380 °C for 1 h, 2 h, 4 h and 10 h, the characteristics of which were used to determine the optimum heat treatment time. Both the ARRm-380 °C_1 h and ARRm-380 °C_2 h featured many fine recrystallized grains, revealing that the fully recovered matrices had equiaxed grains with an average grain size of 5.6 μm and 7.8 μm, respectively. The average grain sizes of ARRm-380 °C_4 h and ARRm-380 °C_10 h significantly grew to 18 μm and 26 μm, respectively. Notably, the tensile properties of ARRm-380 °C_1 h and ARRm-380 °C_2 h showed an obvious improvement in elongation (see Figure 5e and Table 2). According to many previous reports, this improvement is related to the grain boundary sliding (GBS) phenomenon. GBS always occurs in fine-grained microstructures with a grain size smaller than 10 μm [22]. With further increases in treatment time, the time variation significantly altered the microstructure and tensile elongation. The tensile results clearly indicate that the mechanical behavior of ARRm improved with increases in the 380 °C heating duration until 2 h. Prolonging the heating duration of the ARRm heat treatment beyond 4 h led to a serious drop in elongation and yield strength behavior. Obviously, the specimens heat treated at 380 °C for 4 h and 10 h had lower yield strengths (189 MPa and 179 MPa, respectively) and elongations (11.2% and 11.1%, respectively) and displayed the overheating condition, as evidenced by their big grain sizes not being able to trigger GBS. Considering the application requirements, the elongation of the regeneration membrane must be optimized for the clinical pre-bending procedure. In the present cases, the ARRm-380 °C_2 h specimen not only maintained a small grain size, but the brittle $MgZn_2$ particles also dissolved, yielding a synergistic effect that provided the best elongation of 20.2%. Therefore, the heat-treatment parameter of 380 °C for 2 h can be considered as the optimal condition for ARRm, and was named ARRm-H380 to show its representative meaning.

Table 2. Mechanical parameter results of the tensile test (Data presented as mean ± SD, $n = 4$).

Parameter	ARRm	ARRm-380 °C_1 h	ARRm-380 °C_2 h	ARRm-380 °C_4 h	ARRm-380 °C_10 h
UTS (MPa)	268 ± 12	251 ± 9	256 ± 5	224 ± 8	210 ± 5
YS (MPa)	248 ± 8	207 ± 7	200 ± 6	189 ± 6	179 ± 3
EL. (%)	8.5 ± 0.4	19.1 ± 0.6	20.2 ± 0.4	11.2 ± 0.5	11.1 ± 0.3

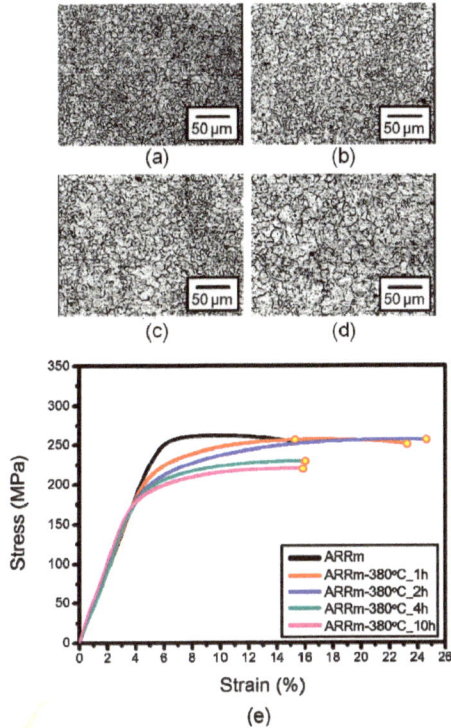

Figure 5. Typical microstructures and tensile stress-strain curves of each groups: (**a**) ARRm-380 °C_1 h; (**b**) ARRm-380 °C_2 h; (**c**) ARRm-380 °C_4 h; (**d**) ARRm-380 °C_10 h, (**e**) stress-strain curves

3.3. Effect of Solid-Solution Treatment and Fluoride Coating on Anti-Corrosion Ability of Mg Regeneration Membrane

Figure 6a,b show the surface morphologies of the fluoride coatings deposited on the ARRm and ARRm-H380 substrate (named ARRm-F24 h and ARRm-H380-F24 h, respectively). The coatings on both surfaces were neat and clean, without obvious coating defects or particle contaminants. The growth reactions of the fluoride conversion coating on the magnesium surface are described in the following:

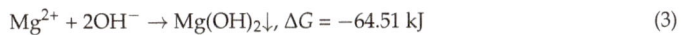

$$Mg \rightarrow Mg^{2+} + 2e^- \tag{1}$$

$$2H_2O + 2e^- \rightarrow H_2 + 2OH^- \tag{2}$$

$$Mg^{2+} + 2OH^- \rightarrow Mg(OH)_2\downarrow, \Delta G = -64.51 \text{ kJ} \tag{3}$$

$$Mg(OH)_2 + 2HF \rightarrow 2H_2O + MgF_2\downarrow, \Delta G = -232.14 \text{ kJ} \tag{4}$$

The EDS spectrums of the ARRm-F24 h and ARRm-H380-F24 h specimen confirm the presence of the MgF_2 on the surface, moreover, the fluorine amount of ARRm-H380-F24 h is significantly higher than ARRm-F24 h. ARRm-F24 h possessed only a 1.6 μm fluoride layer, while ARRm-H380-F24 h possessed a 2.3 μm fluoride layer, as shown in Figure 6e,f, respectively. According to our previous work, the fluoride coating of the Mg-Zn-Zr series alloy is composed of nano-MgF_2 and $MgZn_2$, with the distribution and homogeneity of the latter playing a key role in the coating formation mechanism [20]. Interestingly, the heat-treated specimens offered a 1.44-fold higher coating conversion efficiency, and consequently obtained a thicker coating structure than the non-heat-treated specimens. Generally speaking, a thick and dense coating offers better anti-corrosion performance.

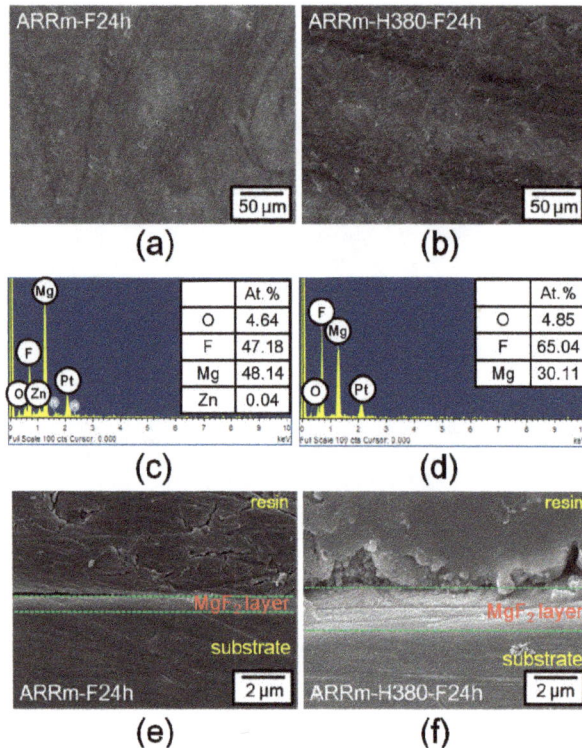

Figure 6. Surface morphologies, elemental analysis and cross-section profiles of the fluoride coated surfaces: (**a**) ARRm-F24 h surface; (**b**) ARRm-H380-F24 h surface; (**c**) EDS spectrum acquired from ARRm-F24 h surface; (**d**) EDS spectrum acquired from ARRm-H380-F24 h surface; (**e**) cross-section profile of ARRm-F24 h; (**f**) cross-section profile of ARRm-H380-F24 h.

In-vitro electrochemical polarization tests are commonly used to evaluate the corrosion resistance of biodegradable metals. For the electrochemical reaction of the Mg alloy, the cathodic reaction mentioned above (Equation (2)) is the water-reduction reaction, which is also closely related to the driving force of hydrogen gas evolution; meanwhile, the anodic reaction represents the oxidation driving force of the α-Mg matrix (Equation (1)). Figure 7 shows the typical polarization curves of each experimental group in the r-SBF solution at 37 °C. The corrosion current density (I_{corr}) is the most important electrochemical parameter, and is often used to calculate polarization resistance (R_p) [23]. I_{corr} can be derived via the intersection point of E_{corr} and the cathodic extrapolation line. Thereafter, R_p can be calculated using the Stern-Geary equation, the results of which are listed in Table 3.

Among the ARRm, ARRm-F24 h, ARRm-H380 and ARRm-H380-F24 h groups, the ARRm unmodified group showed the highest I_{corr} and lowest R_p of 31.6 μA/cm² and 1046 Ω·cm², respectively, which indicates that it might encounter severe oxidation and corrosion in a physiological electrolytic environment. After solid-solution heat treatment, the MgZn$_2$ phases decomposed and dissolved into the α-Mg matrix of ARRm-H380 (fewer micro-galvanic couples), resulting in a lower I_{corr} (21.2 μA/cm^2) than ARRm. Moreover, the significant change in current slope suggests that the ARRm-H380 microstructures had the passivation behavior, which indicates that a protective oxide film formed on the surface [24]. This protective oxide film can inhibit aggressive ions from penetrating and reacting with the inner metal surface, thereby reducing the risk of forming hydrogen cavities and releasing highly-concentrated alkali ions [25]. Both the cathodic and anodic current densities were significantly reduced in the presence of the MgF$_2$ coating; in particular, ARRm-H380-F24 h offered the highest corrosion resistance and lowest anodic current density, and featured a wide passivation window (E_{break} − E_{corr}), indicating that the solid-solution heat treatment could further trigger better coating quality, performance and the overall anti-corrosion ability.

$$R_p = \frac{\beta_a \beta_c}{2.303(\beta_a + \beta_c) I_{corr}}$$

where I_{corr} is the corrosion current density, while β_a and β_c are the anodic and cathodic slopes, respectively, as obtained from the Tafel region.

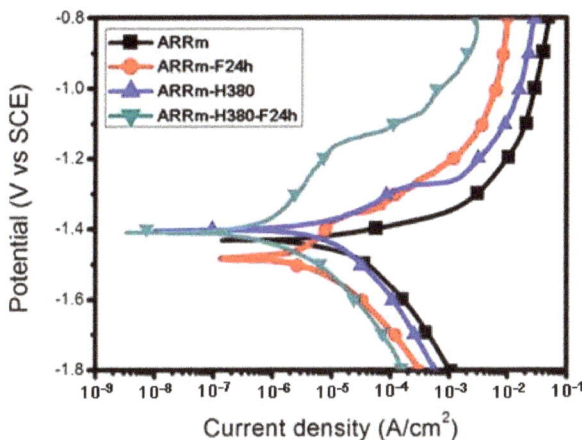

Figure 7. Electrochemical polarization curves obtained in r-SBF solution at 37 °C.

In our previous work, a duplex modified (microstructure modified and surface coated) Mg-Zn-Zr alloy was developed [20]. Although this material exhibited improved anti-corrosion behavior, the importance of the solid-solution heat treatment for fluoride conversion has not been discussed. To further understand the effect of pre-solid-solution heat treatment, the protection efficiency percentage (PE%) is employed to elucidate the importance and contribution of this novel fabrication process.

$$PE(\%) = \frac{I_{corr,sub} - I_{corr,MgF_2}}{I_{corr,sub}} \times 100\%$$

The PE value of the fluoride coating grown on the ARRm substrate was found to be smaller (~80.3%) than that of the ARRm-H380 substrate (~93.8%); this may be due to the latter having superior coating homogeneity and a thicker coating. It is worth noting that the solid-solution heat treatment can balance the potential difference between the MgZn$_2$ particles and Mg matrix. Moreover, owing to hydrogen gas evolution usually occurring at the micro-cathode site, the solid-solution heat treatment can also prevent the formation of a coating-depletion region on un-dissolved MgZn$_2$ particles, resulting in higher coating efficiency and protection ability (as shown in Figures 6 and 7).

Table 3. The resultant electrochemical polarization parameters.

Group	E_{corr} (V)	I_{corr} (μA/cm^2)	$\beta_{cathodic}$ (V/dec)	β_{anodic} (V/dec)	R_p ($\Omega \cdot$cm^2)	PE (%)
ARRm	−1.44	31.6	0.32	0.10	1046	-
ARRm-F24 h	−1.48	6.2	0.25	0.64	12,590	80.3
ARRm-H380	−1.40	21.2	0.23	0.16	1932	-
ARRm-H380-F24 h	−1.40	1.3	0.25	0.53	56,739	93.8

3.4. Effect of Clinical Pre-Bending Procedure on Corrosion Rate and Behavior of Mg Regeneration Membrane

Considering the practical application of the dental GBR procedure, dentists need to bend (inducing strain and residual stress) the regeneration membrane to fit the alveolar ridge [5]. However, Mg-based alloys generally have a known issue, namely the stress-corrosion cracking (SCC) phenomenon, which can accelerate localized corrosion and further cause early failure of the materials [26]. Therefore, the additional effect of bending residual stress on Mg regeneration membranes must be considered and examined before clinical trial. To realize the effect of bending-corrosion behavior, un-bent/r-SBF immersion and pre-bent/r-SBF immersion experiments are discussed in the following.

The un-bent/immersion experiment was carried out at 37 °C in r-SBF for 1, 2 and 4 weeks. Figure 8a shows the corrosion rate data calculated from the normal/immersion test. During the testing period, the corrosion rates of the modified samples were significantly higher than those of the unmodified samples, the corrosion-trend sequence of which from fast to slow corrosion is: ARRm > ARRm-H380 > ARRm-F24 h > ARRm-H380-F24 h. In addition, the corrosion trends of the immersion and electrochemical tests were identical. Figure 8b shows typical corroded surfaces of the different samples tested in r-SBF. The surface corrosion morphologies of the ARRm-H380 and ARRm-H380-F24 h specimens after 1, 3 and 7 days immersion shows the most homogeneous corrosion morphology. By contrast, localized corrosion, characterized by severe oxidation from the surface to the interior of the matrix, can be observed in the ARRm and ARRm-F24 h samples.

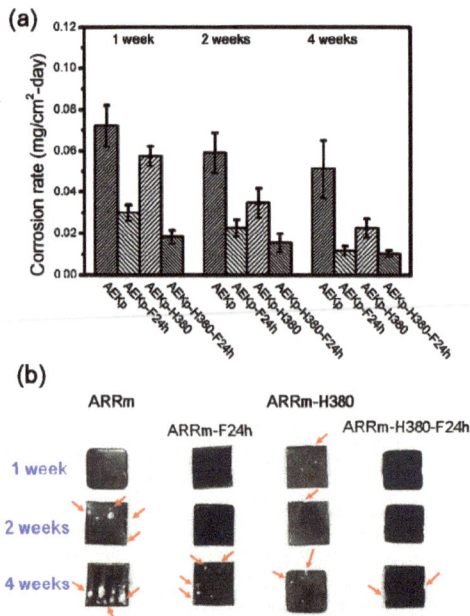

Figure 8. (a) Corrosion rates calculated from immersion test; (b) corrosion macrograph (red arrow: the location of corrosion pits).

Figure 9 presents the corrosion macrograph and micrographs of pre-bent/r-SBF immersion specimens. There are two regions of note on the pre-bent specimens. Firstly, the top surface of the strained area is where large tensile stress is located. Only the ARRm specimen displays poor anti-stress corrosion performance, as compared with the other three specimens. Secondly, due to the heterogeneous contact interface of the nylon screw fixation site, crevice corrosion might be triggered. According to related reports, Ghali et al. stated that crevice corrosion could be initiated due to the different hydrolysis rates between the heterogeneous contact interface of Mg alloys. Accordingly, the formation of $Mg(OH)_2$ could affect the corrosion driving force between the Mg regeneration membrane/screw interface in the crevices [27]. As seen in Figure 9a, the screw fixation interface of the ARRm, ARRm-F24 h, and ARRm-H380 specimens display severe crevice corrosion and accumulation of corrosion-product morphologies; however, the corrosion damage of the heat-treated sample (ARRm-H380) was clearly less than that of the ARRm and ARRm-F24 h samples. Notably, ARRm-H380-F24 h showed a satisfactory crevice corrosion-resistant behavior without significant corrosion damage, confirming that the solid-solution heat treatment can improve crevice-corrosion resistance.

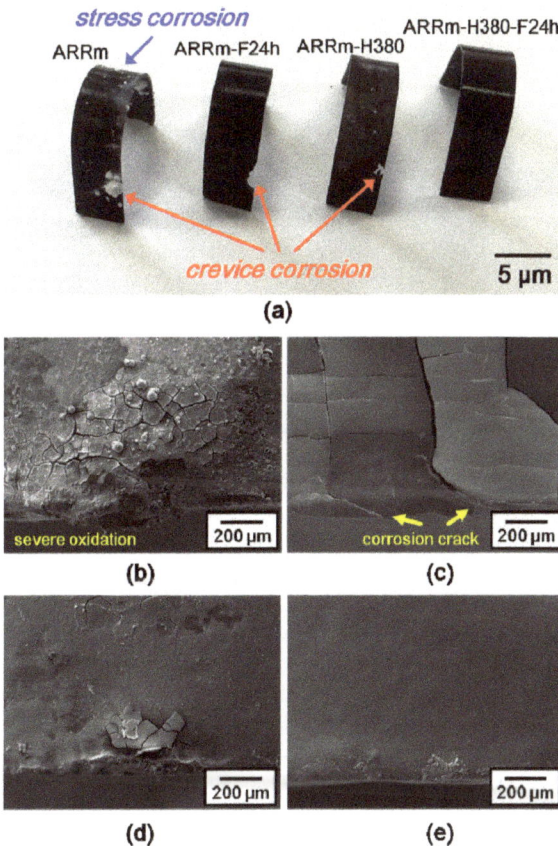

Figure 9. The pre-bent/immersion test examined specimens: (a) macrograph; and micrograph of (b) ARRm; (c) ARRm-F24 h; (d) ARRm-H380; (e) ARRm-H380-F24 h.

After immersion in r-SBF for 1 week, the corrosion trends from the un-bent/immersion and pre-bent/immersion tests were the same (Figure 10); however, the corrosion rates calculated from the pre-bent/immersion test were higher than those calculated from un-bent/immersion test. This can be logically explained by the effects of the bending residual stress and heterogeneous interface (screw fixation site), which can further accelerate the corrosion reaction.

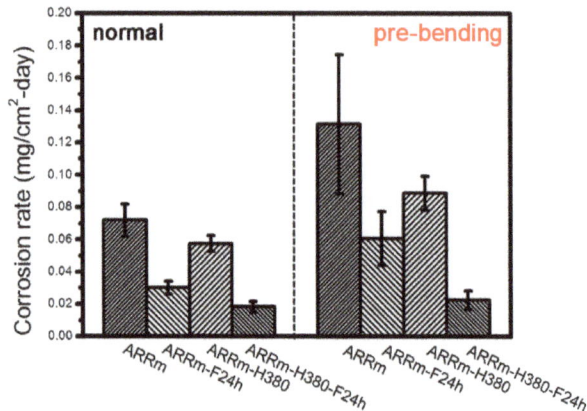

Figure 10. Corrosion rates calculated from un-bent/immersion (marked as normal) and pre-bent/immersion test (marked as pre-bending).

Relatively higher degradation rates for the non-coated series groups (namely ARRm and ARRm-H380) were found, which suffered aggressive corrosion by chlorine and/or other reactive ions. By comparison, after the H380 treatment, the specimens showed relatively lighter corrosion than the untreated ARRm. Moreover, the results also indicate that the fluoride coatings formed on the H380-treated specimens are denser and have fewer defects, thereby providing lower corrosion rates. As such, it is of great significance to reduce the secondary phase of the Mg alloy substrate (the contribution of H380), in order to improve the coating integrity and its protection against corrosion. Song et al. reported similar outcomes, where pre-solid-solution treatment could effectively improve the coating integrity of a Mg-2Zn-Mn-Ca-Ce alloy [28]. Considering that the performance of the heat-treated samples was superior, the following animal experiments only used the ARRm-H380 and ARRm-H380-F24 h alloys to examine the practicability.

3.5. In Vivo Degradation and Bone Healing Situation

During the implantation period, all experimental rats showed good health and wound healing until the end of this research. There were no severe side effects, obvious weight reduction, allergies, rejection and postoperative infection in the rats. Figure 11 shows the hydrogen accumulation phenomenon in the ARRm-H380 group after 4 weeks and 12 weeks of implantation. As can be seen, the photograph shows that ARRm-H380 produced a subcutaneous hydrogen gas cavity, indicating that the degradation amount of ARRm-H380-F24 h was significantly lower than ARRm-H380. The generation of hydrogen gas cavity is inevitable, due to the nature of Mg corrosion. This issue might cause swollen feeling at implanted area. Fortunately, swollen feeling can be minimized by subcutaneous puncture procedure (to leak the hydrogen cavity).

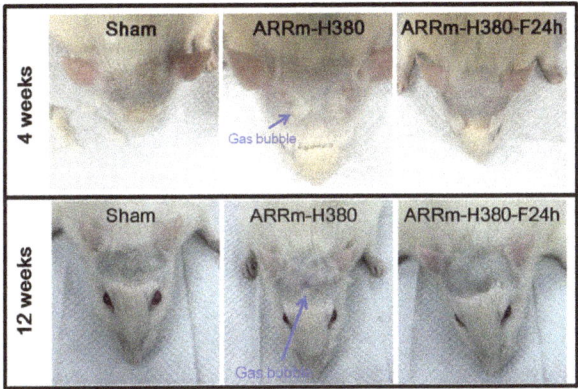

Figure 11. Hydrogen gas cavity and allergy observation.

As shown in Figure 12, the difference in degradation between ARRm-H380 and ARRm-H380-F24 h lies in the corrosion behavior. Whereas the ARRm-H380 sample showed a homogenous corrosion morphology, resulting in an evenly corroded cross-section, the ARRm-H380-F24 h specimen showed a localized corrosion morphology due to the corrosive factors penetrating into the substrate at the weak points in the fluoride coating [29]. Therefore, the weak points corroded first, after which they evolved into localized corrosion pits and/or holes. Furthermore, the thickness and weight retention of the degraded ARRm-H380-F24 h was significantly thicker and heavier than the degraded ARRm-H380, which means that the degradation and amount of released Mg ions of ARRm-H380-F24 h were relatively lower than for ARRm-H380.

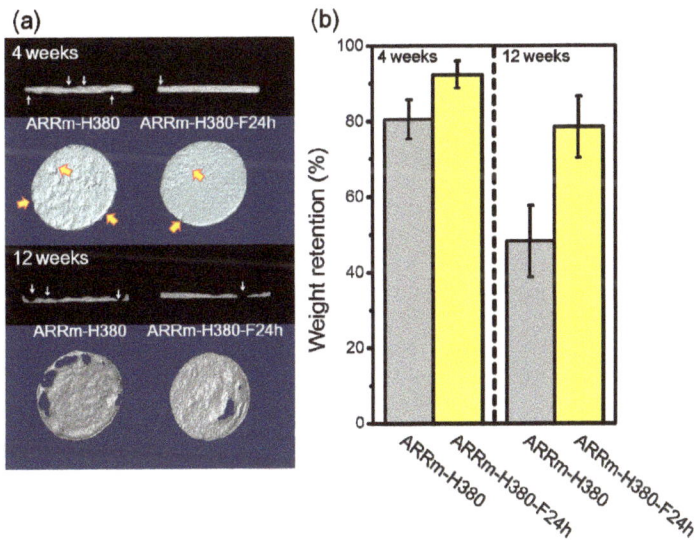

Figure 12. (**a**) 3D images of implanted Mg samples; (**b**) weight retention calculated from CT data. (arrow: the location of initial corrosion pit).

Figure 13 presents the 3D reconstruction images derived from the μ-CT analysis. The sham group showed no significant new bone regeneration; a bone fracture of the 5 mm size-level defect can not be healed because the rat cranial bone lacks blood supply and muscle tissue. However, a remarkable new bone-regeneration phenomenon can be observed in the ARRm-H380 group after 4 weeks and 12 weeks implantation. In comparison, the ARRm-H380-F24 h group showed relatively lower new bone-regeneration capability than with ARRm-H380.

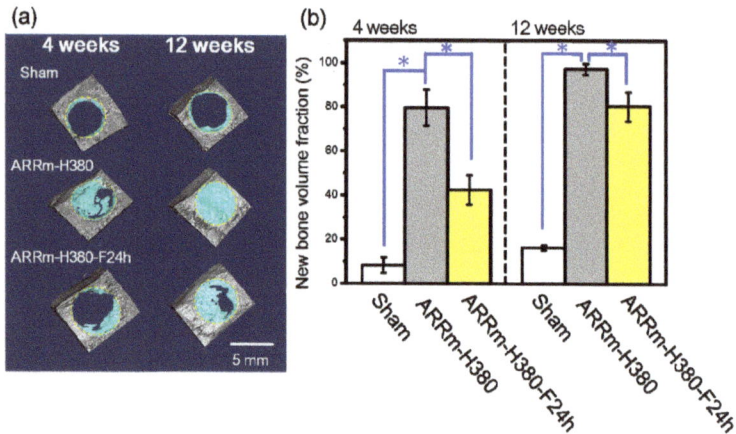

Figure 13. (**a**) 3D images of CSD sites of different experimental groups; (**b**) new bone volume fraction calculated from CT data. (Data presented as mean $\pm$ SD, n = 4 and analyzed using a one-way ANOVA, * p < 0.05).

Interestingly, in comparing Figures 12 and 13, it appears that a higher degradation amount of Mg substrate leads to superior bone regeneration ability. Hence, our results indicate that Mg ions act as an effective factor for bone growth. Moreover, with respect to the statistical data (Figure 13b), it seems clear that the bone-regenerative capability of the Mg-based regeneration membrane accelerated bone tissue formation. The new bone volume fraction with the ARRm-H380 membrane approached almost 100% after 12 weeks implantation. Previous reports have verified that Mg ions act as a stimulator to enhance cell proliferation and migration, and that the functional biochemical stimulation of Mg ions can improve wound healing in vitro and in vivo [7,30]. To the best of our knowledge, this report is the first to discover that Mg-based materials are capable of being applied in dentistry with excellent outcomes. This suitability could be attributed to the excellent mechanical structuring function and appropriate degradation properties of modified ECO505 (especially ARRm-H380 and ARRm-H380-F24 h).

In this study, we demonstrated the first report of the promoting effect for bone healing of the cranial bone in SD rats using modified ECO505 magnesium alloy to acquire the proper magnesium releasing concentration for bone tissue regeneration. According to the referenced articles and the study reported here, we created a representative illustration to demonstrate the promoting mechanism (see Figure 14). These results verified that the ARRm-H380 possessed a proper Mg-releasing ability to stimulate and enhance the regeneration of the bone defect areas, which is one of the key success factors for a GBR material. Although ARRm-H380-F24 h showed relatively lower bone-regeneration ability, its delayed degradation characterization can be effectively applied in older patients who generally require more recovery time. Therefore, this research successfully provides two future therapy solutions for different patients or therapy demands, namely short implantation period with ultra-fast healing effect, and long implantation period with moderate healing effect.

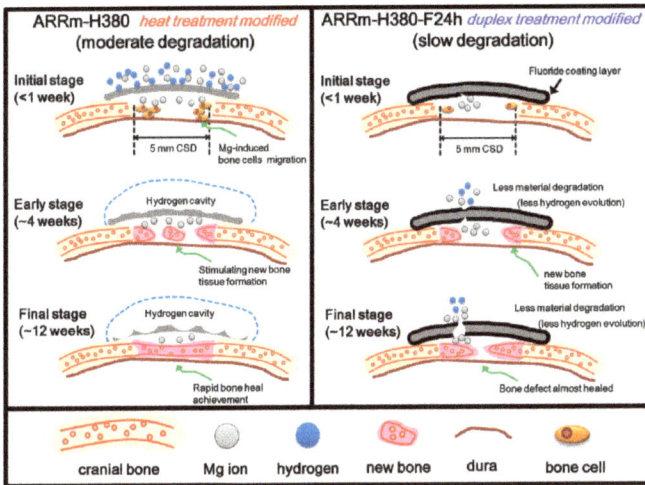

Figure 14. Schematic representation of Mg-based regeneration membrane degradation and new bone regeneration characteristics.

4. Conclusions

1. The environmentally friendly ECO505 material was successfully developed as the raw material of a novel Mg-based regeneration membrane.

2. The H380 solid-solution treatment could lead to the recovery of the matrix, refinement and reduction of the $MgZn_2$ secondary phases while enhancing elongation. Optimization of the solid-solution heat treatment requires the precisely controlled condition of 380 °C_2 h to prevent excessive grain growth in the ARRm material, resulting in ultimate elongation (20.2%). The mechanical properties of the coarse-grained microstructure (ARRm-380 °C_4 h and ARRm-380 °C_10 h) are not reliable for practical application.

3. ARRm-H380 can form a better fluoride coating quality on substrates than non-heat-treated ARRm substrate, thereby showing a higher gain of PE value (93.8%) and corrosion resistance (56 kΩ·cm^2).

4. ARRm-H380 and ARRm-H380-F24 h specimens can effectively enhance corrosion resistance and minimize the effects of stress corrosion and crevice corrosion, and so constitutes promising candidates for regeneration membrane treatment.

5. ARRm-H380 had a significant positive influence on new bone regeneration, where the CSD could heal almost 100% after 12 weeks implantation. In addition, although ARRm-H380-F24 h showed relatively lower bone regeneration ability, it nevertheless exhibited better long-term corrosion resistance than ARRm-H380.

Acknowledgments: The authors are grateful to the Instrument Center of National Cheng Kung University and Southern Taiwan Medical Device Industry Cluster, AZ-10-16-35-105, for their financial support of this research.

Author Contributions: Da-Jun Lin conceived and designed the experiments; Da-Jun Lin and Hung-Pang Lee performed the experiments; Da-Jun Lin and Fei-Yi Hung analyzed the data; Fei-Yi Hung and Ming-Long Yeh contributed reagents/materials/analysis tools; Da-Jun Lin wrote the paper.

Conflicts of Interest: The authors declare no conflict of interest.

References

1. Bottino, M.C.; Thomas, V.; Schmidt, G.; Vohra, Y.K.; Chu, T.M.G.; Kowolik, M.J.; Janowski, G.M. Recent advances in the development of GTR/GBR membranes for periodontal regeneration—A materials perspective. *Dent. Mater.* **2012**, *28*, 703–721. [CrossRef] [PubMed]

2. Mealey, B.L. Periodontal disease and diabetes—A two-way street. *J. Am. Dent. Assoc.* **2006**, *137*, S26–S31. [CrossRef]

3. Page, R.C.; Altman, L.C.; Ebersole, J.L.; Vandesteen, G.E.; Dahlberg, W.H.; Williams, B.L.; Osterberg, S.K. Rapidly progressive periodontitis—A distinct clinical condition. *J. Periodontol.* **1983**, *54*, 197–209. [CrossRef] [PubMed]

4. Yoshikawa, G.; Murashima, Y.; Wadachi, R.; Sawada, N.; Suda, H. Guided bone regeneration (GBR) using membranes and calcium sulphate after apicectomy: A comparative histomorphometrical study. *Int. Endod. J.* **2002**, *35*, 255–263. [CrossRef] [PubMed]

5. Rakhmatia, Y.D.; Ayukawa, Y.; Furuhashi, A.; Koyano, K. Current barrier membranes: Titanium mesh and other membranes for guided bone regeneration in dental applications. *J. Prosthodont. Res.* **2013**, *57*, 3–14. [CrossRef] [PubMed]

6. Witte, F.; Hort, N.; Vogt, C.; Cohen, S.; Kainer, K.U.; Willumeit, R.; Feyerabend, F. Degradable biomaterials based on magnesium corrosion. *Curr. Opin. Solid State Mater. Sci.* **2008**, *12*, 63–72. [CrossRef]

7. Lin, D.J.; Hung, F.Y.; Yeh, M.L.; Lui, T.S. Microstructure-modified biodegradable magnesium alloy for promoting cytocompatibility and wound healing in vitro. *J. Mater. Sci. Mater. Med.* **2015**, *26*, 1–10. [CrossRef] [PubMed]

8. Lin, D.J.; Hung, F.Y.; Yeh, M.L.; Lee, H.P.; Lui, T.S. Development of a novel micro-textured surface using duplex surface modification for biomedical Mg alloy applications. *Mater. Lett.* **2017**, *206*, 9–12. [CrossRef]

9. Xin, Y.; Hu, T.; Chu, P.K. In vitro studies of biomedical magnesium alloys in a simulated physiological environment: A review. *Acta Biomater.* **2011**, *7*, 1452–1459. [CrossRef] [PubMed]

10. Lin, D.J.; Hung, F.Y.; Jakfar, S.; Yeh, M.L. Tailored coating chemistry and interfacial properties for construction of bioactive ceramic coatings on magnesium biomaterial. *Mater. Des.* **2016**, *89*, 235–244. [CrossRef]

11. Cai, S.H.; Lei, T.; Li, N.F.; Feng, F.F. Effects of Zn on microstructure, mechanical properties and corrosion behavior of Mg-Zn alloys. *Mater. Sci. Eng. C* **2012**, *32*, 2570–2577. [CrossRef]

12. Song, Y.W.; Han, E.H.; Shan, D.Y.; Yim, C.D.; You, B.S. The role of second phases in the corrosion behavior of Mg-5Zn alloy. *Corros. Sci.* **2012**, *60*, 238–245. [CrossRef]

13. Lin, D.J.; Hung, F.Y.; Lui, T.S.; Yeh, M.L. Heat treatment mechanism and biodegradable characteristics of ZAX1330 mg alloy. *Mater. Sci. Eng. C* **2015**, *51*, 300–308. [CrossRef] [PubMed]

14. Yan, T.T.; Tan, L.L.; Xiong, D.S.; Liu, X.J.; Zhang, B.C.; Yang, K. Fluoride treatment and in vitro corrosion behavior of an AZ31B magnesium alloy. *Mater. Sci. Eng. C* **2010**, *30*, 740–748. [CrossRef]

15. Bakhsheshi-Rad, H.R.; Idris, M.H.; Kadir, M.R.A.; Daroonparvar, M. Effect of fluoride treatment on corrosion behavior of Mg-Ca binary alloy for implant application. *Trans. Nonferr. Met. Soc. China* **2013**, *23*, 699–710. [CrossRef]

16. Lin, D.J.; Hung, F.Y.; Liu, H.J.; Yeh, M.L. Dynamic corrosion and material characteristics of Mg-Zn-Zr mini-tubes: The influence of microstructures and extrusion parameters. *Adv. Eng. Mater.* **2017**. [CrossRef]

17. Li, Z.; Shizhao, S.; Chen, M.; Fahlman, B.D.; Debao, L.; Bi, H. In vitro and in vivo corrosion, mechanical properties and biocompatibility evaluation of MgF2-coated Mg-Zn-Zr alloy as cancellous screws. *Mater. Sci. Eng. C* **2017**, *75*, 1268–1280. [CrossRef] [PubMed]

18. Ramakrishnan, S.; Koltun, P. Global warming impact of the magnesium produced in china using the pidgeon process. *Resour. Conserv. Recycl.* **2004**, *42*, 49–64. [CrossRef]

19. Ha, W.; Kim, Y.J. Effects of cover gases on melt protection of Mg alloys. *J. Alloy. Compd.* **2006**, *422*, 208–213. [CrossRef]

20. Lin, D.J.; Hung, F.Y.; Yeh, M.L.; Lee, H.P.; Lui, T.S. Correlation between anti-corrosion performance and optical reflectance of nano fluoride film on biodegradable Mg-Zn-Zr alloy: A non-destructive evaluation approach. *Int. J. Electrochem. Sci.* **2017**, *12*, 3614–3634. [CrossRef]

21. Prasad, Y.V.R.K.; Rao, K.P. Effect of homogenization on the hot deformation behavior of cast AZ31 magnesium alloy. *Mater. Des.* **2009**, *30*, 3723–3730. [CrossRef]

22. Watanabe, H.; Mukai, T.; Ishikawa, K.; Higashi, K. Low temperature superplasticity of a fine-grained ZK60 magnesium alloy processed by equal-channel-angular extrusion. *Scr. Mater.* **2002**, *46*, 851–856. [CrossRef]
23. Stern, M.; Geary, A.L. Electrochemical polarization-1. A theoretical analysis of the shape of polarization curves. *J. Electrochem. Soc.* **1956**, *103*, C205.
24. Chang, J.W.; Guo, X.W.; Fu, P.H.; Peng, L.M.; Ding, W.J. Effect of heat treatment on corrosion and electrochemical behaviour of Mg-3Nd-0.2Zn-0.4Zr (wt. %) alloy. *Electrochim. Acta* **2007**, *52*, 3160–3167. [CrossRef]
25. Lee, H.P.; Lin, D.J.; Yeh, M.L. Phenolic modified ceramic coating on biodegradable Mg alloy: The improved corrosion resistance and osteoblast-like cell activity. *Materials* **2017**, *10*. [CrossRef] [PubMed]
26. Winzer, N.; Atrens, A.; Song, G.L.; Ghali, E.; Dietzel, W.; Kainer, K.U.; Hort, N.; Blawert, C. A critical review of the stress corrosion cracking (SCC) of magnesium alloys. *Adv. Eng. Mater.* **2005**, *7*, 659–693. [CrossRef]
27. Ghali, E.; Dietzel, W.; Kainer, K.U. Testing of general and localized corrosion of magnesium alloys: A critical review. *J. Mater. Eng. Perform.* **2004**, *13*, 517–529. [CrossRef]
28. Song, D.; Li, C.; Zhang, L.; Ma, X.; Guo, G.; Zhang, F.; Jiang, J.; Ma, A. Decreasing bio-degradation rate of the hydrothermal-synthesizing coated Mg alloy via pre-solid-solution treatment. *Materials* **2017**, *10*. [CrossRef] [PubMed]
29. Witte, F.; Fischer, J.; Nellesen, J.; Vogt, C.; Vogt, J.; Donath, T.; Beckmann, F. In vivo corrosion and corrosion protection of magnesium alloy LAE442. *Acta Biomater.* **2010**, *6*, 1792–1799. [CrossRef] [PubMed]
30. Schaller, B.; Saulacic, N.; Imwinkelried, T.; Beck, S.; Liu, E.W.Y.; Gralla, J.; Nakahara, K.; Hofstetter, W.; Iizuka, T. In vivo degradation of magnesium plate/screw osteosynthesis implant systems: Soft and hard tissue response in a calvarial model in miniature pigs. *J. Cranio-Maxillofac. Surg.* **2016**, *44*, 309–317. [CrossRef] [PubMed]

Article

Investigation on Mechanical Behavior of Biodegradable Iron Foams under Different Compression Test Conditions

Reza Alavi, Adhitya Trenggono, Sébastien Champagne and Hendra Hermawan *

Department of Mining, Metallurgical and Materials Engineering and Centre Hospitalier Universitaire de Québec Research Center, Laval University, Quebec City, QC G1V 0A6, Canada; reza.alavi.1@ulaval.ca (R.A.); adhitya.trenggono.1@ulaval.ca (A.T.); sebastien.champagne.2@ulaval.ca (S.C.)
* Correspondence: hendra.hermawan@gmn.ulaval.ca; Tel.: +1-418-656-2131 (ext. 5876)

Academic Editor: Eli Aghion
Received: 13 April 2017; Accepted: 30 May 2017; Published: 2 June 2017

Abstract: Biodegradable metal foams have been studied as potential materials for bone scaffolds. Their mechanical properties largely depend on the relative density and micro-structural geometry. In this work, mechanical behavior of iron foams with different cell sizes was investigated under various compression tests in dry and wet conditions and after subjected to degradation in Hanks' solution. Statistical analysis was performed using hypothesis and non-parametric tests. The deformation behavior of the foams under compression was also evaluated. Results show that the mechanical properties of the foams under dry compression tests had a "V-type" variation, which is explained as a function of different geometrical properties by using a simple tabular method. The wet environment did not change the compression behavior of the iron foams significantly while degradation decreased the elastic modulus, yield and compression strengths and the energy absorbability of the specimens. The deformation of open cell iron foams under compression is viewed as a complex phenomenon which could be the product of multiple mechanism such as bending, buckling and torsion.

Keywords: biodegradable metals; iron foam; scaffold; compression; degradation; cell size; open cell foam

1. Introduction

Porous metals or metal foams are used in different applications where altered material properties of the parent metal are beneficial to the quality of the application. In biomedical engineering, metal foams can be used for biodegradable orthopedic implants such as bone scaffolds. Although foaming does not change all the material properties of the cell-wall material, there are properties which depend on the density (and therefore porosity) and micro-structural geometry of cellular materials: "the stiffness, the mechanical strength, the thermal and electrical conductivity as well as acoustic properties" [1]. One should note that the stiffness (elastic modulus) of a cellular structure, e.g., iron foam, depends on the architecture of the structure to a great extent. Therefore, it should not be confused with the elastic modulus of the cell wall material, e.g., iron [1]. As stated in [2], the material properties of metal foams "most directly" depend on the relative density and the properties of the parent material. However, structural properties such as pore sizes, cell types, etc. also influence the material properties of the foams. Bone scaffolds are implanted in the body to serve as a platform on which bone formation takes place. An ideal bone scaffold would resemble the mechanical properties of the natural bone. One advantage of porous metals over solid metals is their lower stiffness, which makes it closer to that of a bone. This would help to reduce stress shielding which can happen in case of using solid metals due to their higher Young's modulus than that of the bone. Porous structure can also contribute to new tissue infiltration and bone formation [3,4]. The structure of the porous scaffold (pore size distribution, interconnectivity and porosity) is an influential factor on the quality of the scaffold as it affects the level of cell penetration, cell growth and material transportation into and out of the scaffold [5].

Different biocompatible porous metals have been studied to serve as scaffolds for orthopedic applications: tantalum, magnesium and its alloys, titanium and its alloys, and iron-foam based materials [5,6]. Magnesium, iron and zinc are considered as biodegradable metals. However, among the available studies on biodegradable systems, the majority of the investigations have been on magnesium based materials due to their non-toxicity and similar mechanical behavior to that of human bone [7]. There have been controversies over biocompatibility of iron due to the emergence of metallosis, local destruction of tissues as a result of mechanical-biological, electro-energetic, and chemical-toxic effects of metal after implantation, of iron implants [8]. Nevertheless, there exist studies that show iron-based materials are possibly suitable for temporary biodegradable implants [9,10]. They provide answers to the two major drawbacks of magnesium-based materials which are high degradation rate that limits the use of such materials on small implants with approximate life span of 6–12 months, and hydrogen evolution during corrosion that can disturb the healing process [7,11–13]. Mechanical properties of iron-based alloys, e.g., strength and ductility, can be easily tailored to meet the criteria for some biomedical applications. They are viewed as good candidates for load bearing biodegradable implants owing to their high mechanical properties, e.g., high strength [13], and biocompatible, non-toxic characteristics [7]. However, a major challenge of using iron as a biodegradable implant is its slow rate of degradation [7,14]. Different approaches have been proposed to improve the corrosion rate of iron such as alloying elements modification, poly(lactic-*co*-glycolic acid) infiltration, and coating [6,15–17]. The objective of this study is to investigate the mechanical behavior of iron-foams as a function of their structural geometry where several iron foam specimens with different structural properties underwent uniaxial compression tests. Results are discussed in terms of the influence of porous properties, i.e., cell size, pore size, number of pores and strut thickness, and influence of environmental conditions, i.e., wet and dry condition as well as degradation on mechanical behavior of iron foam samples. Deformation mechanisms of iron foams under compression were also studied using scanning electron microscope images (SEM) analysis.

2. Materials and Methods

2.1. Iron Foam Specimens

The specimens were open cell pure iron foam with nominal cell diameters of 450 (IF45), 580 (IF58) and 800 (IF80) μm manufactured by Alantum. The foams were produced by the replication of open cell polyurethane (PU) foams [18]. To do so, a thin layer of Ni is sputtered on the PU foam to make the foam conductive for the following electroplating process. Then, iron would be electroplated on the foam to produce the open cell iron foams [18]. Measurements of cell sizes, pore sizes and strut thicknesses of iron foam samples were performed using scanning electron microscope images (SEM, Quanta 250 FEI, Hillsboro, OR, USA). More details on the geometrical measurement are available in Section A of the Supplementary Materials (Figure S1 and Table S1). One should note that these measurements were conducted on 2D images, and the 3D structure of the foams was not considered. Thus, the average measurements should be considered as estimations. The relative density of a foam is defined as ratio of the foam density over the density of the cell-wall material (ρ^*/ρ_s) [19]. To obtain the foam densities, the iron foam sheets were cut into cubic specimens by a stainless-steel scalpel. For each group of iron foams, three specimens were used. The dimensions of the specimens were measured via caliper to obtain the apparent volume (the bulk volume which contains the struts and pores). The mass of each specimen was measured via a sensitive digital scale. The foam densities were calculated as the ratio of masses over apparent volumes. The density of the cell-wall material (iron) was taken as 7.874 (g/cm^3). The calculated average relative density values of IF45, IF58 and IF80 samples were

0.038, 0.027 and 0.025, respectively (Table 1). It should be mentioned that for calculation of the relative densities, it was assumed that the solid structures of the foams contained only iron, and no remainder of PU was present. To prepare the cubic samples for the mechanical tests, they were initially cut from the sheets via a stainless-steel scalpel slightly larger than the final dimensions, and the thicknesses remained unchanged. Then the width and length were reduced to the final dimension by a rotary cutting tool kit. The nominal width and length of iron foam specimens considered for dry compression tests was 10 mm × 10 mm, and the nominal thickness (along the loading direction) for IF45, IF58 and IF80 was 1.7, 2.1 and 2.6 mm, respectively. According to [20], the minimum requirement for the size of all the spatial dimensions of the specimens and for their ratio to the average pore size is 10 mm and 10, respectively. The length and width of the iron foam specimens satisfy both mentioned requirements and have been verified as detailed in Section B of the Supplementary Materials (Figure S2). For wet compression tests and immersion tests, the specimens (only IF80) had similar nominal dimensions as the dry test specimens. To understand the deformation mechanisms, a new set of IF45 specimens were prepared to undergo dry compression tests before SEM observations. They were also prepared in the same fashion as other specimens with similar nominal dimensions. However, because no quantitative analysis was involved in this part, the accuracy of measured dimensions was not critical.

2.2. Mechanical Testing

2.2.1. Compression Test Parameters and Conditions

Figure 1 presents typical stress–strain curves of metallic foams under compression. Different specimen sizes and cross-head speeds were considered initially to choose proper test parameters. Finally, it was decided to use specimens with compression area of 100 mm² ($A = 10 \times 10$ mm²) and the cross-head speed 0.001 mm/s for all the tests and analyses. The compression tests under dry and wet conditions were carried out by Instron machine (ElectroPuls E1000, Instron, Norwood, MA, USA) with a 2 kN load cell. The details of the procedure for choosing the test parameters as well as the dry compression stress–strain curves of all the iron foam samples are available in Section B of the Supplementary Materials (Figures S2–S5).

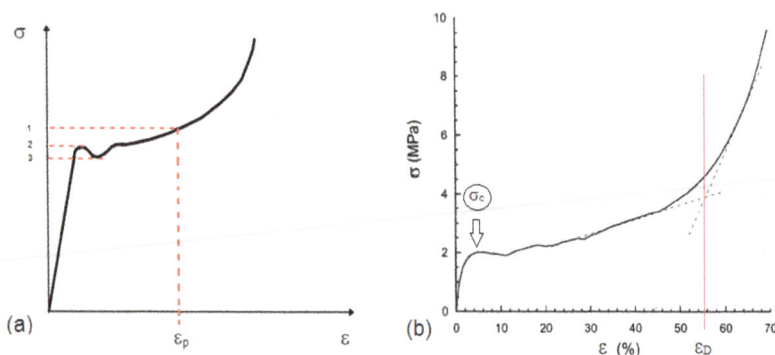

Figure 1. (**a**) A typical stress–strain curve of a foam under compression, numbers indicate different methods to define the compression strength; and (**b**) determination of densification strain. Adapted from [21,22].

Given that the biological environment within the body is not dry, wet compression tests were carried out on IF80 specimens to provide a more realistic condition. The tests were conducted within Hanks' solution (H1387, Sigma Aldrich, Saint Louis, MO, USA) at 37 °C filled into a cylindrical bath (inside diameter = 140 mm, solution fill height = 50 mm) mounted on the Instron machine after the specimen is fixed without gap within the platens (diameter = 50 mm, thickness = 25 mm) and before loading. The configuration allowed the solution to escape during compression and hydrostatic pressure was avoided. In addition, considering that the environment within the body is corrosive, static immersion tests were conducted on IF80 specimens to investigate the effects of degradation on mechanical behavior of iron foams. In the static immersion, the specimens were hung from a wire and submerged in 100 mL of the Hanks' solution then placed in separate incubators (ThermoFisher Scientific, Waltham, MI, USA) for three or seven days at 37 °C and 5% CO_2 atmosphere. The compression test parameters and the specimen nominal sizes were identical to those of the previous dry and wet tests. Finally, a few compression tests were performed on IF45 specimens in order to understand the deformation and failure mechanism of iron foams under compression. To do so, the specimens underwent compression up to different strain levels and then they were observed under SEM.

2.2.2. Compression Properties

In order to assess the mechanical behavior of the iron foams, the following properties were determined using the stress–strain response of the specimens: Elastic modulus (E), yield strength (σ_y), compression strength (σ_c), densification strain (ε_D) and the energy absorbability per volume up to the point of densification (W). The elastic modulus values were approximated by the linear fitting tool of the Quick Fit Gadget provided in OriginPro 2016 software (OriginLab, Northampton, MA, USA). After estimating the slope of the linear elastic regime (E) on each stress–strain curve, the yield strength was approximated by 0.2% offset method [23]. For most of the iron foam specimens investigated in this study, compression strength (σ_c) was taken as the first local maximum after elastic regime (marked by 2 and σ_c in Figure 1a,b, respectively). However, if there was no apparent first local maximum, an arbitrary local maximum in the plateau region was chosen to represent the compression strength stated by Banhart and Baumeister [21]. Densification strain (ε_D) can be defined as a strain at which densification begins. However, because normally there is no abrupt transition from plateau regime to densification regime, ε_D can be defined as the intersection of the tangents to the plateau and densification regimes (marked as ε_D in Figure 1b) [22]. In this study, it was tried to draw the tangents from the midpoint of the plateau regime (where the second derivation is close to zero) and from a part of the densification regime where the slope seems to become stable. The points were selected by eyeballing. A larger ε_D implies that the material undergoes higher strains before the onset of densification. The value of energy absorbed per volume up to the densification strain (W) is equal to the area underneath the stress–strain diagram from $\varepsilon = 0$ to $\varepsilon = \varepsilon_D$, expressed as [22]:

$$W = \int_0^{\varepsilon_D} \sigma(\varepsilon)d\varepsilon \tag{1}$$

The point of $\varepsilon = 0$ was determined as the intersection of the line continuing the linear elastic regime with the strain axis. This method was inspired from the definition of the "zero point for the compressive strain" provided in [20]. High energy absorption capacity indicates the higher level of impact absorption by the material. However, in order to study the impact behavior in a more comprehensive fashion, conducting dynamic impact tests with higher strain rates beside static tests is recommended [24].

2.3. Statistical Analysis

Values of E, σ_y, σ_c, ε_D and W were compared against one another for different groups of specimens via statistical analysis and were shown as mean $\pm$ standard deviation. In order to draw a reliable conclusion, hypothesis tests (One-way ANOVA, *t*-test) along with non-parametric tests (Kruskal–Wallis ANOVA, Two-Sample Kolmogorov–Smirnov Test) were carried out in the OriginLab software. The non-parametric tests were conducted due to the small sample sizes. Unlike parametric hypothesis tests, the normal distribution of population is not assumed when conducting non-parametric tests [25].

3. Results

3.1. Iron Foam Structure

Figure 2 depicts the structure of iron foams with different average cell sizes. It shows that IF80 specimens tend to have thicker struts than those of IF45 and IF58 specimens. Strut thicknesses of the iron foams of different cell sizes are shown in Table 1. The mean values of both branch-strut and end-strut thicknesses of the IF80 specimen are significantly higher than those of the IF45 and IF58 specimens. Definition of cell size, branch-strut and end-strut as well as detail measurement are given in Figure S1 and Table S1 of the Supplementary Materials.

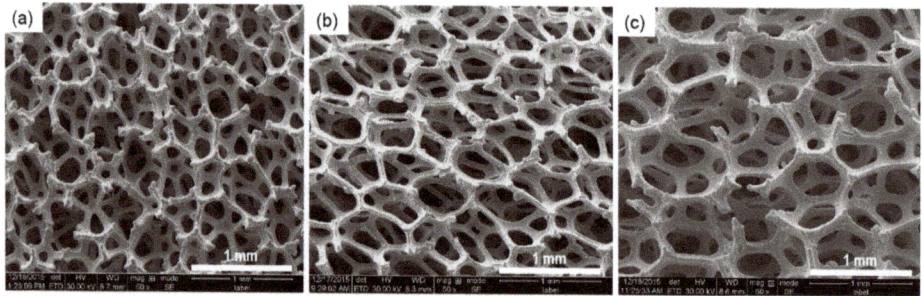

Figure 2. Scanning electron microscope (SEM) images of iron foam structures: (**a**) IF45; (**b**) IF58; and (**c**) IF80.

Table 1. Relative density, cell size, end-strut and branch-strut thicknesses of the iron foams.

Specimen	Relative Density	Cell Size (μm)	Pore Size (μm) *	End-Strut (μm)	Branch-Strut (μm)
IF45	0.038 ± 0.001	461.77 ± 72.26	155.59± 27.94	74.73 ± 10.30	55.52 ± 6.18
IF58	0.027 ± 0.001	617.73 ± 76.08	150.8 ± 29.43	63.62 ± 9.95	59.88 ± 7.55
IF80	0.025 ± 0.001	828.11 ± 79.87	157.33 ± 28.50	97.79 ± 17.54	80.91 ± 12.27

* Between 100 and 200 μm.

3.2. Dry Compression Behavior of the Iron Foams

The stress–strain curves resulted from the dry compression tests on iron foams of different cell sizes are shown in Figure 3. The shifting of the curves within a sample group can be the result of non-identical micro configurations even if they are in the same sample group. The IF45 specimens tend to have higher compression strength than those of IF58 and IF80 specimens. Strain hardening up to the peak followed by a softening is more visible in the stress–strain curves of IF45 and IF80 specimens than those of IF58 specimens. Thus, the local maximum after the linear elastic regime followed by a local minimum in IF58 curves does not stand out as much as it does in IF45 and IF80 curves. The IF58 specimens seem to experience a shorter plateau region than that of IF45 and IF80 specimens. In general, IF45 specimens tend to have the highest strength under compression as they experience the highest level of stress in all regimes. The compression properties of the iron foams are summarized in Table 2. There exists a "V-type" variation of the mechanical properties with respect to the cell sizes, i.e., the mean values of the compression properties of the iron foams of 580 μm nominal cell sizes tend to be lower than those of the specimens with 450 and 800 μm nominal cell size.

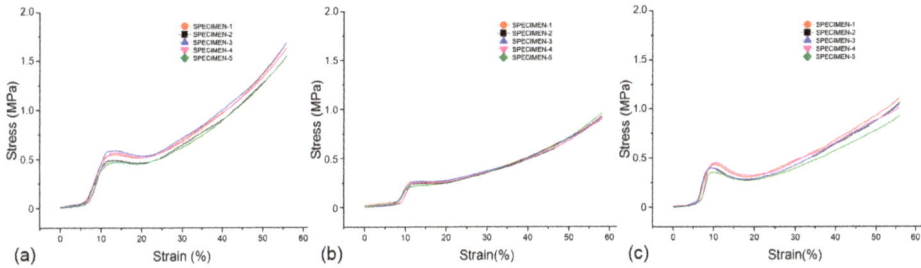

Figure 3. The stress–strain curves of the dry compression tests for different cell sizes: (**a**) IF45; (**b**) IF58; and (**c**) IF80.

Table 2. Compression properties of iron foam specimens tested in dry condition.

Specimen	E (MPa)	σ_y (MPa)	σ_c (MPa)	ε_D (mm/mm)	W (MJ/m^3)
IF45	11.60 ± 1.39	0.48 ± 0.07	0.53 ± 0.05	15.03 ± 0.91	0.054 ± 0.009
IF58	8.24 ± 0.67	0.23 ± 0.03	0.26 ± 0.02	9.94 ± 0.54	0.016 ± 0.003
IF80	17.11 ± 2.3	0.36 ± 0.03	0.41 ± 0.04	12.52 ± 0.48	0.033 ± 0.004

3.3. Compression Behavior of Iron Foams in Hanks' Solution and after Degradation

Elastic modulus, yield and compression strength, densification strain and energy absorbability of five IF80 specimens under wet compression tests were calculated and compared against those of IF80 specimens which had undergone dry compression. The properties of specimens after three- and seven-day immersions were also compared with those which had not undergone any immersion. The stress–strain diagrams and mean values of compression properties are demonstrated in Figure 4 and Table 3, respectively. None of the properties are significantly affected by the wet environment. Differently, the compression behavior of IF80 iron foams was affected after three- and seven-day immersions in Hanks' solution as shown in Figure 4c,d and in Table 3. All the calculated compression strength decreased as a result of degradation. However, these reductions are not significant for densification strain and between the three and seven days after immersion samples.

Table 3. Compression properties of IF80 specimens under dry and wet condition, and after immersion tests.

Specimen	E (MPa)	σ_y (MPa)	σ_c (MPa)	ε_D (%)	W (MJ/m^3)
Wet condition	14.14 ± 1.39	0.33 ± 0.44	0.37 ± 0.04	12.49 ± 0.27	0.030 ± 0.003
No immersion	14.78 ± 2.28	0.39 ± 0.04	0.43 ± 0.04	13.21 ± 0.96	0.039 ± 0.004
3-day immersion	10.48 ± 1.39	0.25 ± 0.03	0.31 ± 0.04	13.17 ± 0.50	0.027 ± 0.005
7-day immersion	10.06 ± 1.49	0.25 ± 0.03	0.30 ± 0.03	13.09 ± 0.73	0.025 ± 0.004

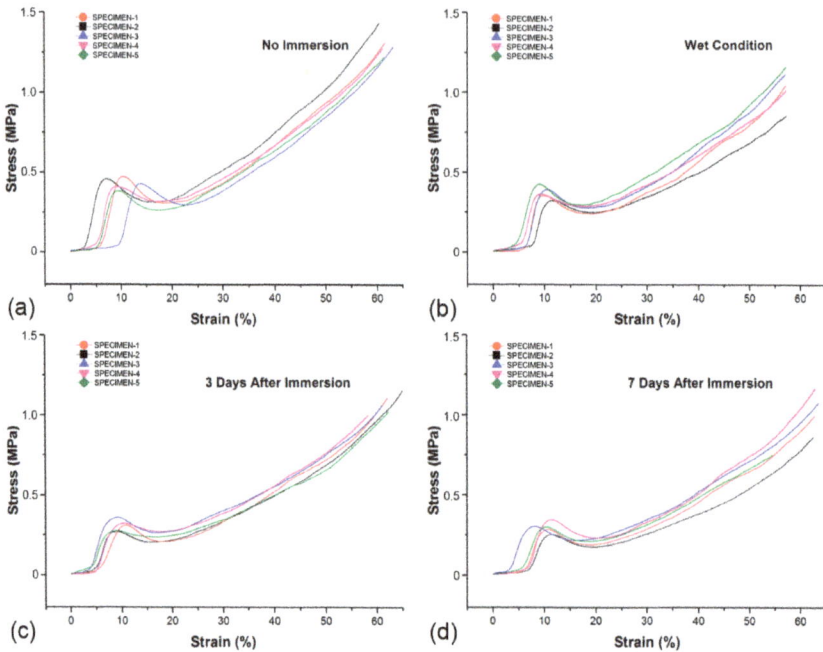

Figure 4. The stress–strain curves of IF80 specimens: (**a**) no immersion; (**b**) wet conditions; (**c**) three days after immersion; and (**d**) seven days after immersion.

3.4. Statistical Analysis

Both hypothesis tests (ANOVA and Tukey) and non-parametric tests were employed to investigate the significance of differences among the values of the elastic and plastic properties of the different iron foams. Figure 5a,b represents the elastic modulus (E) and yield strength (σ_y) box-charts of the specimens under compression, respectively. Although both of the properties represent a V-type variation, an interesting difference is observed between the two: highest elastic modulus is presented by IF80 specimens, while the highest yield strength is presented by IF45 specimens. Indeed, among all the determined properties, it is only the elastic modulus that has a V-type variation in which IF80 specimens present the highest values. Figure 5c depicts the compression strength box-charts of the iron foam specimens of different cell sizes. The figure shows a V-type variation of the strength (Figure 5c). The mean values of compression strengths within IF45, IF58 and IF80 samples were 0.53, 0.255 and 0.41 MPa, respectively. According to the ANOVA and Tukey tests, at $\alpha = 0.05$, all the population means of σ_y, E and σ_c were significantly different from one another. The results obtained from non-parametric tests agreed with those of hypothesis tests: the null hypothesis in Kruskal–Wallis ANOVA test was rejected at significance level of 0.05. Two-sample Kolmogorov–Smirnov (K–S) tests with significance level of 0.05 were carried out within each possible sample pair. The results of K–S tests showed that all the distributions were significantly different from one another, except for the case of the yield strength difference between IF45 and IF80 samples which contradicts the result drawn from the Tukey test. However, given the difference shown by the box-chart of Figure 5b, the Tukey test result seems to be more reliable, i.e., the difference between the yield strength value of IF45 and IF80 populations is significant.

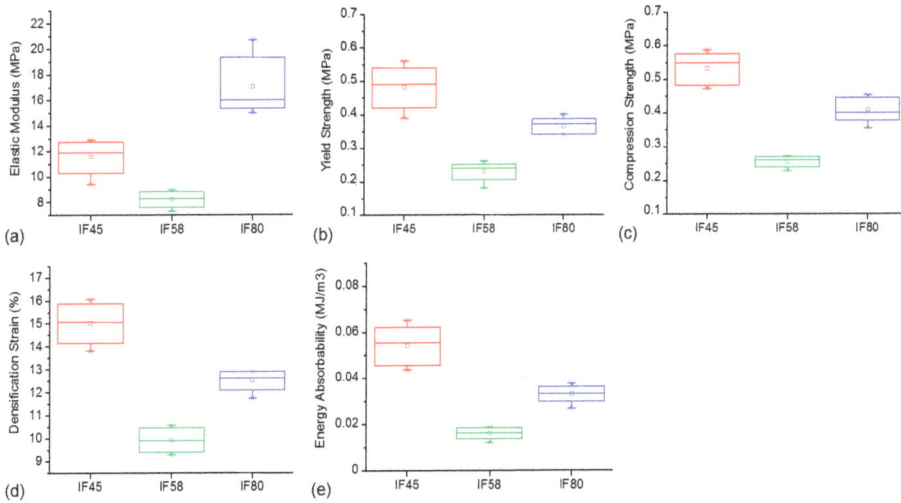

Figure 5. Box charts for iron foam specimens of different cell sizes: (**a**) elastic modulus; (**b**) yield strength; (**c**) compression strength; (**d**) densification strain and (**e**) energy absorbability.

Figure 5d depicts the densification strain (ε_D) box-charts of the iron foam specimens of different cell sizes. Similar to the compression strength variations, IF45 and IF58 samples show the highest and the lowest values of densification strain, respectively (V-type variation). Both One-way ANOVA and non-parametric Kruskal–Wallis ANOVA tests were conducted to evaluate the significance of difference between the populations. Based on the ANOVA and Tukey test results, all the population means were significantly different form one another with a significance level of 0.05. This was in agreement with the result obtained by Kruskal–Wallis ANOVA test at $\alpha = 0.05$, and the three two-sample K–S tests between pairs at $\alpha = 0.05$. Therefore, it can be concluded that IF45 and IF58 samples possess the highest and the lowest level of densification strain, respectively. Figure 5e depicts the amount of absorbed energy per Volume (*W*) up to the densification strain for specimens of different cell sizes which was also compared against one another via one-way ANOVA, Tukey, Kruskal–Wallis ANOVA, and Two-Sample Kolmogorov–Smirnov (K–S) tests. The mean value of energy absorbability within IF45, IF58 and IF80 samples was 0.054, 0.016 and 0.033 MJ/m^3, respectively. According to the Homogeneity of Variance test results, the population variances were significantly different ($\alpha = 0.05$) which violates the equality-of-variance requirement to perform ANOVA test. Although the ANOVA test results stated that at least two populations are significantly different, three two-sample *t*-tests at significance level of 0.01 were performed due to the violation. The results of the *t*-tests suggest that every population is significantly different from the other two ($\alpha = 0.01$). Same conclusion was drawn from performing a Kruskal–Wallis ANOVA test in conjunction with three two-sample Kolmogorov–Smirnov (K–S) tests at significance level of 0.05. Therefore, it is concluded that iron foams of 450 μm and 580 μm average cell sizes have the highest and the lowest level of energy absorbability, respectively.

To compare the results of wet and dry compression tests, both two-sample *t*-test and two-sample non-parametric Kolmogorov–Smirnov test at significance level of 0.05 were utilized. For the case of the elastic modulus, *E*, the *t*-test showed that the difference between the values of the two groups was significant which disagreed with what the Kolmogorov–Smirnov test suggested, i.e., the two distributions were not significantly different. Thus, considering that the values of *E* were just an estimation and that the sample sizes were small, it is not feasible to make a reliable comment on the elastic behavior of the iron foams within the wet environment. However, for the other compression properties the results of the statistical tests were consistent. Although the results of compression tests showed that under wet condition, the mean values of the σ_y, σ_c, ε_D and W tended to be slightly lower than those of the dry tests, the results of all the hypothesis and non-parametric tests showed otherwise: None of the calculated mechanical properties under wet condition were significantly different from those of dry compression. This suggests that the wet environment did not significantly influence the mentioned mechanical properties of the iron foams.

The significance of difference between the mechanical properties of the iron foams after different periods of immersion was investigated via hypothesis and non-parametric tests. Figure 6a–c,e depicts the elastic modulus, yield strength, compression strength and the energy absorbability box charts of IF80 specimens after three- and seven-day immersion tests, respectively. In addition, the box charts samples upon which no immersion test was performed are included. The figure shows that there is a relatively high reduction in the levels of all the four properties after three-day immersion (Figure 6). In addition, the result of one-way ANOVA indicated that ($\alpha = 0.05$), at least two sample population were significantly different, and the results of Tukey test showed that ($\alpha = 0.05$) the mean differences between "three-day immersion" and "seven-day immersion" samples were not significant unlike the other two comparisons, i.e., "no immersion–three-day immersion" and "no immersion–seven-day immersion" samples. Identical conclusions were drawn from conducting Kruskal–Wallis ANOVA tests in conjunction with three Kolmogorov–Smirnov tests within the sample pairs at $\alpha = 0.05$.

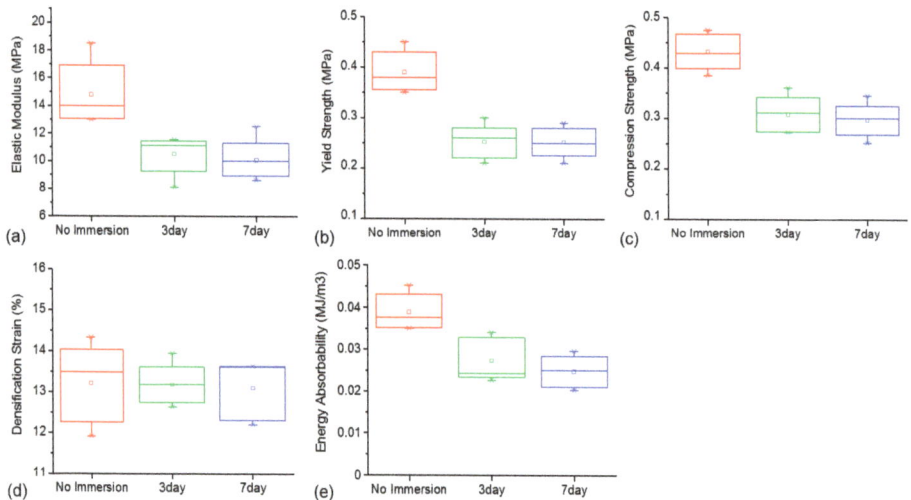

Figure 6. Box charts for iron foam specimens of different immersion time: (**a**) elastic modulus; (**b**) yield strength; (**c**) compression strength; (**d**) densification strain and (**e**) energy absorbability.

Figure 6d depicts the densification strain box-charts of IF80 iron foam specimens after three- and seven-day immersions as well as those on which no immersion test was performed. Looking at the box-charts, the difference between the densification strain levels does not seem to be significant. As the normality of the densification strain values within each sample group was evaluated by Shapiro–Wilk tests, it appeared that the normality of "seven-day immersion" sample population was rejected unlike the other two groups. The result of one-way ANOVA and Kruskal–Wallis ANOVA showed that at $\alpha = 0.05$, the difference between the sample population were not significant. Although the normality requirement for the ANOVA test was violated, the result of the test is expected to be valid given the proximity of strain densification mean values of the samples.

3.5. Deformation of the Iron Foams after Compression

To understand the deformation and failure mechanism of open cell iron foams during compression, various compression tests were conducted on IF45 specimens followed by SEM observation of the compressed specimens. Figure 7 depicts four specimens compressed up to strain levels of 10.8%, 12.8%, 29.8% and 49.5%. The areas marked with red circles depict the struts that experienced bending or plastic deformation. Because the direction of the compression force is perpendicular to the image surfaces, it is difficult to identify buckling in the struts, if it existed. However, considering that the majority of the struts are not completely perpendicular or parallel to the loading direction, it is expected that both bending and buckling contribute to the deformation of some struts.

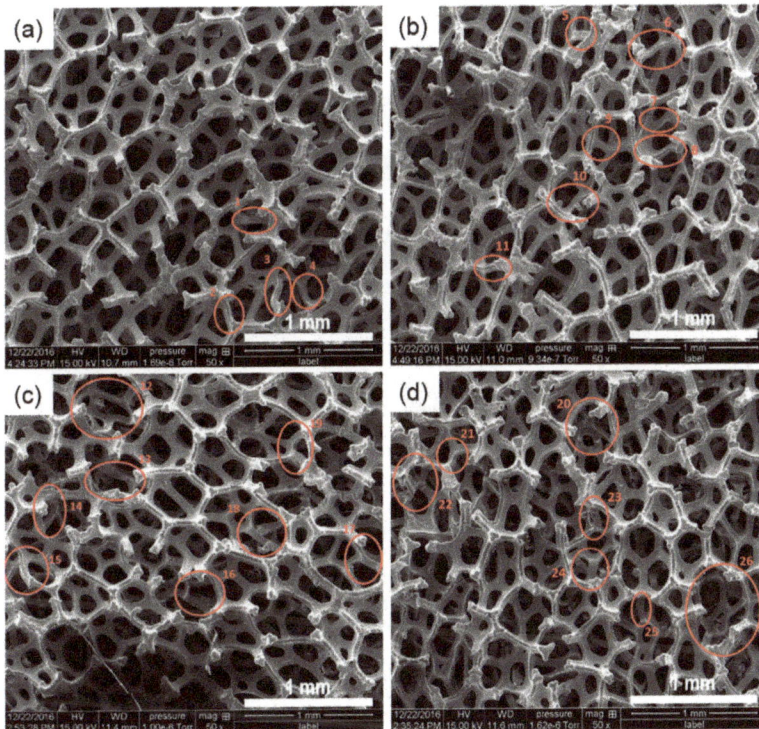

Figure 7. Deformed structure of four IF45 specimens at compression strain of: (**a**) 10.8%; (**b**) 12.8%; (**c**) 29.8%; and (**d**) 49.5%.

Figure 8 depicts the magnified images of some marked regions in Figure 7. It shows the deformation of struts of four specimens at different stages of compression after elastic regime. In samples with the maximum compression strain levels of 10.8%, 12.8% and 29.8%, formation of S-shape plastic hinges in some of the marked regions can be observed, e.g., region 2, 3, 7, 8, and 14 (Figure 8a). C-shape bending is observed in some of the struts of specimens with compression strains of 29.8% and 49.5%, e.g., region 13, 24 and 26 (Figure 8b). Plastic S-shape and C-shape deformation in some struts of open cell aluminum alloy (A356) and 316L stainless steel foams under quasi-static compression has been observed in the previous works by others [26,27]. Deformation bands are present in a few struts of the specimen with the highest level of compression strain (49.5%). These adjacent struts are marked with red circles of number 21, 23 and 27 (Figure 8c).

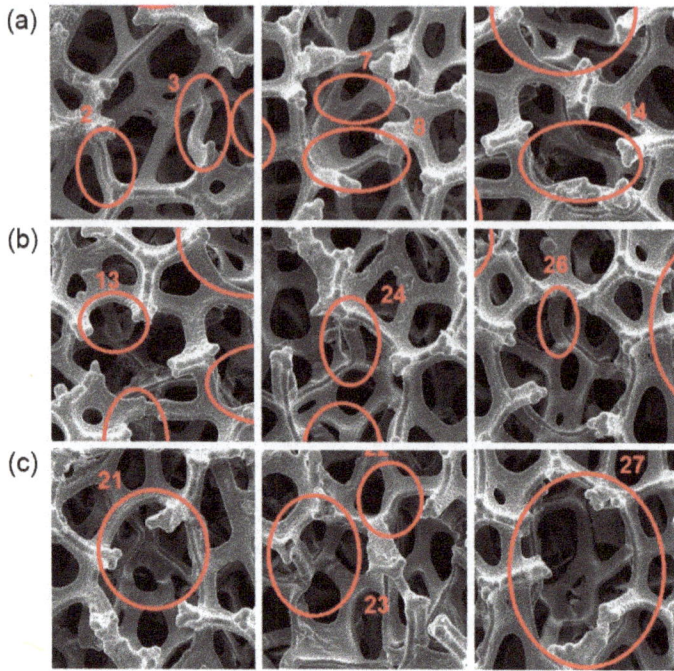

Figure 8. Different shapes formed on plastically deformed struts: (**a**) S-shape; (**b**) C-shape; and (**c**) deformation bands.

3.6. Morphology of the Iron Foams after Immersion Tests

Figure 9 depicts the morphological structure of three specimens, one of which was not immersed and the other two were immersed in Hanks' solution for three and seven days. As shown in the picture, layers of corrosion products were formed on the struts after immersion. The figure suggests that the layers on seven-day immersed specimen are thicker than those on three-day immersed specimens.

Figure 9. Morphology of the IF80 samples: (**a**) no immersion test; (**b**) after three-day immersion; and (**c**) after seven-day immersion.

4. Discussion

4.1. Effect of Structural Properties on Elastic and Plastic Compression Behavior of Iron Foams

As mentioned in the previous sections, stress–strain curves of iron foams under compression are composed of three main regimes: linear elastic, plateau and densification. In the case of metal foams, or other foams made of materials with a plastic yield point, as the stress goes beyond the linear elastic region, plastic collapse takes place. The collapse of cell walls continues until the cells collapse to an extent which further strain requires much higher level of stress, starting the densification part of the compression diagrams [28]. However, the presence of some plastic deformations during elastic regime has been pointed out in compression of some closed-cell aluminum foams [29]. In addition, a significant softening after hardening was observed in many of the iron foam specimens (Figure 3). As explained in [29], softening in closed-cell aluminum foams that underwent compressive loading and unloading was attributed to the collapse of the cells resulted from the plastic collapse of a single deformation band perpendicular to the loading direction. Similarly, plastic collapse of struts/cells may have caused the softening in some of the iron foam specimens. Thus, absence of a significant softening indicates that massive cell collapses do not take place immediately after the beginning of the plateau. Lower densification strain of IF58 specimens (Table 2) could be the result of smaller end-strut thicknesses (Table S1) in IF58 specimens, making the cells weaker especially on the loaded ends, leading them to a faster densification.

In general, the mechanical properties of foams depend on structural properties and the properties of the cell-wall (parent) material [28]. The most important structural properties are relative density (the ratio of the density of the cellular material to the density of the solid material), cell type (open or close) and the level of anisotropy in the cells. The most influential cell wall properties are the solid density (ρ_s), the Young's modulus (E_s), the yield strength (σ_{ys}), the fracture strength (σ_{fs}) and the creep parameters (n_s, $\dot{\epsilon}_{os}$ and σ_{os}) [28]. Because the cell-wall material properties of the samples are not significantly different in this study, it is expected that at least some of the geometrical and structural properties such as the relative densities, cell size, pore size and strut thicknesses influence the mechanical behavior of the iron foams. In the work of Amsterdam et al. [30], it was observed that the relative density influenced the plastic collapse stress values of open cell aluminum foams. In this study, however, the relative density of IF58 and IF80 specimens are very close, so it is expected that other parameters have had more contribution to the variation of the level of the mechanical properties between the two samples. The influence of cell size on mechanical behavior of the foams has been a matter of controversy. As stated in [29], in most cases, mechanical properties of metal foams do not depend on cell size. This was observed in the study of the deformation behavior of the open-cell stainless steel conducted by Kaya and Fleck [27], wherein, at the same relative density, the inhomogeneity in the microstructure was found the influential factor and not the cell size. Investigating the influence of density, cell size and cell shape on the mechanical properties of open cell 6101 aluminum foams, Nieh et al [31] found that, under similar densities, cell size does not have any significant effect on strength while the cell shape had some influences [31]. On the other hand, there have been researchers who found the cell sizes would affect the mechanical properties of the foam or that of a lattice structure [32,33].

In their work, Jian et al. [34] observed that the compressive and fracture strength of porous NiTi alloy samples increased with decreasing mean pore size. The samples of different porosities and pore sizes went under quasi-static compression tests (crosshead speed of 2.4 mm/min) while the range of porosities and pore sizes were 53–55.6% and around 264.8–1026.6 µm, respectively. Therefore, the variation between pore size values seems to be much more considerable than that of the porosities. However, studying the effect of pore size on the mechanical properties of open cell aluminum foams with spherical pores, Bin et al. [35] demonstrated that compressive stress–strain diagrams generally raised as the pore sizes increased. The authors, however, speculated that this was "related to a change in aspect ratio of the wall thickness against the edge length" [35]. Unlike the aforementioned research works wherein increasing pore size resulted in either increasing or decreasing the compression strength, in the study of Xu et al. on biomedical porous NiTi alloys, [36] the strength values varied with pore sizes in a "S type" fashion, i.e., as the pore size increased once, the value of the strength dropped initially, and after the second and third increase of the pore size, the strength increased and decreased, respectively. They attributed this behavior of porous NiTi alloys to the pore size as well as the number of pores both of which represent the same effect on the mechanical properties of the porous alloys, i.e., the increase of the pore size and the number of pores would result in a decrease in the value of mechanical properties including the Rockwell hardness, compression strength and elastic modulus. The similar pattern in variation of the mechanical properties with increasing pore size was observed in this study (V type). According to the information provided by the manufacturer, the nominal pore density of iron foams with nominal cell size of 450, 580, 800 μm is 100–110, 90–100, and 60–70 ppi (the number of pores per linear one inch). Therefore, it is expected that the number of pores in IF80 specimens to be significantly lower than that of IF45 and IF58 specimens around 38% and 32%, respectively. On the other hand, increasing the average pore size from IF58 to IF80 is around 4% for both pore size ranges of 100 < pore size < 200 and pore size > 200 (using the data provided in Table S1). Therefore, it is speculated that the decrease of the number of pores is more influential than the increase of the pore sizes, helping to raise the compression strength from IF58 to IF80 sample. The variation of energy absorbability per volume up to densification strain (W) had a direct relationship with compression strengths. This is expected given the behavior of cellular materials in plateau regime, i.e., an increase in the compression strength is associated with the rise of plateau region which results in increasing the area underneath the stress–strain curve. A linear relationship between the energy absorption capacity and plastic collapse stress (compressive strength) of open-cell 6101 aluminum alloy foams under compression was also observed in the work of Krishna et al. [37]. The same V-type pattern was also observed for variation of the offset yield strength and densification strain values. However, for the case of elastic modulus values under dry compressions, although the V-type variation was still present, an important difference was noticed: IF80 specimens showed the highest values of E unlike the other four properties (σ_y, σ_c, ε_D and W) for which IF45 specimens showed the maximum values. This can be due to the high influence of the end and branch strut thicknesses which are significantly higher for IF80 specimens than those of the other two groups (Table S1). The effect of cell size and relative density on the elastic behavior of an open-cell polymer foam was studied by Maheo et al. [38]. Their experimental results showed that the increase of cell size and relative density led to the increase of the elastic modulus of the foams.

Given that multiple factors potentially control the mechanical behavior of the iron foams, predicting the mechanical behavior of the foams under compression is a complex task. Therefore, a simple tabular method which takes into account multiple parameters, i.e., cell size, pore size, number of pores and strut thicknesses is proposed to explain the V-shape variation of compression strengths of the foams studied in this work. Nonetheless, it should be noted that some of the assumptions may be simplistic. In addition, the method is only proposed as an initial step to develop a model which predicts the variation of the mechanical properties of open cell foams (compression strength for the case of this study) as a function of the geometrical properties. Thus, a more sophisticated model may be developed in the future works. The assumptions of the tabular method, based on the literature review, are as follow:

-   The compression strength has a direct linear relationship with relative density and strut thickness, and it has an inverse relationship with the pore sizes and the number of the pores.
-   All the considered properties have an equal weight to affect the compression strength of the foams.

In the proposed method, for each geometrical property, e.g., relative density, an effect value (EV) which represents the effect of the parameter on the compression strength level of a particular group of open-cell foam is determined. The EV is assigned to each group of iron foams (maximum EV = 100) depending on the average value of the property in the corresponding sample (IF45, IF58, IF80). For each property, the maximum EV of 100 is assigned to the group which would obtain the maximum compression strength, if the corresponding property was the only factor influencing the mechanical strength. For example, for the case of relative density, IF45 sample receive EV of 100, because it has the maximum relative density among all the samples. Therefore, IF45 specimens would have the maximum strength among other cell-size group, if the only determining factor was relative density. Alternatively, for the case of the pore-number factor, IF80 sample is given EV of 100, for IF80 specimens have the lowest number of pores (according to the ppi data). Therefore, if the number of pore was the only parameter to determine the compression strength, IF80 specimens would have the highest strength among all the sample groups of iron foams. Then, after assigning the maximum point for a sample, the EV of the other groups were determined based on their value of the corresponding property relative to that of the sample with maximum EV. For example, for the case of relative density, the EV of the IF58 and IF80 samples was 71.05 and 65.79, respectively, which are resulted from the following equations:

$$100 \times \frac{(\rho * / \rho_S) \text{ IF58}}{(\rho * / \rho_S) \text{ IF45}} = 100 \times \frac{0.027}{0.038} = 71 \tag{2}$$

$$100 \times \frac{(\rho * / \rho_S) \text{ IF80}}{(\rho * / \rho_S) \text{ IF45}} = 100 \times \frac{0.025}{0.038} = 66 \tag{3}$$

More details on the calculations of the points are available in Section C of the Supplementary Materials. All the assigned points are shown in Table 4. Considering the described EV assignment process, a sample with higher total average EV is expected to have a higher compression strength and therefore energy absorbability. Please note that the final assigned EV associated with pore size are the result of taking the average of the EVs for pore sizes larger than 200 and those of between 100 and 200 μm using the data of Table S1. Similarly, the assigned EVs for each group with respect to the strut thickness were the average of the EVs calculated for end-strut and branch-strut thicknesses. That is why a maximum EV of 100 is not shown in those columns. As expected, IF45 and IF58 groups have the highest and the lowest average total EV, respectively (V-type variation).

Table 4. Assigned EVs of each sample group for the corresponding geometrical properties.

Specimen	$\rho * / \rho_S$	Cell Size	Pore Size	Number of Pores	Strut Thickness	Average Total Point
IF45	100	100	98.46	61.9	72.52	86.43
IF58	71.05	74.75	87.27	68.42	69.54	74.21
IF80	65.79	55.76	83.66	100	100	81.04

4.2. Effect of Environmental Conditions on Compression Behavior of Iron Foams

The statistical analysis of the mechanical responses under wet and dry compression tests suggests that the wet environment did not significantly influence the mechanical behavior of the iron foams. This could be due to the presence of open cells which let the fluid, which had a low viscosity, to escape as the compression applied. Moreover, because the compression was applied shortly after immersion of the specimens in Hanks' solution, no degradation effect is expected. Under the condition of higher strain rate and presence of a more viscous fluid inside the cells, an increase in the strength of the foams would have been expected: As an open cell containing a fluid is compressed, more work is needed to act against the viscosity, and a faster deformation of the foam requires more work [28]. However, for the case of this study, the strain rate and the viscosity were so low that their effects on the mechanical properties were negligible.

Comparing the compression behavior of IF80 non-immersed specimens with those of the specimens which were immersed for three- and seven-days showed that the elastic modulus, yield and compression strengths and energy absorption significantly decreased after immersion. This may be explained by degradation of iron due to the corrosion, resulting in lowering the level of the mechanical properties. However, the differences between the three-day-immersed and seven-day-immersed samples were not statistically significant which can be justified by decreasing the degradation rate after a few days. Such degradation behavior was observed in some of the pure iron samples after 10- or 15-day dynamic immersion test in Hanks' solution in the work of Mariot et al. [39]. This was potentially attributed to the formation of corrosion products on the surface such as iron phosphate which impeded additional degradation "by hindering oxygen diffusion" [39]. In addition, in [40], presence of hydroxide layer on the surface of electroformed iron was pointed as a cause of slowing down the degradation process in the static immersion test. As it appears in Figure 9, layers of corrosion products formed on the struts after seven-days are thicker than those formed after three-day immersion. Thus, this may have contributed to the reduction of corrosion rate.

4.3. Deformation and Failure Mechanism

As pointed out previously, three regimes exist during compression of foam: linear elastic, plateau and densification. In open cell foams with low relative densities ($\rho^*/\rho_s \leq 0.1$), cell wall bending mainly controls the linear elastic regime. Plastic collapse of the cells during compression gives rise to the plateau of the metal foams which results in formation of plastic hinges in the cell structure. It should be noted that formation of plastic hinges during plateau occurs in the foams made by materials that experience plastic yielding such as metals or rigid polymers [28]. Plastic hinges were formed in iron foam specimens as shown in Figure 8a. Finally, densification takes place "when the cells have almost completely collapsed" and further strains require much higher level of stress when cell walls being adjacent to each other [28]. Figure 10 represents a simple model of an open cell foam under linear elastic deformation and the formation of plastic hinges during plastic collapse.

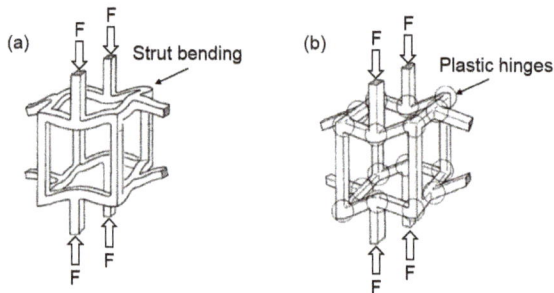

Figure 10. A simple model of an open cell foam experiencing: (**a**) linear elastic deformation; and (**b**) formation of plastic hinges during plastic collapse. Adapted from [28].

Different modes of plastic deformation of struts, including S-shape and C-shape deformations nd deformation bands in IF45 specimens are shown in Figure 8. Although Gibson's report [29] noted that the buckling of cells takes place during softening of aluminum foams, in this study, the presence or absence of buckling cannot be verified with a high level of certainty as the SEM images show only the surfaces on which compression applied. In other words, the struts whose directions are along the loading axis are not clearly visible. However, given that many of the struts are inclined to the loading axis, it is expected that buckling partially contributes to many of the strut deformatins. In Kaya and Fleck's work [27], both bending and buckling were addressed as two different deformation mechanisms appeared in different struts. In Daxner's work [41], both buckling and bending are pointed out as "dominating deformation mechanism" in the struts of "open-cell metallic foams". In the work of Schuler et al. [26], it is stated that C-shape and S-shape deformations are the product of bending and torsion. Therefore, given that the strut orientation relative to the loading direction is an influential factor on deformation behavior of the foams [37], it is fair to state that the deformation of an open cell iron foam under compression is a complex mechanism which could be the product of different mechanisms in conjunction with each other such as bending, buckling and torsion. However, some of the struts may experience only one form of deformation. The complexity of the strut deformations has been acknowledged in [26] as well.

5. Conclusions

To investigate the influence of geometrical properties and environmental condition on mechanical behavior of porous iron, iron foams of three different cell sizes underwent dry and wet compression tests. In addition, mechanical properties of iron foam specimens after static immersion were assessed and compared with those of non-immersed specimens. Using the stress–strain responses obtained from compression tests, elastic and plastic mechanical properties were analyzed: elastic modulus, yield strength, compression strength, densification strain and energy absorbability per volume up to the point of densification. Two major groups of statistical tests were carried out to analyze the significance of difference between the values of properties: hypothesis tests including ANOVA in conjunction with Tukey post hoc test and *t*-test, and non-parametric tests including Kruskal–Wallis ANOVA and two-sample K–S tests. In most, though not all, cases, hypothesis and non-parametric tests led to similar conclusions.

It was observed that the mechanical properties of the foams under dry compression tests had a "V-type" variation. Comparing the values of elastic modulus revealed that IF80 specimens had a the highest level of stiffness while for other properties, i.e., yield strength, compression strength, densification strain and energy absorbability, IF45 specimens possessed the highest level under dry compression. A simple tabular method was proposed to explain the variation in compression strength of the iron foams of different geometrical properties with respect to each other. Wet environment generally did not alter the mechanical behavior of the iron foams significantly while degradation decreased the elastic modulus, yield and compression strength and the energy absorbability of the samples. The deformation behavior of the foams under compression was also evaluated via SEM images and different deformation modes were identified. It was speculated that the deformation of open cell iron foams under compression is a complex phenomenon that could be the product of multiple mechanism such as bending, buckling and torsion. However, further studies are needed to understand the failure mechanism of iron foams.

Supplementary Materials: The following are available online at http://www.mdpi.com/2075-4701/7/6/202/s1. Figure S1: Measurement of cell size, pore size and different strut thicknesses on an IF58 specimen, Figure S2: Experimental design and parameters with the number of specimens used in each group, Figure S3: Stress-strain curves of IF45 samples under compression: (a) compression area A_1 and cross-head speed S_1, (b) compression area A_2 and cross-head speed S_1, (c) compression area A_1 and cross-head speed S_2, (d) compression area A_2 and cross-head speed S_2, Figure S4: Stress-strain curves of IF58 samples under compression: (a) compression area A_1 and cross-head speed S_1, (b) compression area A_2 and cross-head speed S_1, (c) compression area A_1 and cross-head speed S_2, (d) compression area A_2 and cross-head speed S_2, Figure S5: Stress-strain curves of IF80 samples under compression: (a) compression area A_1 and cross-head speed S_1, (b) compression area A_1 and cross-head speed S_1, (c) compression area A_2 and cross-head speed S_1, (d) compression area A_2 and cross-head speed S_2, Table S1: Results of the measurement.

Acknowledgments: This work is supported by the Natural Sciences and Engineering Research Council of Canada (NSERC) via the Discovery Grant.

Author Contributions: Hendra Hermawan and Adhitya Trenggono conceived and designed the experiments; Reza Alavi, Adhitya Trenggono and Sébastien Champagne performed the experiments; Reza Alavi and Hendra Hermawan analyzed the data; Reza Alavi wrote the paper; Hendra Hermawan revised and approved the paper.

Conflicts of Interest: The authors declare no conflict of interest.

References and Note

1. Pippan, R.; Motz, C.; Kriszt, B.; Zettl, B.; Mayer, H.; Stanzl-Tschegg, S.; Simancik, F.; Kovacik, J. Material Properties. In *Handbook of Cellular Metals: Production, Processing, Applications*; Degischer, H.P., Kriszt, B., Eds.; Wiley-VCH: New York, NY, USA, 2002.

2. Ashby, M.F.; Evans, A.G.; Fleck, N.A.; Gibson, L.J.; Hutchinson, J.W.; Wadley, H.N.G. *Properties of Metal Foams in Metal Foams: A Design Guide*; Butterworth-Heinemann: Oxford, UK, 2000; pp. 40–54.

3.   Wu, S.; Liu, X.; Yeung, K.W.K.; Liu, C.; Yang, X. Biomimetic porous scaffolds for bone tissue engineering. *Mater. Sci. Eng. R Rep.* **2014**, *80*, 1–36. [CrossRef]
4.   He, J.; He, F.L.; Li, D.W.; Liu, Y.L.; Yin, D.C. A novel porous Fe/Fe–W alloy scaffold with a double-layer structured skeleton: Preparation, in vitro degradability and biocompatibility. *Colloids Surf. B Biointerfaces* **2016**, *142*, 325–333. [CrossRef] [PubMed]
5.   Alvarez, K.; Nakajima, H. Metallic scaffolds for bone regeneration. *Materials* **2009**, *2*, 790–832. [CrossRef]
6.   Wen, Z.; Zhang, L.; Chen, C.; Liu, Y.; Wu, C.; Dai, C. A construction of novel iron-foam-based calcium phosphate/chitosan coating biodegradable scaffold material. *Mater. Sci. Eng. C* **2013**, *33*, 1022–1031. [CrossRef] [PubMed]
7.   Čapek, J.; Vojtěch, D.; Oborná, A. Microstructural and mechanical properties of biodegradable iron foam prepared by powder metallurgy. *Mater. Des.* **2015**, *83*, 468–482. [CrossRef]
8.   Eliaz, N. Biodegradable Metals. In *Degradation of Implant Materials*; Springer Science and Business Media: New York, NY, USA, 2012; p. 94.
9.   Peuster, M.; Wohlsein, P.; Brügmann, M.; Ehlerding, M.; Seidler, K.; Fink, C.; Brauer, H.; Fischer, A.; Hausdorf, G. A novel approach to temporary stenting: Degradable cardiovascular stents produced from corrodible metal-results 6–18 months after implantation into New Zealand white rabbits. *Heart* **2001**, *86*, 563–569. [CrossRef] [PubMed]
10. Oriňáková, R.; Oriňák, A.; Bučková, L.M.; Giretová, M.; Medvecký, L.; Labbanczová, E.; Kupková, M.; Hrubovčáková, M.; Kova, M. Iron based degradable foam structures for potential orthopedic applications. *Int. J. Electrochem. Sci.* **2013**, *8*, 12451–12465.
11. Zberg, B.; Uggowitzer, P.J.; Löffler, J.F. MgZnCa glasses without clinically observable hydrogen evolution for biodegradable implants. *Nat. Mater.* **2009**, *8*, 887–891. [CrossRef] [PubMed]
12. Vojtěch, D.; Kubásek, J.; Šerák, J.; Novák, P. Mechanical and corrosion properties of newly developed biodegradable Zn–based alloys for bone fixation. *Acta Biomater.* **2011**, *7*, 3515–3522. [CrossRef] [PubMed]
13. Vojtěch, D.; Kubásek, J.; Čapek, J.; Pospíšilová, I. Comparative mechanical and corrosion studies on magnesium, zinc and iron alloys as biodegradable metals. *Mater. Technol.* **2015**, *49*, 877–882. [CrossRef]
14. Kraus, T.; Moszner, F.; Fischerauer, S.; Fiedler, M.; Martinelli, E.; Eichler, J.; Witte, F.; Willbold, E.; Schinhammer, M.; Meischel, M. Biodegradable Fe–based alloys for use in osteosynthesis: Outcome of an in vivo study after 52 weeks. *Acta Biomater.* **2014**, *10*, 3346–3353. [CrossRef] [PubMed]
15. Hermawan, H.; Alamdari, H.; Mantovani, D.; Dubé, D. Iron–manganese: New class of metallic degradable biomaterials prepared by powder metallurgy. *Powder Metall.* **2008**, *51*, 38–45. [CrossRef]
16. Yusop, A.H.M.; Daud, N.M.; Nur, H.; Kadir, M.R.A.; Hermawan, H. Controlling the degradation kinetics of porous iron by poly(lactic-co-glycolic acid) infiltration for use as temporary medical implants. *Sci. Rep.* **2015**, *5*, 11194. [CrossRef] [PubMed]
17. Hermawan, H.; Dubé, D.; Mantovani, D. Degradable metallic biomaterials: Design and development of Fe–Mn alloys for stents. *J. Biomed. Mater. Res. Part A* **2010**, *93*, 1–11. [CrossRef] [PubMed]
18. Oh, K.; Lee, E.; Bae, J.S.; Jang, M.J.; Poss, R.; Kieback, B.; Walther, G.; Kloeden, B. Large scale production and applications of alloy metal foam. *Metfoam 2011 Proc.* **2011**, *S*, 601–606.
19. Gibson, L.J.; Ashby, M.F. The structure of cellular solids. In *Cellular Solids: Structure and Properties*, 2nd ed.; Cambridge University Press: Cambridge, UK, 1997.
20. International Organization for Standardization. ISO13314: 2011(E), Mechanical testing of metals_Ductility testing_Compression test for porous and cellular metals, Switzerland, 2011-12-15.
21. Banhart, J.; Baumeister, J. Deformation characteristics of metal foams. *J. Mater. Sci.* **1998**, *33*, 1431–1440. [CrossRef]
22. Paul, A.; Ramamurty, U. Strain rate sensitivity of a closed-cell aluminum foam. *Mater. Sci. Eng. A* **2000**, *281*, 1–7. [CrossRef]
23. Beer, F.P.; Johnston, E.R.; Dewolf, J.T.; Mazurek, D.F. *Mechanics of Materials*, 6th ed.; McGraw-Hill: New York, NY, USA, 2012.
24. Montanini, R. Measurement of strain rate sensitivity of aluminium foams for energy dissipation. *Int. J. Mech. Sci.* **2005**, *47*, 26–42. [CrossRef]
25. OriginLab. Nonparametric Tests. Available online: http://www.originlab.com/index.aspx?go=Products/Origin/Statistics#Nonparametric_Tests_PRO (accessed on 31 March 2017).

26. Schüler, P.; Fischer, S.F.; Bührig-Polaczek, A.; Fleck, C. Deformation and failure behaviour of open cell Al foams under quasistatic and impact loading. *Mater. Sci. Eng. A* **2013**, *587*, 250–261.

27. Kaya, A.C.; Fleck, C. Deformation behavior of open-cell stainless steel foams. *Mater. Sci. Eng. A* **2014**, *615*, 447–456. [CrossRef]

28. Gibson, L.J.; Ashby, M.F. The mechanics of foams: Basic results. In *Cellular Solids: Structure and Properties*, 2nd ed.; Cambridge University Press: Cambridge, UK, 1997.

29. Gibson, L.J. Mechanical behavior of metallic foams. *Annu. Rev. Mater. Sci.* **2000**, *30*, 191–227. [CrossRef]

30. Amsterdam, E.; Vries, J.H.B.; Hosson, J.T.M.; Onck, P.R. The influence of strain-induced damage on the mechanical response of open-cell aluminum foam. *Acta Mater.* **2008**, *56*, 609–618. [CrossRef]

31. Nieh, T.G.; Higashi, K.; Wadsworth, J. Effect of cell morphology on the compressive properties of open-cell aluminum foams. *Mater. Sci. Eng. A* **2000**, *283*, 105–110. [CrossRef]

32. Stephani, G.; Andersen, O.; Göhler, H.; Kostmann, G.; Kümmel, K.; Quadbeck, P.; Reinfried, M.; Studnitzky, T.; Waag, U. Iron Based Cellular Structures—Status and Prospects. *Adv. Eng. Mater.* **2006**, *8*, 847–852. [CrossRef]

33. Xiao, L.; Song, W.; Wang, C.; Liu, H.; Tang, H.; Wang, J. Mechanical behavior of open-cell rhombic dodecahedron Ti–6Al–4V lattice structure. *Mater. Sci. Eng. A* **2015**, *640*, 375–384. [CrossRef]

34. Jian, Y.T.; Yang, Y.; Tian, T.; Stanford, C.; Zhang, X.P.; Zhao, K. Effect of Pore Size and Porosity on the Biomechanical Properties and Cytocompatibility of Porous NiTi Alloys. *PLoS ONE* **2015**, *10*, e0128138. [CrossRef] [PubMed]

35. Jiang, B.; Wang, Z.; Zhao, N. Effect of pore size and relative density on the mechanical properties of open cell aluminum foams. *Scr. Mater.* **2007**, *56*, 169–172. [CrossRef]

36. Xu, J.L.; Bao, L.Z.; Liu, A.H.; Jin, X.F.; Luo, J.M.; Zhong, Z.C.; Zheng, Y.F. Effect of pore sizes on the microstructure and properties of the biomedical porous NiTi alloys prepared by microwave sintering. *J. Alloys Compd.* **2015**, *645*, 137–142. [CrossRef]

37. Krishna, B.V.; Bose, S.; Bandyopadhyay, A. Strength of open-cell 6101 aluminum foams under free and constrained compression. *Mater. Sci. Eng. A* **2007**, *452–453*, 178–188. [CrossRef]

38. Maheo, L.; Viot, P.; Bernard, D.; Chirazi, A.; Ceglia, G.; Schmitt, V.; Mondain-Monval, O. Elastic behavior of multi-scale, open-cell foams. *Compos. Part B Eng.* **2013**, *44*, 172–183. [CrossRef]

39. Mariot, P.; Leeflang, M.A.; Schaeffer, L.; Zhou, J. An investigation on the properties of injection-molded pure iron potentially for biodegradable stent application. *Powder Technol.* **2016**, *294*, 226–235. [CrossRef]

40. Moravej, M.; Purnama, A.; Fiset, M.; Couet, J.; Mantovani, D. Electroformed pure iron as a new biomaterial for degradable stents: In vitro degradation and preliminary cell viability studies. *Acta Biomater.* **2010**, *6*, 1843–1851. [CrossRef] [PubMed]

41. Daxner, T. Deformation Mechanisms and Yielding in Cellular Metals. In *Plasticity of Pressure-Sensitive Materials*; Altenbach, H., Öchsner, A., Eds.; Springer: Berlin/Heidelberg, Germany, 2014; pp. 165–166.

Review

Biodegradable Metallic Wires in Dental and Orthopedic Applications: A Review

Mohammad Asgari [1,2], Ruiqiang Hang [1,3], Chang Wang [4,5], Zhentao Yu [4,5], Zhiyong Li [1,2,*] and Yin Xiao [1,2,6,*]

[1] The Institute of Health and Biomedical Innovation, Queensland University of Technology, Brisbane, QLD 4059, Australia; m.asgari@hdr.qut.edu.au

[2] School of Chemistry, Physics & Mechanical Engineering, Science & Engineering Faculty, Queensland University of Technology, Brisbane, QLD 4000, Australia

[3] Research Institute of Surface Engineering, Taiyuan University of Technology, Taiyuan 030024, China; hangruiqiang@tyut.edu.cn

[4] Northwest Institute for Nonferrous Metal Research, Shaanxi Key Laboratory of Biomedical Metal Materials, Xi'an 710016, China; cch_wang@163.com (C.W.); yzt@c-nin.com (Z.Y.)

[5] China-Australia Joint Research Centre of Biomedical Metallic Materials, Shaanxi Key Laboratory of Biomedical Metal Materials, Xi'an 710016, China

[6] The Australia-China Centre for Tissue Engineering and Regenerative Medicine (ACCTERM), Queensland University of Technology, Brisbane, QLD 4059, Australia

* Correspondence: zhiyong.li@qut.edu.au (Z.L.); yin.xiao@qut.edu.au (Y.X.); Tel.: +61-7-3138-5112 (Z.L.); +61-7-3138-6240 (Y.X.); Fax: +61-7-3138-8381 (Z.L.); +61-7-3138-6030 (Y.X.)

Received: 30 January 2018; Accepted: 22 March 2018; Published: 26 March 2018

Abstract: Owing to significant advantages of bioactivity and biodegradability, biodegradable metallic materials such as magnesium, iron, and zinc and their alloys have been widely studied over recent years. Metallic wires with superior tensile strength and proper ductility can be fabricated by a traditional metalworking process (drawing). Drawn biodegradable metallic wires are popular biodegradable materials, which are promising in different clinical applications such as orthopedic fixation, surgical staples, cardiovascular stents, and aneurysm occlusion. This paper presents recent advances associated with the application of biodegradable metallic wires used in dental and orthopedic fields. Furthermore, the effects of some parameters such as the surface modification, alloying elements, and fabrication process affecting the degradation rate as well as biocompatibility, bioactivity, and mechanical stability are reviewed in the most recent works pertaining to these materials. Finally, possible pathways for future studies regarding the production of more efficient biodegradable metallic wires in the regeneration of bone defects are also proposed.

Keywords: magnesium; zinc; iron; biodegradable materials; wire drawing; bone regeneration; bone tissue engineering

---

1. Introduction

Metallic wires have been widely used in a wide range of industries in different applications. Drawing is a conventional metal forming process which has been applied to fabricate metallic wire. Briefly, this process continuously decreases the cross-sectional area of a specimen by pulling it through a single (in single-pass drawing) or series (in the multi-pass drawing) of the conical die(s) [1]. Figure 1 shows the schematic of die geometry in single-pass wire drawing process. d_i, d_o and α are the input diameter, the output diameter and die angle respectively. Due to imparting of a huge cold working happened during the drawing process, production of thin metallic wire with considerable strength is potentially achievable [2]. Equation (1) represents the percentage of cold work that could be achieved by single-pass wire drawing process [3]:

$$Cold\ work\ (\%) = 1 - \left(\frac{d_o}{d_i}\right)^2 \tag{1}$$

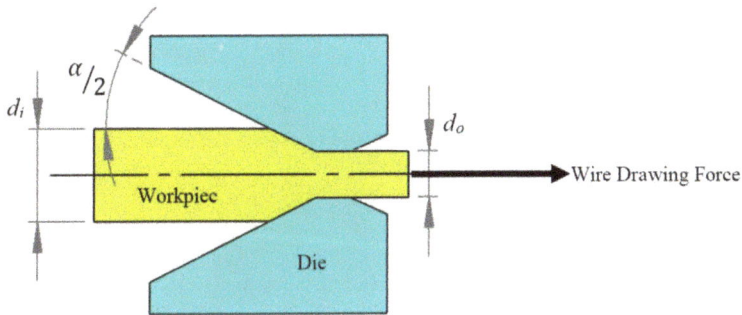

Figure 1. Schematic representation of a typical pass in the wire drawing process.

Moreover, different intermediate and/or post heat treatments coinciding with precise die(s) design and manufacturing could lead to the fabrication of the fine wire with acceptable ductility for different applications [3].

Over recent decades, the application of the wire drawing process in the production of biomedical metallic wires has been investigated. Drawn wires from titanium (Ti) and its alloys, stainless steels (SSs), and nitinol have been used in different medical devices such as cardiovascular stents [4], coronary guide catheter [5], coil occlusion of the aneurysm [6], Kirschner wires (K-wire) [7], orthodontic archwires [8], ligature wire [9], surgical sutures and versus staples [10]. Based on the application, such devices might be implanted either for a short or long time in the human body and regardless of the duration time, after finishing their mission, they should be removed from the body by a secondary operation. Because they cannot be degraded by human body fluids and, in some cases, if they are kept for more time, they will lead to complications such as harsh inflammations [11].

To decrease the clinical expenses and increase the quality of life of patients by reducing any secondary surgeries for removal of metal implants as well as minimizing any negative consequences, a new generation of metallic biomaterials has been introduced as a temporary support. These kinds of materials, called biodegradable metals (BMs), are strong enough during the target tissue healing process and can be gradually resorbed in the human body [12,13]. The degradation of BMs mainly ascribes to electrochemical corrosion, during which process the metal atoms lose their free electrons, become cations and enter body fluids. The content and the releasing allowance of those ions should not be toxic to the human body, especially for the surrounding host tissues [14]. To that aim, up to now different biodegradable alloys have been introduced. Among them, magnesium (Mg), iron (Fe), zinc (Zn), and their alloys as biodegradable implant metals are the three main categories that are considered more than other beneficial elements such as calcium (Ca) and strontium (Sr) [13].

In addition to the structural role of Mg, Fe, Zn, and their alloys as an implant device, they play important roles in many metabolic reactions and biological mechanisms of the body and consequently reduce the healing duration [15]. That is the reason they are named nutrient metals (NMs) [14,16]. Up to now, many studies have been conducted to increase the performance of these NMs by enhancing their initial mechanical properties and mitigating their potential drawbacks such as imperfect degradation rate to convert them from mechanical replacement devices to nutrient implants.

As an implant device, BMs in the shape of wire were firstly applied in 1878 by Edward C. Huse. He used pure Mg wire as a ligature to stop bleeding vessels. Although Huse reported the first clinical application of Mg wire, the Austrian surgeon Prof. Dr. Erwin Payr is known as the pioneer in the field of biodegradable Mg implants. His valuable results and consequently his inspiring articles published from 1892 to 1905 led to the utilization of biodegradable metallic wires in both human and animal models by other researchers [17]. Andrews in 1917 [18], and Seelig in 1924 [19] were early followers who investigated the animal in vivo experiments of pure Mg wire. From that time, a considerable number of feasibility studies have been done to propose Mg, Fe and more recently Zn and their alloys as ideal candidates in interventional clinical devices such as cardiovascular stents [20], and orthopedic implants [1,21]. However, there are still some challenges including improvement of both the surface and bulk properties such as corrosion resistance, bioactivity, strength, ductility, and toxicity. These properties need to be modified by more investigations to approve the application of BMs and their alloys in different medical operations without any negative consequences about the health of patients.

The present review mainly focuses on the recent progress related to the usage of biodegradable metallic wires in the orthopedic and dental studies. To that aim, the positive and any possible negative effects of Mg, Fe, and Zn on the bone biological mechanisms and bone healing process are generally presented first. Then, the history of clinical application development of each of these metals and their alloys in the shape of wires is reviewed. Furthermore, any other related scientific results to in vitro and in vivo studies of the biodegradable metallic wires are briefly summarized. Finally, the remaining problems in this research field that could be addressed in the future work are discussed and some possible options that are beneficial to the improvement of the clinical performance of these wires in the bone applications are also proposed.

2. Biological and Mechanical Aspects of Biodegradable Metallic Materials for Bone Regeneration Applications

2.1. Mg and Its Alloys

2.1.1. Advantages of Mg and Its Alloys

Mg as a mineral is the fourth most abundant cation element in the body [22]. It has a pivotal role in many body metabolisms including enzymatic reaction, the formation of apatite, and bone cells adsorption [23]. About half of the total amount of Mg in our body is stored in bone tissue [24]. Wu et al. reported that Mg ions (Mg^{2+}) govern a stimulating effect on the new bone formation by increasing the proliferation and differentiation of osteogenic (stem) cells via osteogenesis-related signaling pathways in vitro [25]. The release of Mg^{2+} is related to the chemical reaction of Mg in the human body fluids based on the following reaction which produces Mg hydroxide ($Mg(OH)_2$) and hydrogen gas (H_2) evolution [26]:

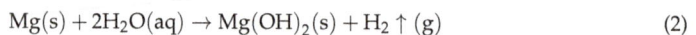

$$Mg(s) + 2H_2O(aq) \rightarrow Mg(OH)_2(s) + H_2 \uparrow (g) \tag{2}$$

Zheng et al. [27] presented a degradation mechanism for Mg in the physiological environment by the following anodic dissolution and cathodic reduction reactions:

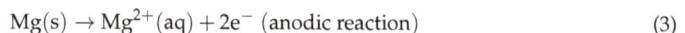

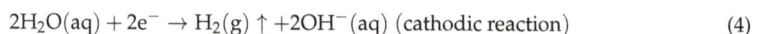

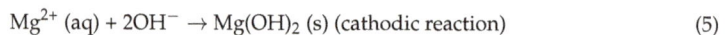

$$Mg(s) \rightarrow Mg^{2+}(aq) + 2e^- \text{ (anodic reaction)} \tag{3}$$

$$2H_2O(aq) + 2e^- \rightarrow H_2(g) \uparrow +2OH^-(aq) \text{ (cathodic reaction)} \tag{4}$$

$$Mg^{2+}(aq) + 2OH^- \rightarrow Mg(OH)_2(s) \text{ (cathodic reaction)} \tag{5}$$

It is clear that Mg^{2+} originates from the anodic reaction. Then, some of them take part in a reaction with hydroxide ions (OH^-) to produce $Mg(OH)_2$, while others enter the solution.

Mg(OH)$_2$, which is called milk of magnesia, includes both water and MgO solid particles. Hence, some part of it can convert to MgO and vice versa (see Equation (6)):

$$Mg(OH)_2(aq) \leftrightarrow MgO(s) + H_2O(aq) \tag{6}$$

Both of Mg(OH)$_2$ and MgO are beneficial inorganic materials to the body [28]. In addition to the conventional medical application of Mg(OH)$_2$ which has been widely used as an antacid to neutralize stomach acid and a laxative, it also possesses numerous valuable properties including excellent biocompatibility, nontoxic antibacterial activity, and high drug loading ability [29]. In bone regeneration application, as a major degradation product from any magnesium alloys, Mg(OH)$_2$ can enhance the bone formation and temporarily decrease the bone resorption resulting in a higher bone mass [30,31]. Therefore, it could be a potential protective coating material for Mg-based implants to reduce the corrosion rate and improve bone ingrowth simultaneously [31].

Regardless of the wide range of industrial applications, MgO similar to Mg(OH)$_2$ is applied as both an antacid and a laxative. It is also used as a drying agent because it rapidly reacts with water and creates Mg(OH)$_2$ [28]. MgO in shape of nanoparticles has been introduced as a novel antibacterial material to enhance tissue regeneration process and reduce bacterial infection during orthopedic implantations [32]. Huang et al. [33] claimed that the toxicity of MgO against bacteria increased by reducing its particle size. Hickey et al. [32] fabricated nanocomposites using MgO nanoparticles to mineralize poly(L-lactic acid) (PLLA) and applied in traumatic brain injury(TBI)-related orthopedic tissue regeneration. They reported that all the seeded bacteria were killed by PLLA/MgO nanocomposite. Therefore, MgO, as one of the corrosion products of Mg and its alloys play an antibacterial role for orthopedic tissue engineering applications.

In terms of mechanical aspects, Mg and its biocompatible alloys are more advantageous in bone applications in comparison with other metallic materials, ceramics, and biodegradable polymers (BPs). Because the mechanical properties of Mg are more similar to those of natural cortical bone compared to other materials which can decrease the stress shielding effect and prevent consequent bone resorption in the implanted sites [34,35], that makes Mg the most attractive biodegradable material in the orthopedic devices and implants as a suitable substitution for the non-bioresorbable one. Table 1 summarizes the information about the typical mechanical properties of cortical bone and commonly used metallic biomaterials.

Table 1. Comparison of typical mechanical properties of cortical bone and commonly used metallic biomaterials, data from [35–37].

Tissue/Material	Young's Modulus, E (GPa)	Yield Tensile Strength (MPa)	Ultimate Tensile Strength (MPa)	Density (g/cm^3)
Cortical bone	5–23	104.9–114.3	35–283	1.8–2.0
Pure Mg, wrought	41–45	100	180	1.74
Pure Zn, as cast	90	10	20	7.13
Pure Fe	211.4	50	540	7.87
Ti-6Al-4V, as cast	114	760–880	830–1025	4.43
316L stainless steel	193	200–300	450–650	8
Co-Cr alloy	240	500–1500	900–1540	8.3

2.1.2. Disadvantages of Mg and Its Alloys

The standard electrode potential for Mg(s) $\leftrightarrow$ Mg^{2+}(aq) + 2e$^-$ reaction is very low (-2.37 V) [38], making Mg very reactive in aqueous solutions such as body fluids. Although there is a gap between in vitro and in vivo results regarding the corrosion mechanism of Mg [39], commonly for both in vitro and in vivo the oxide film formed on the surface of Mg is porous rather than dense, thus cannot effectively prevent direct contact between metallic Mg and solution, leading to the rapid degradation of Mg and its alloys. The reduction in the mechanical stability due to the stress corrosion cracking is the first negative consequence related to such high corrosion rate of Mg and its alloys, risking the adequate load shielding over the tissue regeneration duration, especially in load-bearing applications [13,40]. Therefore, it is vitally important that there should be a balance between the degradation time and the healing period of damaged tissue to prevent implantation mechanical failure. It is the point that should be considered in the geometry design, material selection, and fabrication process of all BMs and their alloys generally, and more critical for Mg and its alloys.

Hydrogen gas (H_2) evolution is another challenging issue in the application of Mg and its alloys in the body. Based on Equation (2), 1 g of pure Mg could produce 1 liter (L) of H_2 gas [41]. Rely on previous in vivo experiments, some of this gas is soluble in the blood. The solubility of H_2 depends on the amount of the blood flow in the implantation site. The remained H_2 bubbles are accumulated in the local host tissue and create gas cavity endangering the cell adhesion to implants [42]. In some applications such as cardiovascular stents, due to the convective transport phenomena, the evolution of H_2 is not as concerning as orthopedic applications; because bone inherently is a poorly vascularized tissue leading to the formation of H_2 cavities potentially [26]. Some animal studies have shown that it might be dangerous [43]. In 2016, Noviana et al. [41] investigated the effect of H_2 evolution from porous pure-magnesium implants on the mortality of adult rats. They reported that the cavity formation was not tolerable for all the rats and the survival rate at Day 18 post-implantation reached zero percent (see Figure 2). Hence, such amount of post-implantation mortality is a warning sign for the risk of excessive H_2 accumulation in the large implants such as fixation plates. Nonetheless, due to the very light mass and negligible surface area of fine magnesium wire having sub-millimeter diameter there should not be that much of concern regarding the risk of H_2 gas evolution. However, it is still an estimation that needs to be investigated precisely by more in vitro and in vivo studies.

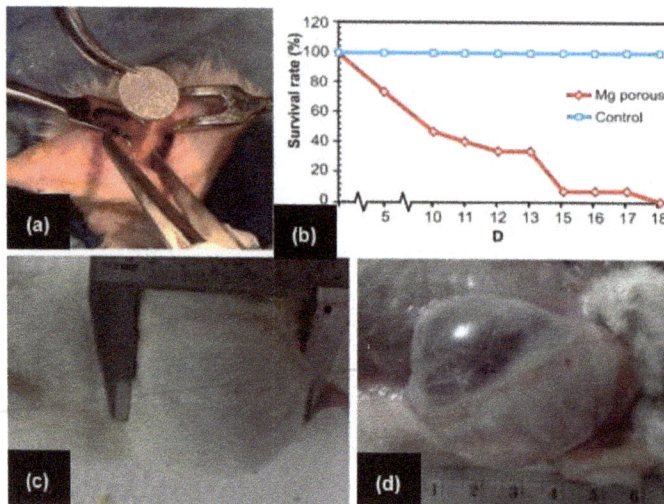

Figure 2. (a) Porous pure-magnesium implantation; (b) survival rate of the group of rats with magnesium implants compared with control group; (c) gas cavity formation; and (d) subcutaneous opening of the implantation area at Day 7, reproduced with permission from [41], Elsevier, 2016.

2.2. Fe and Its Alloys

2.2.1. Advantages of Fe and Its Alloys

Iron (Fe) as the fourth most abundant element on earth is significantly necessary for the human body and takes part in many enzymatic systems. The hemoglobin uses the largest part of total Fe (about 60–70%). Whereas, only about 10% of it is stored in bone marrow [44]. Fe has an essential role in the function of red blood cells and oxygen transport throughout the body. Fe deficiency may decrease the number of healthy oxygen-carrying red blood cells, thus has a negative effect on physical activity and increases susceptibility to infections [45]. Fe has a crucial role in bone metabolism by participating in two main biological processes including collagen synthesis and vitamin D activation and deactivation [46]. The clinical observations have shown that both Fe overload and deficiency (with or without anemia) could happen for the patients who suffered from osteoporosis [45,46]. The link between Fe and bone metabolism, especially in the homeostasis of bone ageing to maintain a balance between bone formation and resorption, is still in a need of further investigation, especially for elderly people to increase the quality of their life.

Almost all of the previous in vivo studies in animal models have been mainly focused on the application of Fe and its alloys as suitable degradable implant materials in the fabrication of cardiovascular stents [47]. While there are few studies in literature related to the bone application of Fe and its alloys. Kraus et al. [48] followed up the degradation behavior of Fe-based wires implanted into the femur of 38 rats for 52 weeks. Regardless of some results [48] showing that Fe ions could release from the implanted wires into the surrounding tissues, there is no clear information about its benefits on bone regeneration process.

2.2.2. Disadvantages of Fe and Its Alloys

The standard electrode potential for the reaction of $Fe(s) \leftrightarrow Fe^{2+}(aq) + 2e^-$ is relatively high (-0.444 V) [49]. Furthermore, owing to including a high concentration of CO_2 gas in blood causing to create a uniform, compact and dense manganese carbonate crystals the degradation rate of Fe in vivo is slower than in vitro [50] and in most of the cases is longer than clinical needs [51]. Additionally, because of the low rate of blood circulation near the implants, utilization of Fe and its alloys for bone reconstruction purposes would be definitely lower than their corrosion rate. Adding some alloying elements to pure Fe has been reported to be one of the most effective methods to accelerate the degradation rate of Fe to an acceptable level for practical clinical usage [12,44]. Among all of the Fe-based alloys, Fe–Mn alloys have been proved to be the most promising ones in accelerating the corrosion rate of Fe with similar mechanical properties (i.e., Young's modulus and ultimate tensile strength) to 316L SS [51]. Moreover, Mn can play an important role in promoting the growth of new bone and connective tissues as well as reducing bone loss [52,53] which could increase its bioactivity in orthopedic and dental applications. However, Fe and its alloys have much higher mechanical properties than those of the human cortical bone (see Table 1), which keeps the stress shielding effect as a challenging issue to select them as a suitable candidate in the load-bearing locations, especially as bulk temporary implants.

2.3. Zn and Its Alloys

2.3.1. Advantages of Zn and Its Alloys

Zinc (Zn) as the second most abundant transition metal element in the human body is found in all organs, tissues, fluids, and body secretions [54,55]. It is considered to be one of the vital mineral elements participating in a variety of fundamental biological functions, including signal transduction, apoptosis regulation as well as nucleic acid metabolism, DNA and RNA polymerase, and organic ligands interactions [56–58]. Approximately 85% of the Zn content of body exists in bone and muscle [14] which serves a crucial role in the improvement of cell proliferation and osteo-related gene expressions of osteoblasts affecting bone in growth [59,60]. Li et al. [56] investigated the biological effects of the Zn-1X(Mg, Ca and Sr) binary alloys using thin pin shape implants inserted into the bone tunnel created along the axis of the femoral shaft from the distal femur of mice. They reported that the Zn-1X binary alloys with nutrient alloying elements improved new bone formation around the pins effectively. Other in vivo and in vitro studies have shown that Zn affects the osteoclastic bone resorption metabolism [61,62]. The Zn content in the bone matrix of patients suffering from skeletal diseases such as osteoporosis (commonly in elderly people) and bone cancer is less than that of normal people [56,63]. In summary, having a healthy bone with sufficient density is correlated with the balancing of Zn content in the human body.

It has been well established that Zn ions (Zn^{2+}) show antibacterial property [64–66]. Recently, Zhu et al. [59] introduced plasma immersion ion implantation (PIII) process as a promising method for Zn-incorporation on Ti alloy, which is beneficial to both osteogenic and antibacterial abilities to enhance osseointegration while decreases implant-associated infections. Zn^{2+} are released directly from the corrosion of Zn and its possible alloys when in contact with the human body fluids through electrochemical corrosion. Drelich et al. proposed [67] the hypothetical mechanism happening during the degradation of Zn in the physiological environment (Figure 3). The reaction between some of Zn^{2+} with hydroxyl ions (OH^-) forms Zn oxide (ZnO). In addition to the wide range of dental and orthopedic applications of ZnO, due to its desirable antibacterial and osteogenic abilities, the anti-cancer property of ZnO has been also approved as a beneficial material to cure osteosarcoma [68,69].

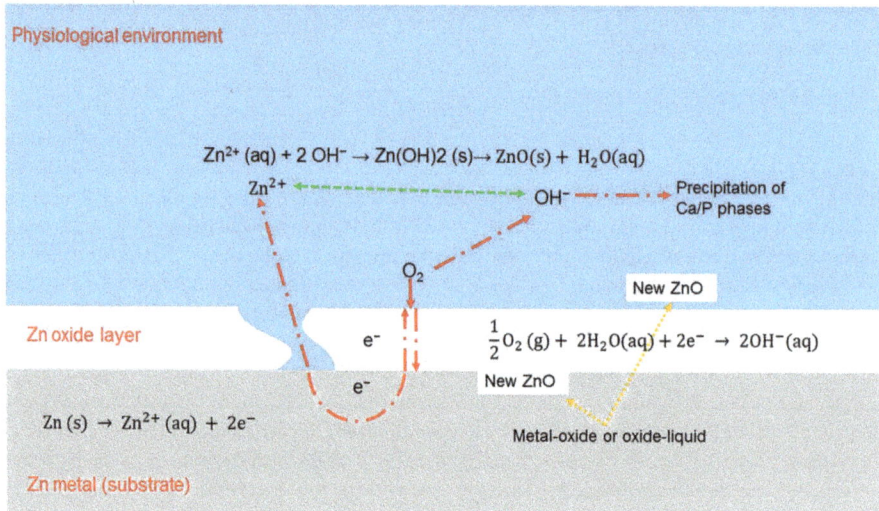

Figure 3. Schematic diagram of degradation mechanism of Zn in a physiological environment.

The standard potential for the cathodic reaction of Zn is -0.762 V [49], which is between that of Mg and Fe. Therefore, the moderate degradation rate of Zn is faster than Fe but slower than Mg, which could be an ideal option for clinical applications, especially in terms of mechanical stability [55]. More recently, Yang et al. [57] implanted pure Zn stents into the abdominal aorta of rabbits for 12 months. They reported that the stent was capable to serve mechanical stability for the first 6 months and degraded $41.75 \pm 29.72\%$ of the stent volume after 12 months of implantation without any severe inflammations, platelet aggregation, and thrombosis formation. Furthermore, in the matter of accumulation of H_2 during the degradation, there is less concern in the application of Zn and its alloys in bone applications than that of Mg and its alloys. Because it seems that the corrosion rate of Zn is low enough to prepare a time for the dissolving of H_2 in the body without the creation of any gas cavities.

2.3.2. Disadvantages of Zn and Its Alloys

Generally, pure Zn is soft, brittle with low strength (Table 1). Therefore, it seems impossible to apply pure zinc in load-bearing applications. The addition of alloying elements has been recognized as one of the most effective methods to improve the mechanical properties of pure Zn [56]. Even adding a little amount of alloying elements can improve its mechanical properties considerably. Briefly, depending on the composition of the Zn alloys the ultimate tensile strength (UTS) has been achieved in a wide range of 87 to 399 MPa till now [55,70]. Moreover, other parameters such as using different metalworking processes such as rolling, extortion, and post-processing heat treatments can definitely affect the strength, ductility, and degradation behavior of Zn-based alloys. However, Young's modulus (E) of Zn is about twice as that of Mg, which indicates that the stress shielding effect is more challenging in Zn-based implants.

Non-uniform corrosion behavior of Zn compared to uniform corrosion of both Mg and Fe is another problem. Chen et al. [15] compared the corrosion behavior of Zn, Fe and Mg in a long-term course in vitro. They observed that by increasing immersion time the corrosion rate of Zn developed faster than Fe and Mg, which is justified by the occurrence of localized corrosion in the surface of Zn samples. Similar behavior was reported in vivo [57]. The localized corrosion causes to lose the strength uniformity of an implant which makes its failure behavior more ambiguous. Drelich et al. [67] claimed that coating of Zn with a passive layer of ZnO film could be an effective approach to delay the degradation rate and decrease the corrosion non-uniformity. However, Törne et al. [71] claimed that the corrosion behavior of Zn is affected by the types of examined solution. They proposed that whole blood or real plasma are suitable solutions for short-term in vitro studies which can resemble in vivo situation, while Ringer's solution is more suitable for long-term degradation studies. Collectively, more in vivo and in vitro investigations should be conducted for the better understanding of the corrosion behavior of Zn in bone applications to address those contradicting results.

3. Current Orthopedic and Dental Implant Applications of Metallic Wires

The skeletal system of the human body as a complex three-dimensional structure plays two main roles, supporting the many body organs and their related tissues and attaching the numerous muscle groups which are needed for body movements. Bone as the hard part of the skeleton is a natural two- phase organic-inorganic ceramic composite. The organic phase gives bone its flexibility, while the inorganic phase provides bone with its structural rigidity [72]. The combination of organic and inorganic phases in the matrix provides bone with the unique mechanical properties such as toughness, strength, and stiffness. These properties give the ability to withstand against various mechanical loads applied during the normal and intense physical activities. Despite the remarkable mechanical and structural characteristics, the bone may fracture due to various loads resulting from sudden injuries and repeated cyclic loads. An effective solution in restoring the functionality of damaged bone is surgical implantation of artificial biomaterials. Several types of bone fixation devices in the form of screws, plates, nails, staples, and wires are produced for orthopedic and dental implant applications. The selection of biomaterials for these devices is highly dependent on the specific medical application. The new biomaterials which were introduced and developed for implant applications should have excellent biocompatibility, comparable strength to natural bone and produce no cytotoxicity effects [73]. The present review paper focuses on the bone application of wires; therefore, in this section, bone fixation devices which are made from wires are introduced first and then the evolution of the biomaterials to be compatible with natural bone are explained.

3.1. K-Wires

Martin Krischner (1879–1942) was a German surgeon who developed a device that allowed him to insert chromed piano wire from 0.7 to 1.5 mm in diameter into fractured bones to be used as a traction anchor instead of the large Steinmann pins [74]. Krischner wires (K-wires) were never used by their inventor for fracture fixation and orthopedic implants [75]. Otto Lowe [74] in 1932 published the first paper describing the use of K-wires for fracture fragment stabilization. Nowadays, these wires with different sizes have been used to hold bone fragments together (pin fixation) or to provide an anchor for skeletal traction. K-wires can also serve as guide pins for the placement of cannulated screws and can be placed in bone either by hand or at low speed with a power drill [75,76]. Figure 4a shows radiographic picture of K-wires used to repair a fracture of the medial epicondyle in the elbow.

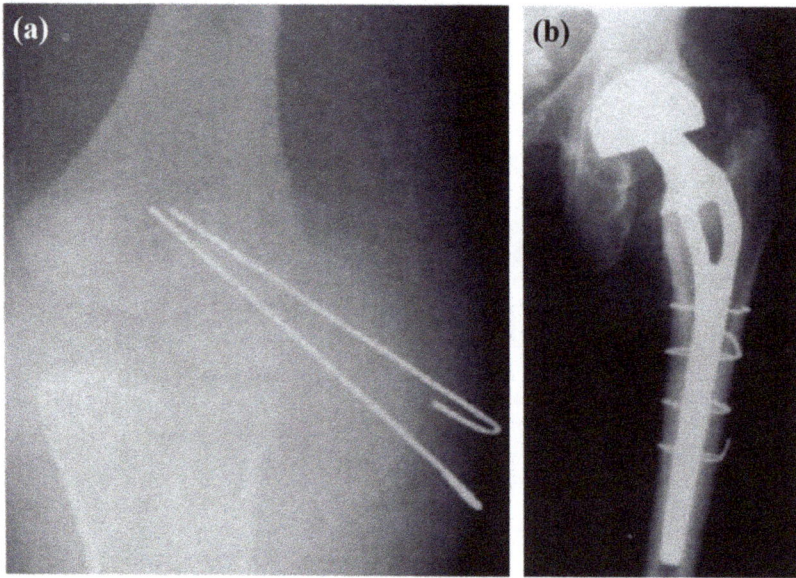

Figure 4. (**a**) Radiographic picture of K-wires used to repair a fracture of the medial epicondyle in the elbow and (**b**) radiograph shows cerclage wiring used to contain bone fragments in a fracture around the stem of a femoral prosthesis, reproduced with permission from [76], The Radiological Society of North America (RSNA®), 1991.

3.2. Cerclage Wires

The other application of thinner orthopedic wire is as a means of stabilizing fracture fragments. Cerclage wiring refers to encircling fragments and tensioning the wire to hold fragments in alignment [76]. A single or double strand of the wire is placed around the bone, and then the ends are twisted together. Cerclage wires are usually used in combination with other types of fixation devices such as plates or nails, but can also be used alone in special situations [76]. The radiograph in Figure 4b shows cerclage wiring used to contain bone fragments in a fracture around the stem of a femoral prosthesis.

3.3. Tension-Band Wires

The tension band technique converts a tensile force to the compression one. Tension banding is particularly useful in the setting of fractures where a muscle pull produces distraction of the fracture fragments, such as fracture of the patella [76]. Parallel K-wires are placed to provide rotational stability and reduce shearing forces between the fragments. Figure 5 presents an application of tension band wiring in repairing of an olecranon fracture.

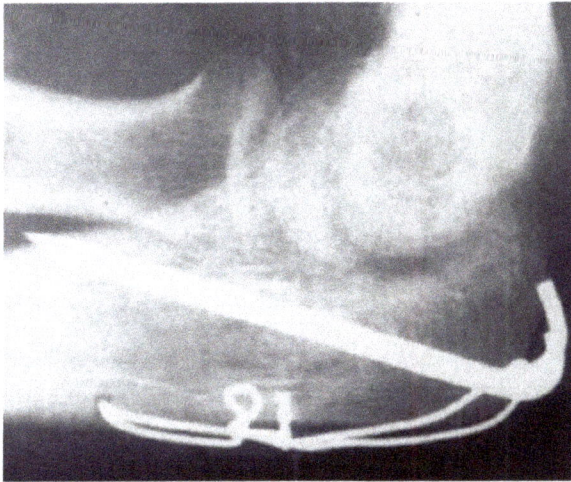

Figure 5. The radiograph shows tension band wiring used to repair an olecranon fracture, reproduced with permission from [76], The Radiological Society of North America (RSNA®), 1991.

3.4. Orthodontic Archwires

An orthodontic archwire is a wire conforming the dental arch and exerts forces to correct irregularities in the position of teeth. An archwire can be used with dental braces as a source of force. Orthodontic wires release the energy stored upon its placement by applying forces and torque to the teeth through the appliances placed on them [77]. The forces are applied to move the teeth to a targeted position. Another application of these wires is maintaining the existing dental position; in this case, they have a retentive purpose [77].

3.5. Ligature Wires

Ligature wires are small flexible and twistable ties or rings, which are used widely in dentistry to hold or ligate the archwires to the brackets [9]. This kind of wires is mostly made from SS and elastomer [78].

3.6. Staples

The bone staple is essentially simple in design and use. This device is made up two or more points of entry into bone that are attached to each other. They can be fabricated by double bending a pin or wire that is biased cut to create the sharp ends or tips [10]. The bone staples are fixed into the bone to help stabilize a fracture to promote bone healing. Bone staples can be used as adjuncts to other forms of fixations or as single or multiple bone staples at one site.

3.7. Sutures

Metallic sutures are available in many sizes and in two forms: monofilament or multi-strand. Most surgeons find the multi-strand wires are easier to handle since they are pliable and ties like silk [79].

4. The evolution of Metallic Wires in Bone Fixation Devices

K-wires as a versatile tool in the hand of orthopedic surgeons have been used for temporary or definitive osteosynthesis, temporary joint transfixation or for the guidance of other implants such as cannulated screws. K-wires are present in different forms and designs. Traditional K-wires have a circular cross-section and two tip designs (diamond and trocar tips). Diamond tip K-wires require the least axial force to penetrate the bone and consequently generate low heat during drilling, while the trocar tip wires need a greater force and generate higher temperature on insertion [80]. According to a surgical instrumentation catalogue, the diameter of standard K-wires based on their applications varies between 0.4 and 3.5 mm. Besides the strengthening of K-wires by enhancing the mechanical properties of their materials, the configuration of them has been studied in some researches [81,82]. Different K-wire fixation techniques contained two or four K-wires with different diameter were investigated and it was observed that the type of configuration was an effective parameter in increasing of the rigidity and stability of the fractured bone [81,82].

The first usage of wires in fracture fixation was reported in 1775 when Fe wires were used in a bone surgery [27]. In 1906, Lambotte used a combination of a Fe wire cerclage, Mg plate and steel screws to stabilize the fracture of the lower leg [17]. The occurrence of an electrochemical reaction between Mg and Fe caused extensive subcutaneous gas cavities, local swelling and pain one-day post-surgery [27]. By observing these symptoms, Lambotte learned that to avoid galvanic corrosion of the Mg, it should not be implanted with other metals [17]. Lambotte also used pure Mg nails alone to fix supracondylar fractures of four children. He observed the formation of gas cavities due to the high corrosion rate of magnesium in vivo [37]. Mainly, because of this problem, the application of Mg wires was restricted, and SS wires were introduced instead. The biometals used in wire bone fixation devices are classified into two categories, permanent and biodegradable metals. Among permanent metallic wires, surgical SS, cobalt-chromium (CoCr) alloys and Ti alloys are the most common ones. In the following sections, the application of these two groups was described in wire bone fixation devices.

4.1. Permanent Metallic Wires

In 1920, the first type of SS namely 18-8 was used as a bone implant. This alloy showed higher corrosion resistant and mechanical strength in comparison with that of pure Fe. Since then, various types of SSs were developed and some of them were used as a proper material for K-wires and cerclage wires. In 1962, the American Society for Testing and Materials (ASTM) has formed committee F04 on medical and surgical materials and devices to standardized metals for medical applications. Table 2 shows the differently registered biomaterials for bone wires in the ASTM standard.

In 1988, Stark et al. were the first researchers to recommend the use of SS K-wires in all grafted fractures, since they were easy to accomplish and required short operating time [83]. They also achieved 97% union of long-standing ununited fractures of the scaphoid by examining 151 patients through 22 years, which was a great achievement with K-wires [83]. SS K-wires provide strong fixation but can be difficult to manipulate and often requires a secondary operation to remove the wire. Due to the mismatch between the mechanical properties of the wires and the natural bone, mechanical forces and loads are retained by the wires and are not transferred to the damaged bone. Consequently, stress shielding happens and results in bone resorption and wire loosening. In addition to stress shielding effect as well as need for the second surgery, K-wires may cause percutaneous pin tract infection and wire migration [84,85].

Table 2. Common permanent biomaterials used for bone wires, data from [86].

Chemical Composition	Alloy	ASTM Number	Application
Fe-18Cr-14Ni-2.5Mo (C < 0.03)	316 L	F138-08/F139-08/F1350-08	Kirschner wires, cerclage wires
Fe-19Cr-10Ni-2Mn	302	A313/F899	guidewires
Fe-20Cr-10.5Ni-2Mn	-	A313/F899	Guidewires, orthodontic wires
Fe-20Cr-10.5Ni-2Mn (C < 0.03)	-	A580	Guidewires, Orthodontic wires
35Co-35Ni-20Cr-10Mo (Ti = 0.7)	MP35N	F562	Orthopedic wires
35Co-35Ni-10Mo (Ti = 0.01)	35NLT	F562	Orthopedic wires
40Co-20Cr-16Fe-15Ni-7Mo	Elgiloy	F1058-08	Surgical wires
Ti-6Al-4V	-	F136	Kirschner-wires
Ti-Ni	Nitinol	F2063	Archwires, Kirschner-wires

The most basic materials used in K-wires are SSs, Co-Cr alloys, Ti and its alloys and nitinol. It should be noted that the nitinol is an alloy of nickel and titanium elements where the two elements are present in roughly equal atomic percentages. Since 1930, it has been known that Co-based alloys due to their excellent wear resistance which is higher than SS, are proper candidates for bone repair devices [87]. These alloys possess excellent mechanical properties. CoCr alloys also have one order of magnitude greater corrosion resistance in comparison with SSs. Mainly, due to these properties, cobalt and its alloys specially, MP35N, modified 35NLT and Elgiloy, have achieved wide attention for fabrication of K-wires, cerclage wires and guide wires [86].

The performance of medical grade Ti alloys is superior to that of SS and cobalt alloys, due to the 50% greater strength to weight ratio of the former. These alloys have a lower modulus, higher corrosion resistance and better biocompatibility in comparison with SSs and cobalt-based alloys. These advantages make the Ti and its alloys a better-suited alternative for bone applications with higher loading rates [88]. As compared to commercial unalloyed Ti, Ti-6Al-4V alloy has greater strength, it has been selected as a proper material for K-wires and cerclage wires. The application of this alloy has been approved by the ASTM for wire bone fixation. Therefore, various medical device manufacturers utilize Ti-6Al-4V alloy for producing K-wires in a wide range of diameters. Clauss et al. have treated 135 toe deformities using both SS K-wires and Ti K-wires for one year. They observed that the Ti K-wires represent less recurrence of deformity, pain, and biofilm formation than SS k-wires. Hence, the authors suggested utilization of Ti K-wires instead of SS ones for transfixation of toe deformities [89]. Despite the lower density of Ti alloys, their wear resistance and bending strength are lower than cobalt alloys [87].

Nitinol has proper mechanical stability, lower stiffness, and thermo-elasticity as well as acceptable corrosion resistance. These properties make it a suitable substitution for SS implants [88]. Fracture fixation by SS K-wires provides more bending stiffness in comparison with the nitinol wires, while nitinol wires are more durable. Additionally, nitinol typically costs about 4 times as much as SS with all else equal.

4.2. Biodegradable Polymeric (BP) Wires

In order to eliminate the secondary surgery and resolve the shortcomings of permanent metallic wires, the fracture fixation with biodegradable wires made of polymeric materials was suggested and applied in the clinical trials [90,91]. The wires made of biodegradable synthetic polymers were applied in the fracture fixation of the knee [91], wrist [92], elbow [84] olecranon and patella [93] and small bones of hand [90] and their usage were compared with the metallic K-wires. It was concluded that in both cases the healing time was the same and polymeric wires were suitable alternatives for the metallic ones [84,90,91,93]. Despite the high cost of polymeric biomaterials, the total expenses of treatment by polymeric wires was estimated lower than the metallic ones, since there is no need for the second surgery [93]. In the fixation of wrist fracture, treatment by BP wires was not recommended, because of high complication rate in patients [92]. In other cases, it was suggested to use polymeric wires as an adjunct to metal fixations [94,95]. It has been reported that BP materials have a common disadvantage in load-bearing applications according to their innate mechanical properties. Considering a proper mechanical strength, biodegradable metals are promising candidates in load-bearing situations, where a high mechanical strength and a suitable Young's modulus are required.

5. Recent Developments of BM Wires in Bone Applications

Although the idea of deployment of BM wires has been presented more than a century ago, their development has been retarded until nearly a decade ago. In this section, the evolution of BM wires has been introduced with the only focus on Mg, Fe, and Zn and their alloys.

The first experimentally employed Mg wires were used as sutures to anchor nerves and muscles. During the early attempts, the lack of ductility prevented the use of Mg wires as suture [19]. Although the pure Mg as a suture wire presented an adequate degradation rate [19], rapid degradation was an obstacle to develop the usage of Mg as an implant in bone surgery. In orthopedic applications, implants usually undertake a certain load during the healing of injured bone. Due to the rapid degradation, the mechanical strength and integrity of the Mg-based implants will be deteriorated seriously. Hence, the implants lose their load-bearing ability during their service time. Pits or cracks created by the rapid corrosion might result in sudden failure, even at the initial stage after implantation [96]. Therefore, it can be concluded that as a bone fixation wire material, two main drawbacks of Mg could be low ductility and high degradation rate.

Recent studies on Mg-based BMs have focused on the improvement of mechanical properties and corrosion resistance of Mg and its alloys. In this regard, various methods such as selection on alloying elements [97–100], microstructural adjustment [101–104] and surface modification [96,105] have been applied. All the mentioned methods have been widely investigated for examining the in vivo and in vitro characteristics of common implant devices such as screws, plates, and pins [106–108]. In addition to the clinical trials of McBride and Verbrugge [109], at the first half of the 20th century, to employ Mg screws, plates and nails, recently, a few researchers [110,111] developed Mg-based implant devices and utilized them in patients'0 body to fix the bone fractures. Although plenty of studies has investigated the application of BP materials for manufacturing K-wires and cerclage wires, less attention has been paid to develop BMs for these wire bone fixation devices. By analyzing the plenty of publications about biodegradable Mg-based materials, it has been found that just a few pieces of research have investigated the development of the Mg alloys in the form of wires for bone fraction surgeries. As it was mentioned, the diameter of K-wires and cerclage wires vary between 0.4 and 3.5 mm. Hence, in this section, only the wires in this range of diameters were considered.

Various alloying elements were added to pure Mg to reduce the corrosion rate and concurrently keep the biocompatibility of the alloy for bone surgery. Among these alloys, Mg-calcium (Ca) alloys were biocompatible but degraded quite rapidly [97]. The combination of Mg with rare earth metals decreased the corrosion rate but also created cytotoxic effects [97,99]. Tie et al. [112] introduced a novel Mg alloy containing 2% silver (Mg2Ag) and showed that the alloy possessed appropriate mechanical properties and a rather low degradation rate in vitro. In order to assess the in vivo behavior of the alloy, Jahn et al. [113] implanted Mg2Ag intramedullary wires into mice with and without a femoral shaft fracture. The wires with the final diameter of 0.8 mm were produced via the cold rolling following the hot extrusion. To avoid the loss of ductility, heat treatment between the mentioned metal forming processes were employed. As a matter of biological effects, Mg2Ag alloy does not have any side effects on the growth rate of the bone or on inner organ morphology that represents the high biocompatibility of this alloy. They have examined the degradation rate of the alloy both in vitro and in vivo. Although the corrosion rate of Mg2Ag alloy in vivo was three times faster than in vitro, the alloy still showed sufficient rate of biodegradability which maintained proper mechanical stability while supporting fracture healing. Figure 6 illustrates the X-ray images of fractured bone fixed by steel and Mg2Ag wires during 133 days in mice [113]. Finally, they have concluded that Mg2Ag might be a promising material that has a potential application in musculoskeletal medicine.

In another attempt, Bian et al. introduced Mg-based BMs with dietary trace element germanium (Ge) for orthopedic implant application [114]. They added various mass percentages of Ge to Mg but in a limited range because of biosafety concerns. The MgGe samples with a diameter of 2.2 mm were prepared by hot rolling and machining. The highest strength and elongation were reported for Mg3Ge alloy with the yield tensile strength (YTS) of 135 MPa, UTS of 236 MPa and 17.7% elongation. It could be concluded that in comparison to other Mg-X alloy (X represents for essential/possibly essential element in human health) Mg-3Ge exhibits the highest strength and significantly improves elongation and serves the lowest in vivo corrosion rate. Figure 7 gives a comparative diagram to assess the strength and elongation of various biodegradable Mg-X alloys manufactured in different process conditions [114].

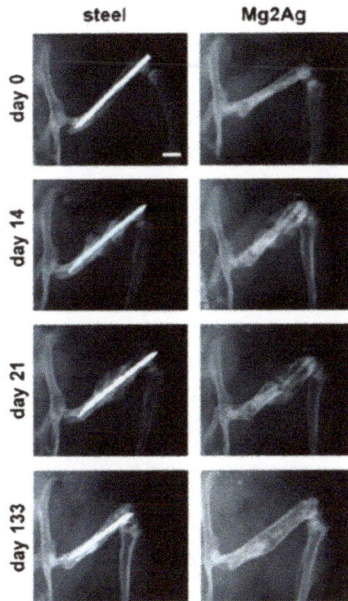

Figure 6. X-ray images to determine degradation of intramedullary Mg2Ag wires during 133 days in mice compared to a steel pin (scale bar = 5 mm), reproduced with permission from [113], Elsevier, 2016.

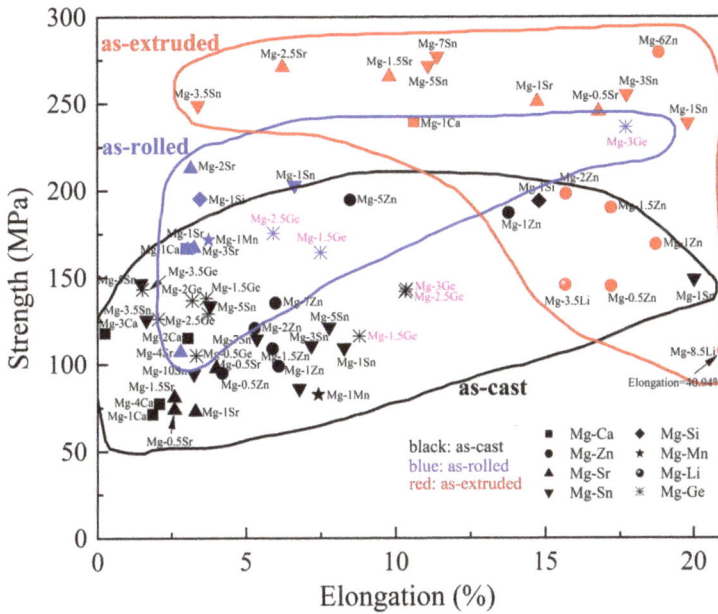

Figure 7. Comparison of the mechanical properties among various biodegradable Mg-X alloys (X: essential/possibly essential element in human health). Different symbols represent different alloy systems and symbol color reflects the process condition, reproduced with permission from [114], Elsevier, 2017.

Drawing and extrusion are the most common industrial processes for the fabrication of wires based on the desired diameters. Bian et al. [114] have produced the Mg3Ge wires by machining following hot rolling process which is not a proper method for manufacturing wires. It is clear from Figure 7 that extruded parts for most of the Mg alloys have a higher ductility and strength in comparison with rolled and casted parts. Considering the last two points, one can conclude that more ductility and higher strength would be achieved by the fabrication of Mg3Ge wires via wire extrusion.

The relevance between the manufacturing process and the corrosion rate for Mg alloys have been investigated by Koo et al. [115]. They fabricated as-cast Mg-Zn-Mn and as-extruded Mg-Zn-Mn wires with a diameter of 1.6 mm. Both the alloys were examined in vivo and in vitro. It was observed that as-cast Mg-Zn-Mn alloy had higher corrosion rate with severely localized pattern in comparison with the extruded one which had the uniformly localized pattern. This observation has been attributed to the larger grains boundary and the lower mechanical strength of the as-cast alloy compared to the extruded one which had smaller grain size and higher mechanical strength [115]. This research highlighted the effect of the manufacturing process on the mechanical properties and the corrosion rate of the BM alloys. Therefore, it can be stated that by imposing higher strains during the manufacturing process of wires, finer microstructure could be achieved, which in turn improves the mechanical properties and corrosion rate of BMs. Despite the limited studies about Mg wires applied in bone fixation devices, it could be stated that Mg wires have a high potential for utilization as K-wires and guide wires in orthopedic surgeries.

In order to overcome the low corrosion rate of pure Fe, various alloying elements such as Mn, Co, Al, W, Sn, B, C, S and Si have been introduced to accelerate its degradation rate [116–118]. After adding Mn and Si into pure Fe (Fe30Mn6Si), its corrosion rate was increased [118], while adding Co, Al, W, Sn, B, C and S have no appreciable influences on the degradation rate of pure Fe [116]. All the mentioned elements can increase the strength of the base metal except Sn which decreases the mechanical properties [116]. Fe-based alloys also have high strength and high ductility, making them suitable candidates for utilization as wire fixation devices. Additionally, the superior mechanical properties of these alloys are favored for making fine wires even in sub-micron sizes which could be beneficial for the plenty of clinical operations. However, the high Young's modulus of these alloys offers an inadequate match to those of natural bone and potentially causes stress shielding.

Hermawan et al. [119] reported the corrosion rate of Fe-Mn alloy which was finally one order of magnitude lower than that of Mg and its alloys. To achieve high degradation rates, Fe-Mn-Pd alloys were produced and showed higher corrosion rate compared to pure Fe and Fe-Mn alloys [51]. It is worth mentioning that these researchers obtained the corrosion rate of those alloys in vitro, which in fact are different from in vivo results. Kraus et al. [48] examined Fe-Mn-Pd alloy wires in vivo to perform a feasibility study of these alloys in osteosynthesis applications. They produced the wires with a diameter of 1.6 mm by casting and machining processes. The degradation rate of Fe-Mn-Pd wires was not significantly higher than that of Fe and Fe-Mn alloys (see Figure 8).

Attempting to find a proper Fe-based alloy with suitable degradation rate, some researchers achieved novel Fe alloys such as Fe-Pd [120], Fe-Mn-C-S [121] and Fe-Ga [122] which showed higher corrosion rate compared to pure Fe in vitro. Capek et al. [120] stated that the preparation route significantly influenced material properties such as mechanical strength and corrosion rate. In spite of numerous studies for improving the degradation rate of Fe-based alloys have been reported, only one research in the literature [48] has applied Fe-based wires in vivo experiments in the rats' long bone.

As discussed in Section 2.3, Zn and its alloys have moderate corrosion rate and biologically beneficial corrosion products thus have been considered as one of the first priorities in the material selection for the regeneration of bone defects. However, there is no in vivo studies in the literature about the utilization of Zn-based wires in the orthopedic or dental applications. Actually, the contribution of Zn-based wires in vivo experiments has been mostly limited to the cardiovascular stent application which has different biological environment compared to bone tissues. Anyway, the alloying elements, process conditions for wire drawing, mechanical properties, and the corrosion behavior of Zn-based stents could be an initial guidance for the future use of this kind wires in the bone regeneration purposes.

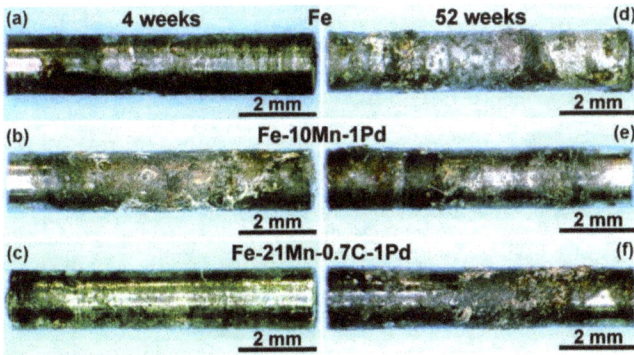

Figure 8. Optical micrographs of the wires after implantation for (**a–c**) 4 weeks and (**d–f**) 52 weeks of (**a,d**) pure Fe, (**b,e**) Fe–10Mn–1Pd and (**c,f**) Fe–21Mn–0.7C–1Pd, reproduced with permission from [48], Elsevier, 2014.

5.1. Medical Sutures and Staples

In 1849, Sims successfully used a fine structure of drawn silver wires by a jeweler to close a large vesicovaginal fistula [123]. In 1947 Babcock et al. [123] published their experimental results for utilization of tantalum (Ta) and SS wires as sutures. They found that both the materials were suitable candidates for suturing but the 18-8 SS which had a reasonable price could be more applicable [123]. For many years and until now, SS sutures have gained a lot of attentions. Permanent metals adopted for utilization as sutures were mentioned in Table 2. Since the beginning of the 2nd world war, BP materials were introduced. The BP materials could be classified in two groups, synthetic polymers (nylon, polyester and propylene) and natural polymers (catgut and collagen) [124]. Materials that were introduced for suture applications should have high mechanical strength and integrity and should be biocompatible. It has been reported that polymers cannot withstand high loads securely and exhibit a low creep and stress relaxation resistance which limits their range of application [13]. Metallic sutures possess, in general, the higher strength and more advantageous creep and relaxation behavior than polymers, and they also show, in most of cases, good biocompatibility. Therefore, metallic sutures are preferred to polymeric ones especially when a biodegradable metallic suture is applied. Mg, Fe and Zn as biodegradable metals could attract more attention for suture applications.

The first usage of Mg as a suture material dates to the beginning of the 20th century. There were some attempts to use Mg wires for osteosynthesis and in vascular surgeries. The low ductility of pure Mg due to its close-packed hexagonal structure has prevented their usage as suture materials. In 1924, Seeling produced high purity Mg in a distillation process to overcome the low ductility of the material [13]. The high purity Mg did not exhibit adequate mechanical properties to allow a defect-free kinking and knotting of the wires. Seeling also tried to increase the ductility of Mg by adding some alloying elements [13]. The corrosion rate of Mg as suturing material was reported to be adequate. Hence, Seeling focused on the improvement of its mechanical properties. As it was mentioned earlier, Mg and its alloys have high corrosion rate which is inappropriate for most of the applications. Therefore, decreasing the degradation rate of Mg by some methods such as adding alloying elements were carried out. Researches in this field were retarded until the beginning of 21th century.

Seitz et al. [125,126] were the first researchers produced Mg wires for suture applications. They selected four Mg alloys, ZEK100, MgCa0.8, AL36 and AX30, which were previously reported to be biocompatible. Mg suture wires of 0.3 to 0.5 mm in diameter were fabricated by hot extrusion [125] and their diameter was further decreased via multiple drawing [126]. Despite low mechanical properties of the extruded wires, they could meet the parameters required for a surgical suture. As an extruded suture, the AL36 wire had the greatest ductility which allowed for tightening knots (Figure 9). With the aid of multiple wires drawing process, the Mg wires were reduced to monofilament wire possessing diameters between 0.5 and 0.1 mm and they also twisted into polyfilament sutures using stranding. Compared to monofilament strands, polyfilament alloy materials possessed significantly lower tensile strengths but higher fracture strains [126]. More ductility along with the adequate strength of polyfilament alloy materials made them suitable candidates for suture applications.

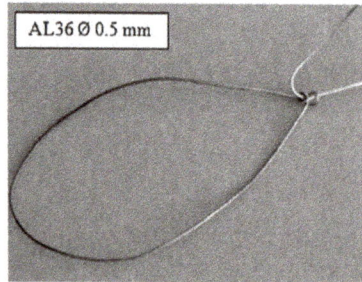

Figure 9. Knotted noose made of the alloy AL36, reproduced with permission from [125], WILEY, 2010.

The previously mentioned studies have investigated pure Mg in combination with aluminum (Al) as the alloying element. It has been shown that it could be detrimental to the body. In high doses, Al has been shown to cause neurotoxicity [26]. Considering the neurotoxicity of element Al, Mg-Al-based alloys are not recommended candidates for biodegradable implant material [27]. However, recently, Al-free Mg alloys with better biological performance such as Mg-RE (rare earth element) alloys have been developed and investigated in vivo and in vitro [98,99,127]. Bai et al. [127] employed hot extrusion and cold drawing with intermediate annealing process to prepare fine wires with final diameters less than 0.4 mm of three kinds of ternary Mg-4%RE (Gd/Y/Nd)-0.4%Zn alloys. High yield strength along with moderate and adequate ductility was reported for all the fine wires. The in vitro degradation performance of the finished fine wires was evaluated, and the results indicated that Mg-4Gd-0.4Zn and Mg-4Nd-0.4Zn wires showed a similar good corrosion resistance and a uniform corrosion behavior in simulated body fluid (SBF) solution. In contrast, Mg-4Y-0.4Zn fine wires exhibited a relatively high degradation rate and pitting corrosion behavior [127]. All the studies dedicated to wires with the diameter of less than 1 mm have been reviewed in this section. Generally, one can conclude that Mg-based fine wires might be good candidates for suture applications with adequate mechanical properties and adequate degradation rate. Another application of fine wires is as staples. Staples made of pure Mg [128,129] and its alloys [130] were recently investigated for gastric anastomosis. Although the examined staples have exhibited enough closure strength, homogeneous corrosion behavior, and desirable biodegradation rate both in vivo and in vitro [128–130], their usage as a bone staple were not evaluated in any researches yet.

In the 18th century, Icart successfully employed Fe wires for the first time to support the healing of fractured human bone [13]. At that time, Fe was not considered to be a degradable implant material and was selected only due to its sufficient mechanical properties. It was just at the beginning of the 21st century that Fe attracted researchers' attention as a biodegradable and more importantly nutrient implant material in bone surgeries. However, the use of pure Fe and Fe-based alloys as a nutrient suture material were not reported in the literature. Table 3 summarizes all the applications of biodegradable metallic wires.

Table 3. Biodegradable metallic wires in bone applications.

Base Metal	Alloy	Wire Diameter (mm)	Fabrication Process	Application	Type of Biological Experiments	Published Time
Mg	ZEK100 [125]	0.3–0.5	Hot extrusion	Suture	-	2010
	MgCa0.8 [125]	0.3–0.5	Hot extrusion	Suture	-	2010
	AL36 [125]	0.3–0.5	Hot extrusion	Suture	-	2010
	AX30 [125]	0.3–0.5	Hot extrusion	Suture	-	2010
	ZEK100 [126]	0.274^{m}–0.822^{P}	Hot extrusion + Cold drawing + Intermediate annealing	Suture	-	2011
	MgCa0.8 [126]	0.274^{m}–0.822^{P}	Hot extrusion + Cold drawing + Intermediate annealing	Suture	-	2011
	AL36 [126]	0.274^{m}–0.822^{P}	Hot extrusion + Cold drawing + Intermediate annealing	Suture	-	2011
	AX30 [126]	0.274^{m}–0.822^{P}	Hot extrusion + Cold drawing + Intermediate annealing	Suture	-	2011
	Mg–4Gd–0.4Zn [127]	0.3	Hot extrusion + Cold drawing + Intermediate annealing	Suture	in vitro	2014
	Mg–4Y–0.4Zn [127]	0.3	Hot extrusion + Cold drawing + Intermediate annealing	Suture	in vitro	2014
	Mg–4Nd–0.4Zn [127]	0.3	Hot extrusion + Cold drawing + Intermediate annealing	Suture	in vitro	2014
	Mg2Ag [113]	0.8	Hot extrusion + Cold drawing + Intermediate annealing	Guide wire-thin pin	in vitro-in vivo (mouse)	2016
	Mg3Ge [114]	2.2	Hot rolling + Machining	Guide wire-thin pin	in vitro-in vivo (rabbit)	2017
	Mg–Zn–Mn [115]	1.6	Hot extrusion	Guide wire-thin pin	in vitro-in vivo (mouse)	2017
Fe	Fe–10Mn–1Pd [48]	1.6	Casting + Machining	Guide wire-thin pin	in vitro-in vivo (rat)	2014
	Fe–21Mn–0.7C–1Pd [48]	1.6	Casting + Machining	Guide wire-thin pin	in vitro-in vivo (rat)	2014

^{m}: Monofilament, ^{P}: polyfilament.

5.2. Bone Tissue Engineering

Bone grafting is known as an operation aiming to replace diseased or injured bones with grafting materials. The bone graft could be taken from the patient (autograft) or donors (allograft) or utilized synthetic materials [131]. Even though both of autograft and allograft have some significant advantages, owing to some clinical problems such as the necessity of performing additional surgery on the donor site as well as size and geometry limitations of autograft and inconsistency with host tissue, and the possibility of disease transmission from donated allograft, a growing demand for synthetic materials as bone substitutes is received from clinicians [132]. Rely on the diamond concept for the bone healing process, synthetic bone materials could play an important role as a scaffold because it can keep mechanical stability. In addition, a combination of these materials with osteogenic cells and growth factors could accelerate the new bone formation [132,133].

Among the wide range of synthetic materials, calcium phosphate cement (CPC) is known to be the most resemble synthetic material to natural bone thus is clinically suitable for the repair of bone defects in a variety of orthopedic and dental applications [134,135]. Although CPCs are biodegradable, biocompatible, osteoconductive and have proper compression strength, because of their low bending and tensile strength and low fracture toughness their applications are limited to non-load bearing areas such as small cranial and maxillo-facial surgeries [136,137]. Generally, the mechanical strength of CPC is similar to that of trabecular bone, or one-fifth of that of cortical bone [138]. Reinforcement of CPC with fibers is one of the most promising approaches for its application in thin and large bone defects as well as stress-bearing locations [135]. The fiber reinforced calcium phosphate cement (FRCPC) concept was firstly introduced by Gonten et al. in 2000 [139]. They utilized a fiber-mesh made from poly (galectin) to reinforce CPC. Their results showed that the knitted fiber-mesh could enhance the flexural and toughness of CPCs. Then, in 2002, Xu et al. [140] incorporated two types of resorbable fibers into CPC with a fiber volume fraction of 25%. Their FRCPC increased the flexural strength threefold, and work-of-fracture toughness nearly 100 times than those of unreinforced CPC. They also claimed that resorption of the fiber during the regeneration process could facilitate vacuolization and consequently rapid bone-ingrowth. In a same period of time, Dos Santos et al. [138] followed up this field using non-degradable fibers including carbon, nylon which were not suitable for clinical applications. They just tried to enhance the mechanical properties which CPC is weak at.

In the next generation of FRCPC, biocompatible and biodegradable polymeric fibers such as poly (lactic-co-glycolic) acid (PLGA) [141] and biopolymers such as chitosan and gelatin [142] were used. Owing to the lower Young's modulus and strength of these fibers than those of the cement matrix, they are not successful enough as a reinforcement element for the strengthening of CPC. The bending strength of 40–45 MPa is the best mechanical properties have been reported for FRCPC achieved by PLGA fibers up to now. Furthermore, in physiological environment, fast degradation rate of PLGA fibers results in the concern of mechanical stability only after 3–4 weeks [143]. In 2013, the new generation of FRCPC was introduced by Kruger et al. [144]. They utilized AZ31 and ZEK100 Mg wires with 0.6 and 0.228 mm in diameter respectively. It was for the first time that BM wires were used as a reinforcement element for the strengthening of CPC which was successful to improve the bending strength of FRCPC up to 139 ± 41 MPa. They stated that regarding the fracture of the FRCPC initiated by shearing of the matrix at or near the biodegradable metallic wire surface, hybrid reinforcement of CPC with both metal and fine polymer fibers such as polylactic acid (PLA) could postpone the crack initiation and further enhance the mechanical properties of FRCPC (Figure 10). Additionally, in vitro tests with osteoblasts validated cytocompatibility of wire-reinforced CPC composites with BM.

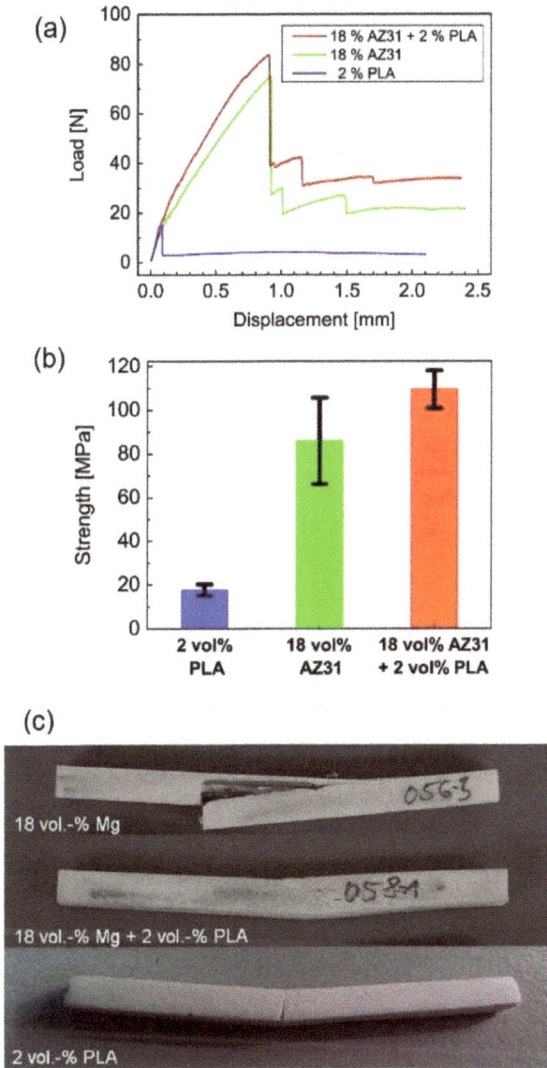

Figure 10. Comparison between three different compositions of FRCPC including AZ31, PLA and hybrid (both AZ31 and PLA). (**a**) Load-displacement curves; (**b**) Bending strength and (**c**) Fracture images of the composites, reproduced with permission from [114], Elsevier, 2013.

However, there is only one pilot study for the use of FRCPC composites. Further investigations need to be done to optimize the mechanical properties of the BM wire reinforced CPC composites by applying a vast range of diameters, plenty types of biodegradable alloys and different fiber dispositions. Moreover, assessment of the mechanical stability should be conducted in much longer time in vitro (several months or even a year), and finally in vivo animal tests should be performed to examine all of the positive and negative biological effects and structural changing during the degradation of these kind of composites before any clinical applications for the grafting of human bone defects in the load-bearing locations.

6. Future Work

Generally, for BM wires and especially for Mg and its alloys, controlling the degradation rate for implant applications is quite important. It becomes more critical for the application of fine BM wires in stress-bearing areas, because the mechanical properties of such wires can be significantly influenced by high corrosion rate. Therefore, the first priority in the application of BM wires is controlling the corrosion rate to a logical rate, at least in the initial stage after implantation. In this regard, methods such as selection of alloying elements, microstructure adjustments and surface modification will be effective. It is worth mentioning that these corrosion controlling methods have some intrinsic differences. Various surface modification techniques create a protective layer on the material which resist against corrosion. In some cases, this layer also increases the osteoinductivity and osteoconductivity of the material while in other cases only controls the corrosion rate on the surface. In general, surface modification methods affect the surface corrosion. Selecting proper alloying elements in combination with the base metal could adjust the volume degradation rate of the material and concurrently could enhance mechanical properties of an alloy. The third corrosion controlling method, by employing a proper fabrication process, creates a fine and uniform microstructure which in turn influences the mechanical properties of the material. Hence, it is obvious that selecting alloying elements and refining microstructure methods will have another beneficial effect on mechanical characteristics of the materials. These two mentioned techniques, except in some special cases, are preferred to surface modification methods.

6.1. Alloying Elements

6.1.1. Magnesium

Common alloying elements used for pure Mg are mentioned in Table 4. Biological and mechanical advantages and disadvantages of the elements are summarized in this table. Since Al and Cu may cause neurotoxicity in the body, their usage in recent studies has been limited. It is obvious from the table that Cu also accelerates the corrosion rate of Mg [26] and hence, it is not suitable for implants. Adding Ca, Si and Li elements could deteriorate the degradation rate of Mg [97]. Therefore, these elements cannot play their role as a corrosion controlling element. It can be seen that other alloying elements, such as Zn, Zr, Mn and RE improve the corrosion resistance of Mg and simultaneously increase the strength and ductility of the obtained alloy. Additionally, for the mentioned elements in standard dosage, harmful side effects were not reported [97]. In recent literature it can be found that newly introduced biocompatible Mg alloys mostly contain these elements [145–148]. Therefore, in designing and developing of new Mg biomaterials the potentials of these alloying elements should be considered.

6.1.2. Iron

Alloying elements such as Mn, Co, Al, W, Sn, B, C, S and Si were added into Fe-based alloys to accelerate their degradation rate. It has been found that adding Co, Al, W, Sn, B, C and S have no influence on the degradation behavior of pure Fe [116], while the alloy Fe30Mn6Si had a relatively more corrosion rate compared to pure Fe [118]. There are two main criteria regarding increase in the corrosion rate of Fe-based alloys which affect the corrosion susceptibility of this kind of alloys: (1) the addition of less noble alloying elements within the solubility limit in Fe to cause the matrix more susceptible to corrosion; and (2) the addition of noble alloying elements to generate small and finely dispersed intermetallic phases (IMPs). Mn and Pd have been shown to be suitable alloying elements according to both of the mentioned approaches [51]. Fe-Mn-Pd alloy has investigated and suggested as a Fe-based alloy which could improve the degradation rate of the base metal [48]. Inspecting the effect of alloying elements on the corrosion behavior of Fe-based alloys is still a challenging and ongoing research area.

Table 4. Biological and mechanical advantages and disadvantages of alloying elements for Mg.

Alloying Element	Advantages	Disadvantages
Al	Increase the strength and ductility and decrease the corrosion rate [26].	Neurotoxicity and accumulation in bone [87].
Ca	Most abundant mineral and mainly stored in bone and teeth; activator or stabilizer of enzymes [87].	Calcium metabolism disorder; kidney stones; increasing corrosion rate [97].
Cu	Increase the ductility [97].	Neurodegenerative diseases including Alzheimer's, Menkes, and Wilson disease; accelerate corrosion and decrease the strength [26].
Zn	Essential trace element; appear in all enzyme classes; increase the strength and improve corrosion resistance [87].	Euro toxic and hinder bone development at higher concentration [87].
Mn	Improve corrosion resistance, strength and ductility; essential trace element; activator of enzyme; Mn deficiency is related to osteoporosis, diabetes mellitus, and atherosclerosis [87].	Excessive Mn results in neurotoxicity [87].
Si	Cross linking agent of connective tissue basement membrane structures; necessary for bone calcification [87].	Excessive SiO_2 causes lung diseases [87]; decreasing the ductility and increasing corrosion rate [97].
Li	Used in the treatment of manic-depressive psychoses [87].	Decrease the strength and increasing corrosion rate [97].
Zr	Increase the strength and ductility [97].	High concentration in liver and gall bladder [87].
RE	Compound of drugs for treatment of cancer; improve the mechanical properties and corrosion resistance [87].	Accumulation in bone and liver [87].

6.1.3. Zinc

Zinc as a newly introduced biodegradable metal with a moderate degradation rate has attracted some attention. Alloying elements of Zn have the role of strengthening the matrix since Zn-based alloys suffer from low mechanical properties. To improve the mechanical properties of pure Zn, some elements such as Mg, Ca, Sr, Sn and Fe were added into Zn-based alloys [56,70,117,149]. Vojtech et al. [70] developed as cast Zn-Mg alloys and studied their mechanical properties and corrosion behavior. Their results showed a significantly improvement in the tensile strength of Zn modifying by 1% Mg. The elongation of Zn-1% Mg alloy was much greater than that of pure Zn but was only 2%, which is quite low for medical usage [70]. Combination of Mg as an alloying element with Zn-based alloys especially about 1% is newly introduced and investigated by several researchers [54,55]. Shen et al. [150] produced Zn-1.2% Mg alloy and achieved the high ultimate strength of 362 MPa and elongation of 21%, which is the highest ductility obtained for biodegradable Zn alloys. The combination of Zn with other alloying elements is still a challenging topic for researchers.

6.2. Surface Modification

6.2.1. Magnesium

Numerous studies have proved that the functional and mechanical behaviors as well as chemical and physical properties of biomaterials can be affected by different surface modification techniques [151]. Although plenty of coating methods such as electrodeposition [152], sol-gel and dipping [153], chemical conversion [154], anodization [155], vapor deposition [156], spin [157], and alkali treatment [158] have been proposed to inhibit rapid corrosion of Mg and its alloys, just few of them are effective and flexible enough for the formation of a uniform coating with desired thickness on the non-flat surfaces such as wires. Simultaneously, the adhesion between the substrate and the coating film is another concern that should be considered. Finally, prevention from deteriorating the mechanical properties of the substrate as much as possible is the third consideration in the selection of an adequate coating method. To summarize, it is claimed that only sol-gel preparation combined with dip coating and alkali treatment can meet all of the three mentioned requirements which make them suitable and cost-effective methods for the coating of Mg-based wires [151,159].

The composition of the coating is another critically important parameter. An ideal coating material for bone applications should be able to enhance the biocompatibility and osteointegration properties of Mg-based alloys and simultaneously increase the corrosion resistance. As discussed in Section 2.1.1, coating with $Mg(OH)_2$ [159,160] could enhance bone formation and passivate the Mg-based wires to decrease the corrosion rate as well. The synthetic calcium phosphate coatings [151,161,162] can definitely facilitate the new bone formation. The biomimetic coating with pure collagen [163] and collagen-based composites [164] is another promising approach that can be applied in the future studies of Mg-based wires in bone tissue regeneration.

6.2.2. Iron

Despite numerous investigations focused on the surface treatment methods for Mg-based alloys, only few research can be found for surface treatment of pure Fe. Chen et al. [165] suggested using micro-patterned Au disc arrays to accelerate degradation of pure Fe. The proposed approach relied on the galvanic corrosion caused by the Au array [165]. In another research, Fe-O thin films were prepared on pure Fe by plasma immersion ion implantation and deposition in order to improve the biodegradability and biocompatibility [166]. The prepared thin films decreased the degradation rate and enhanced the corrosion behavior [166]. Sandblasting method has been proposed by Zhou et al. [167] as a successful surface treatment for increasing the degradation rate of pure Fe in SBF. The reasons for increasing the corrosion rate were the change of surface composition (in the early stage), high roughness and high density of dislocations [167]. The limited number of researches in the surface treatment methods for Fe-based BM wires indicates that more studies is required to introduce and reveal the potentials of various surface treatments approaches.

6.3. Fabrication Processes

As stated before, microstructure adjustment is one of the corrosion control methods. In this regard, selection of a proper fabrication process could effectively influence the microstructure. In case of Mg-Zn-Mn alloy, it has been reported that the extruded alloy compared to casted material had fine grain size and accordingly slower degradation rate. Therefore, selection of the manufacturing process of the alloy needs more attention. There are several common manufacturing processes such as rolling, extrusion, forging, drawing, and continuous casting. Recently severe plastic deformation (SPD) processes such as equal channel angular extrusion (ECAP), multi-axial forging (MAF) and high-pressure torsion (HPT) were introduced. These SPD methods by imposing large shear strains or huge compressive strains decrease the grain size of the materials. Dobatkin et al. [168] employed three different SPD processes (ECAP, MAF, and rotary swaging (RS)) to investigate the effects on the microstructure and corrosion behavior of WE43 Mg alloy. They observed that all the processes could decrease the grain size and create ultra-fine grained (UFG) microstructure. The biodegradation rate of the UFG WE43 was examined in vitro and it was concluded that ECAP and MAF caused some deceleration of biodegradation rate by slowing down the gas formation in the biological fluid and improved the biocompatibility of the WE43 alloy [168]. This research shows the potential of SPD method in controlling the corrosion of Mg alloys which can be spread to other biodegradable metals as well.

7. Conclusions

Application of biodegradable metals (magnesium, iron, and zinc) as promising biodegradable materials is spreading in various biomedical devices especially for dental and orthopedic applications. Wire is one of the most popular forms of these materials that can be easily formed by drawing and other methods. Although currently biodegradable metallic wires have some drawbacks such as unsatisfactory degradation rate and mechanical properties when used in bone surgeries, many methods can be adopted to control their corrosion behavior and tailor their mechanical properties. Owing to the proper mechanical properties, controllable degradation rate, and the release of nutrient ions, which will significantly be beneficial to bone cells and bone tissue regeneration, we firmly believe that in the near future BM wires will be used broadly in dental and orthopedic fields.

Author Contributions: Mohammad Asgari, Zhiyong Li and Yin Xiao conceptualized the paper and gave instructions; Mohammad Asgari wrote the paper; Ruiqiang Hang, Zhiyong Li and Yin Xiao edited the paper; Mohammad Asgari, Zhiyong Li and Yin Xiao revised the paper; Chang Wang and Zhentao Yu gave idea and advice; Zhiyong Li and Yin Xiao gave directions.

Conflicts of Interest: The authors declare no conflict of interest.

References

1. Griebel, A.J.; Schaffer, J.E. Absorbable filament technologies: Wire-drawing to enable next-generation medical devices. In *Magnesium Technology 2016*; Springer: Berlin, Germany, 2016; pp. 323–327.
2. Sardeshmukh, A.; Reddy, S.; Gautham, B.; Joshi, A.; Panchal, J. A data science approach for analysis of multi-pass wire drawing. In Proceedings of the ASME 2017 International Design Engineering Technical Conferences and Computers and Information in Engineering Conference, Cleveland, OH, USA, 6–9 August 2017; American Society of Mechanical Engineers: New York, NY, USA, 2017; p. V001T002A071. [CrossRef]
3. Griebel, A.J.; Schaffer, J.E.; Hopkins, T.M.; Alghalayini, A.; Mkorombindo, T.; Ojo, K.O.; Xu, Z.; Little, K.J.; Pixley, S.K. An in vitro and in vivo characterization of fine WE43B magnesium wire with varied thermomechanical processing conditions. *J. Biomed. Mater. Res. Part B Appl. Biomater.* **2017**. [CrossRef] [PubMed]

4. Duerig, T.; Pelton, A.; Stöckel, D. An overview of nitinol medical applications. *Mater. Sci. Eng. A* **1999**, *273*, 149–160. [CrossRef]
5. Welch, J.M.; Sutton, G. Coronary Guide Catheter. Google Patents WO2017020012 A1, 29 July 2016.
6. Henkes, H.; Bose, A.; Felber, S.; Miloslavski, E.; Berg-Dammer, E.; Kühne, D. Endovascular coil occlusion of intracranial aneurysms assisted by a novel self-expandable nitinol microstent (neuroform). *Int. Neuroradiol.* **2002**, *8*, 107–119. [CrossRef] [PubMed]
7. Tan, L.; Sun, D.-H.; Yu, T.; Wang, L.; Zhu, D.; Li, Y.-H. Death due to intra-aortic migration of kirschner wire from the clavicle: A case report and review of the literature. *Medicine* **2016**, *95*, e3741. [CrossRef] [PubMed]
8. Bravo, L.A.; de Cabañes, A.G.; Manero, J.M.; Rúperez, E.; Gil, F.J. NiTi superelastic orthodontic archwires with polyamide coating. *J. Mater. Sci. Mater. Med.* **2014**, *25*, 555–560. [CrossRef] [PubMed]
9. Chakravorty, B. Configured ligature wire for quick conventional and overtie lingual archwire ligations. *APOS Trends Orthod.* **2017**, *7*, 108–110. [CrossRef]
10. Iavazzo, C.; Gkegkes, I.D.; Vouloumanou, E.K.; Mamais, I.; Peppas, G.; Falagas, M.E. Sutures versus staples for the management of surgical wounds: A meta-analysis of randomized controlled trials. *Am. Surg.* **2011**, *77*, 1206–1221. [PubMed]
11. Bowen, P.K.; Shearier, E.R.; Zhao, S.; Guillory, R.J.; Zhao, F.; Goldman, J.; Drelich, J.W. Biodegradable metals for cardiovascular stents: From clinical concerns to recent Zn-alloys. *Adv. Healthc. Mater.* **2016**, *5*, 1121–1140. [CrossRef] [PubMed]
12. Li, H.; Zheng, Y.; Qin, L. Progress of biodegradable metals. *Prog. Natl. Sci. Mater. Int.* **2014**, *24*, 414–422. [CrossRef]
13. Seitz, J.-M.; Durisin, M.; Goldman, J.; Drelich, J.W. Recent advances in biodegradable metals for medical sutures: A critical review. *Adv. Healthc. Mater.* **2015**, *4*, 1915–1936. [CrossRef] [PubMed]
14. Wang, C.; Yang, H.T.; Li, X.; Zheng, Y.F. In vitro evaluation of the feasibility of commercial Zn alloys as biodegradable metals. *J. Mater. Sci. Technol.* **2016**, *32*, 909–918. [CrossRef]
15. Chen, Y.; Zhang, W.; Maitz, M.F.; Chen, M.; Zhang, H.; Mao, J.; Zhao, Y.; Huang, N.; Wan, G. Comparative corrosion behavior of Zn with Fe and Mg in the course of immersion degradation in phosphate buffered saline. *Corros. Sci.* **2016**, *111*, 541–555. [CrossRef]
16. Li, H.; Yang, H.; Zheng, Y.; Zhou, F.; Qiu, K.; Wang, X. Design and characterizations of novel biodegradable ternary Zn-based alloys with IIA nutrient alloying elements Mg, Ca and Sr. *Mater. Des.* **2015**, *83*, 95–102. [CrossRef]
17. Witte, F. Reprint of: The history of biodegradable magnesium implants: A review. *Acta Biomater.* **2015**, *23*, S28–S40. [CrossRef] [PubMed]
18. Andrews, E.W. Absorbable metal clips as substitutes for ligatures and deep sutures in wound closure. *J. Am. Med. Assoc.* **1917**, *69*, 278–281. [CrossRef]
19. Seelig, M. A study of magnesium wire as an absorbable suture and ligature material. *Arch. Surg.* **1924**, *8*, 669–680. [CrossRef]
20. Hou, L.-D.; Li, Z.; Pan, Y.; Sabir, M.; Zheng, Y.-F.; Li, L. A review on biodegradable materials for cardiovascular stent application. *Front. Mater. Sci.* **2016**, *10*, 238–259. [CrossRef]
21. Li, X.; Chu, C.; Zhou, L.; Bai, J.; Guo, C.; Xue, F.; Lin, P.; Chu, P.K. Fully degradable PLA-based composite reinforced with 2D-braided Mg wires for orthopedic implants. *Compos. Sci. Technol.* **2017**, *142*, 180–188. [CrossRef]
22. Staiger, M.P.; Pietak, A.M.; Huadmai, J.; Dias, G. Magnesium and its alloys as orthopedic biomaterials: A review. *Biomaterials* **2006**, *27*, 1728–1734. [CrossRef] [PubMed]
23. Zhao, N.; Zhu, D. Application of Mg-based alloys for cardiovascular stents. *Int. J. Biomed. Eng. Technol.* **2013**, *12*, 382–398. [CrossRef]
24. Liu, A.; Sun, M.; Shao, H.; Yang, X.; Ma, C.; He, D.; Gao, Q.; Liu, Y.; Yan, S.; Xu, S. The outstanding mechanical response and bone regeneration capacity of robocast dilute magnesium-doped wollastonite scaffolds in critical size bone defects. *J. Mater. Chem. B* **2016**, *4*, 3945–3958. [CrossRef]
25. Wu, L.; Feyerabend, F.; Schilling, A.F.; Willumeit-Römer, R.; Luthringer, B.J. Effects of extracellular magnesium extract on the proliferation and differentiation of human osteoblasts and osteoclasts in coculture. *Acta Biomater.* **2015**, *27*, 294–304. [CrossRef] [PubMed]
26. Persaud-Sharma, D.; McGoron, A. Biodegradable magnesium alloys: A review of material development and applications. *J. Biomim. Biomater. Tissue Eng.* **2012**, *12*, 25–39. [CrossRef] [PubMed]

27. Zheng, Y.F.; Gu, X.N.; Witte, F. Biodegradable metals. *Mater. Sci. Eng. R Rep.* **2014**, *77*, 1–34. [CrossRef]

28. Pilarska, A.A.; Klapiszewski, Ł.; Jesionowski, T. Recent development in the synthesis, modification and application of Mg(OH)$_2$ and MgO: A review. *Powder Technol.* **2017**, *319*, 373–407. [CrossRef]

29. Guo, M.; Muhammad, F.; Wang, A.; Qi, W.; Wang, N.; Guo, Y.; Wei, Y.; Zhu, G. Magnesium hydroxide nanoplates: A pH-responsive platform for hydrophobic anticancer drug delivery. *J. Mater. Chem. B* **2013**, *1*, 5273–5278. [CrossRef]

30. Janning, C.; Willbold, E.; Vogt, C.; Nellesen, J.; Meyer-Lindenberg, A.; Windhagen, H.; Thorey, F.; Witte, F. Magnesium hydroxide temporarily enhancing osteoblast activity and decreasing the osteoclast number in peri-implant bone remodelling. *Acta Biomater.* **2010**, *6*, 1861–1868. [CrossRef] [PubMed]

31. Weizbauer, A.; Kieke, M.; Rahim, M.I.; Angrisani, G.L.; Willbold, E.; Diekmann, J.; Flörkemeier, T.; Windhagen, H.; Müller, P.P.; Behrens, P.; et al. Magnesium-containing layered double hydroxides as orthopaedic implant coating materials—An in vitro and in vivo study. *J. Biomed. Mater. Res. Part B Appl. Biomater.* **2016**, *104*, 525–531. [CrossRef] [PubMed]

32. Hickey, D.J.; Ercan, B.; Chung, S.; Webster, T.J.; Sun, L.; Geilich, B. Mgo nanocomposites as new antibacterial materials for orthopedic tissue engineering applications. In Proceedings of the 2014 40th Annual Northeast Bioengineering Conference (NEBEC), Boston, MA, USA, 25–27 April 2014; pp. 1–2.

33. Huang, L.; Li, D.-Q.; Lin, Y.-J.; Wei, M.; Evans, D.G.; Duan, X. Controllable preparation of Nano-MgO and investigation of its bactericidal properties. *J. Inorg. Biochem.* **2005**, *99*, 986–993. [CrossRef] [PubMed]

34. Haghshenas, M. Mechanical characteristics of biodegradable magnesium matrix composites: A review. *J. Magn. Alloys* **2017**, *5*, 189–201. [CrossRef]

35. Gu, X.-N.; Zheng, Y.-F. A review on magnesium alloys as biodegradable materials. *Front. Mater. Sci. China* **2010**, *4*, 111–115. [CrossRef]

36. Yusop, A.; Bakir, A.; Shaharom, N.; Abdul Kadir, M.; Hermawan, H. Porous biodegradable metals for hard tissue scaffolds: A review. *Int. J. Biomater.* **2012**, *2012*, 641430. [CrossRef] [PubMed]

37. Zhao, D.; Witte, F.; Lu, F.; Wang, J.; Li, J.; Qin, L. Current status on clinical applications of magnesium-based orthopaedic implants: A review from clinical translational perspective. *Biomaterials* **2017**, *112*, 287–302. [CrossRef] [PubMed]

38. Haynes, W.M. *CRC Handbook of Chemistry and Physics*; CRC Press: Boca Raton, FL, USA, 2014.

39. Feyerabend, F.; Wendel, H.-P.; Mihailova, B.; Heidrich, S.; Agha, N.A.; Bismayer, U.; Willumeit-Römer, R. Blood compatibility of magnesium and its alloys. *Acta Biomater.* **2015**, *25*, 384–394. [CrossRef] [PubMed]

40. Harandi, S.E.; Banerjee, P.C.; Easton, C.D.; Singh Raman, R.K. Influence of bovine serum albumin in hanks' solution on the corrosion and stress corrosion cracking of a magnesium alloy. *Mater. Sci. Eng. C* **2017**, *80*, 335–345. [CrossRef] [PubMed]

41. Noviana, D.; Paramitha, D.; Ulum, M.F.; Hermawan, H. The effect of hydrogen gas evolution of magnesium implant on the postimplantation mortality of rats. *J. Orthop. Transl.* **2016**, *5*, 9–15. [CrossRef]

42. Tang, J.; Wang, J.; Xie, X.; Zhang, P.; Lai, Y.; Li, Y.; Qin, L. Surface coating reduces degradation rate of magnesium alloy developed for orthopaedic applications. *J. Orthop. Transl.* **2013**, *1*, 41–48. [CrossRef]

43. Kraus, T.; Fischerauer, S.F.; Hänzi, A.C.; Uggowitzer, P.J.; Löffler, J.F.; Weinberg, A.M. Magnesium alloys for temporary implants in osteosynthesis: In vivo studies of their degradation and interaction with bone. *Acta Biomater.* **2012**, *8*, 1230–1238. [CrossRef] [PubMed]

44. Zheng, Y.; Xu, X.; Xu, Z.; Wang, J.; Cai, H. Development of Fe-based degradable metallic biomaterials. *Metallic Biomater. New Direct. Technol.* **2017**, 113–160.

45. Zimmermann, M.B.; Hurrell, R.F. Nutritional iron deficiency. *Lancet* **2007**, *370*, 511–520. [CrossRef]

46. Toxqui, L.; Vaquero, M.P. Chronic iron deficiency as an emerging risk factor for osteoporosis: A hypothesis. *Nutrients* **2015**, *7*, 2324–2344. [CrossRef] [PubMed]

47. Francis, A.; Yang, Y.; Virtanen, S.; Boccaccini, A.R. Iron and Iron-based alloys for temporary cardiovascular applications. *J. Mater. Sci. Mater. Med.* **2015**, *26*, 138. [CrossRef] [PubMed]

48. Kraus, T.; Moszner, F.; Fischerauer, S.; Fiedler, M.; Martinelli, E.; Eichler, J.; Witte, F.; Willbold, E.; Schinhammer, M.; Meischel, M. Biodegradable Fe-based alloys for use in osteosynthesis: Outcome of an in vivo study after 52 weeks. *Acta Biomater.* **2014**, *10*, 3346–3353. [CrossRef] [PubMed]

49. Huang, T.; Zheng, Y.; Han, Y. Accelerating degradation rate of pure iron by Zinc Ion implantation. *Regener. Biomater.* **2016**, *3*, 205–215. [CrossRef] [PubMed]

50. Mouzou, E.; Paternoster, C.; Tolouei, R.; Chevallier, P.; Biffi, C.A.; Tuissi, A.; Mantovani, D. CO_2-rich atmosphere strongly affects the degradation of Fe-21Mn-1C for biodegradable metallic implants. *Mater. Lett.* **2016**, *181*, 362–366. [CrossRef]

51. Schinhammer, M.; Hanzi, A.C.; Loffler, J.F.; Uggowitzer, P.J. Design strategy for biodegradable Fe-based alloys for medical applications. *Acta Biomater.* **2010**, *6*, 1705–1713. [CrossRef] [PubMed]

52. Keen, C.L.; Zidenberg-Cherr, S. Manganese A2—Caballero, benjamin. In *Encyclopedia of Food Sciences and Nutrition*, 2nd ed.; Academic Press: Oxford, UK, 2003; pp. 3686–3691.

53. Dermience, M.; Lognay, G.; Mathieu, F.; Goyens, P. Effects of thirty elements on bone metabolism. *J. Trace Elements Med. Biol.* **2015**, *32*, 86–106. [CrossRef] [PubMed]

54. Sikora-Jasinska, M.; Mostaed, E.; Mostaed, A.; Beanland, R.; Mantovani, D.; Vedani, M. Fabrication, mechanical properties and in vitro degradation behavior of newly developed Zn Ag alloys for degradable implant applications. *Mater. Sci. Eng. C* **2017**, *77*, 1170–1181. [CrossRef] [PubMed]

55. Katarivas Levy, G.; Goldman, J.; Aghion, E. The prospects of zinc as a structural material for biodegradable implants—A review paper. *Metals* **2017**, *7*, 402. [CrossRef]

56. Li, H.F.; Xie, X.H.; Zheng, Y.F.; Cong, Y.; Zhou, F.Y.; Qiu, K.J.; Wang, X.; Chen, S.H.; Huang, L.; Tian, L.; et al. Development of biodegradable Zn-1X binary alloys with nutrient alloying elements Mg, Ca and Sr. *Sci. Rep.* **2015**, *5*, 10719. [CrossRef] [PubMed]

57. Yang, H.; Wang, C.; Liu, C.; Chen, H.; Wu, Y.; Han, J.; Jia, Z.; Lin, W.; Zhang, D.; Li, W.; et al. Evolution of the degradation mechanism of pure Zinc stent in the one-year study of rabbit abdominal aorta model. *Biomaterials* **2017**, *145*, 92–105. [CrossRef] [PubMed]

58. Liu, X.; Sun, J.; Yang, Y.; Zhou, F.; Pu, Z.; Li, L.; Zheng, Y. Microstructure, mechanical properties, in vitro degradation behavior and hemocompatibility of novel Zn–Mg–Sr alloys as biodegradable metals. *Mater. Lett.* **2016**, *162*, 242–245. [CrossRef]

59. Zhu, H.; Jin, G.; Cao, H.; Qiao, Y.; Liu, X. Influence of implantation voltage on the biological properties of Zinc-implanted titanium. *Surf. Coat. Technol.* **2017**, *312*, 75–80. [CrossRef]

60. Seo, H.-J.; Cho, Y.-E.; Kim, T.; Shin, H.-I.; Kwun, I.-S. Zinc may increase bone formation through stimulating cell proliferation, alkaline phosphatase activity and collagen synthesis in osteoblastic MC3T3-E1 cells. *Nutr. Res. Pract.* **2010**, *4*, 356–361. [CrossRef] [PubMed]

61. Moonga, B.S.; Dempster, D.W. Zinc is a potent inhibitor of osteoclastic bone resorption in vitro. *J. Bone Min. Res.* **1995**, *10*, 453–457. [CrossRef] [PubMed]

62. Bhardwaj, P.; Rai, D.V.; Garg, M.L. Zinc inhibits ovariectomy induced microarchitectural changes in the bone tissue. *J. Nutr. Intermed. Metab.* **2016**, *3*, 33–40. [CrossRef]

63. Yamaguchi, M. Role of nutritional zinc in the prevention of osteoporosis. *Mol. Cell. Biochem.* **2010**, *338*, 241–254. [CrossRef] [PubMed]

64. Pasquet, J.; Chevalier, Y.; Pelletier, J.; Couval, E.; Bouvier, D.; Bolzinger, M.-A. The contribution of zinc ions to the antimicrobial activity of zinc oxide. *Colloids Surf. A Physicochem. Eng. Asp.* **2014**, *457*, 263–274. [CrossRef]

65. Boyd, D.; Li, H.; Tanner, D.A.; Towler, M.R.; Wall, J.G. The antibacterial effects of zinc ion migration from Zinc-based glass polyalkenoate cements. *J. Mater. Sci. Mater. Med.* **2006**, *17*, 489–494. [CrossRef] [PubMed]

66. McCarthy, T.J.; Zeelie, J.J.; Krause, D.J. The antimicrobial action of Zinc ion/antioxidant combinations. *J. Clin. Pharm. Ther.* **1992**, *17*, 51–54. [CrossRef] [PubMed]

67. Drelich, A.J.; Bowen, P.K.; LaLonde, L.; Goldman, J.; Drelich, J.W. Importance of oxide film in endovascular biodegradable Zinc stents. *Surf. Innov.* **2016**, *4*, 133–140. [CrossRef]

68. Moon, S.-H.; Choi, W.J.; Choi, S.-W.; Kim, E.H.; Kim, J.; Lee, J.-O.; Kim, S.H. Anti-cancer activity of ZnO chips by sustained Zinc Ion release. *Toxicol. Rep.* **2016**, *3*, 430–438. [CrossRef] [PubMed]

69. Choi, S.-W.; Choi, W.J.; Kim, E.H.; Moon, S.-H.; Park, S.-J.; Lee, J.-O.; Kim, S.H. Inflammatory bone resorption and antiosteosarcoma potentials of Zinc Ion sustained release ZnO chips: Friend or foe? *ACS Biomater. Sci. Eng.* **2016**, *2*, 494–500. [CrossRef]

70. Vojtech, D.; Kubasek, J.; Serak, J.; Novak, P. Mechanical and corrosion properties of newly developed biodegradable Zn-based alloys for bone fixation. *Acta Biomater.* **2011**, *7*, 3515–3522. [CrossRef] [PubMed]

71. Törne, K.; Larsson, M.; Norlin, A.; Weissenrieder, J. Degradation of Zinc in saline solutions, plasma, and whole blood. *J. Biomed. Mater. Res. Part B Appl. Biomater.* **2016**, *104*, 1141–1151. [CrossRef] [PubMed]

72. Katz, E.P.; Li, S.-T. Structure and function of bone collagen fibrils. *J. Mol. Biol.* **1973**, *80*, 1–15. [CrossRef]

73. Tan, L.; Yu, X.; Wan, P.; Yang, K. Biodegradable materials for bone repairs: A review. *J. Mater. Sci. Technol.* **2013**, *29*, 503–513. [CrossRef]

74. Huber, W. Historical remarks on martin kirschner and the development of the kirschner (K)-wire. *Indian J. Plast. Surg.* **2008**, *41*, 89–92. [CrossRef] [PubMed]

75. Harasen, G. Orthopedic hardware and equipment for the beginner: Part 1. Pins and wires. *Can. Vet. J.* **2011**, *52*, 1025–1026. [PubMed]

76. Ricbard, M.; Slone, M.M.H.; Vander Griend Robert, A.; William, J. Montgomery Orthopaedic fixation devices. *RadioGraphics* **1991**, *11*, 823–847.

77. Rudolph, D.J.; Willes, M.G.; Sameshima, G.T. A finite element model of apical force distribution from orthodontic tooth movement. *Angle Orthod.* **2001**, *71*, 127–131. [PubMed]

78. Henriques, J.F.C.; Higa, R.H.; Semenara, N.T.; Janson, G.; Fernandes, T.M.F.; Sathler, R. Evaluation of deflection forces of orthodontic wires with different ligation types. *Braz. Oral Res.* **2017**, *31*, e49. [CrossRef] [PubMed]

79. Cowley, L.L. Wire sutures: Braided or monofilament? *Am. J. Surg.* **1967**, *113*, 472–474. [CrossRef]

80. Karmani, S.; Lam, F. The design and function of surgical drills and k-wires. *Curr. Orthop.* **2004**, *18*, 484–490. [CrossRef]

81. Massengill, J.B.; Alexander, H.; Parson, J.R.; Schecter, M.J. Mechanical analysis of kirschner wire fixation in a phalangeal model. *J. Hand Surg.* **1979**, *4*, 351–356. [CrossRef]

82. Viegas, S.F.; Ferren, E.L.; Self, J.; Tencer, A.F. Comparative mechanical properties of various kirschner wire configurations in transverse and oblique phalangeal fractures. *J. Hand Surg.* **1988**, *13*, 246–253. [CrossRef]

83. Stark, H.H.; Rickard, T.; Zemel, N.; Ashworth, C. Treatment of ununited fractures of the scaphoid by iliac bone grafts and kirschner-wire fixation. *J. Bone Jt. Surg. Am.* **1988**, *70*, 982–991. [CrossRef]

84. Hope, P.G.; Williamson, D.M.; Coates, C.J.; Cole, W.G. Biodegradable pin fixation of elbow fractures in children. A randomised trial. *J. Bone Jt. Surg. Br. Vol.* **1991**, *73*, 965–968. [CrossRef]

85. Leppilahti, J.; Jalovaara, P. Migration of kirschner wires following fixation of the clavicle—A report of 2 cases. *Acta Orthop. Scand.* **1999**, *70*, 517–519. [CrossRef] [PubMed]

86. Gbur, J.L.; Lewandowski, J.J. Fatigue and fracture of wires and cables for biomedical applications. *Int. Mater. Rev.* **2016**, *61*, 231–314. [CrossRef]

87. Chen, Q.; Thouas, G.A. Metallic implant biomaterials. *Mater. Sci. Eng. R Rep.* **2015**, *87*, 1–57. [CrossRef]

88. Prasad, K.; Bazaka, O.; Chua, M.; Rochford, M.; Fedrick, L.; Spoor, J.; Symes, R.; Tieppo, M.; Collins, C.; Cao, A.; et al. Metallic biomaterials: Current challenges and opportunities. *Materials* **2017**, *10*, 884. [CrossRef] [PubMed]

89. Clauss, M.; Graf, S.; Gersbach, S.; Hintermann, B.; Ilchmann, T.; Knupp, M. Material and biofilm load of k wires in toe surgery: Titanium versus stainless steel. *Clin. Orthop. Relat. Res.* **2013**, *471*, 2312–2317. [CrossRef] [PubMed]

90. Jensen, C.; Jensen, C. Biodegradable pins versus kirschner wires in hand surgery. *J. Hand Surg. Br. Eur. Vol.* **1996**, *21*, 507–510. [CrossRef]

91. Plaga, B.; Royster, R.; Donigian, A.; Wright, G.; Caskey, P. Fixation of osteochondral fractures in rabbit knees. A comparison of kirschner wires, fibrin sealant, and polydioxanone pins. *Bone Jt. J.* **1992**, *74*, 292–296. [CrossRef]

92. Casteleyn, P.P.; Handelberg, F.; Haentjens, P. Biodegradable rods versus kirschner wire fixation of wrist fractures. A randomised trial. *J. Bone Jt. Surg. Br. Vol.* **1992**, *74*, 858–861. [CrossRef]

93. Juutilainen, T.; Patiälä, H.; Rokkanen, P.; Törmälä, P. Biodegradable wire fixation in olecranon and patella fractures combined with biodegradable screws or plugs and compared with metallic fixation. *Arch. Orthop. Trauma Surg.* **1995**, *114*, 319–323. [CrossRef] [PubMed]

94. Lurate, B.; Mukherjee, D.; Kruse, R.; Albright, J. Fixation of osteochondral fractures with absorbable pins. In Proceedings of the 1995 Fourteenth Southern Biomedical Engineering Conference, Shreveport, LA, USA, 7–9 April 1995; pp. 57–58.

95. Maruyama, T.; Saha, S.; Mongiano, D.O.; Mudge, K. Metacarpal fracture fixation with absorbable polyglycolide rods and stainless steel k wires: A biomechanical comparison. *J. Biomed. Mater. Res. Part A* **1996**, *33*, 9–12. [CrossRef]

96. Tian, P.; Liu, X. Surface modification of biodegradable magnesium and its alloys for biomedical applications. *Regen. Biomater.* **2015**, *2*, 135–151. [CrossRef] [PubMed]

97. Witte, F.; Hort, N.; Vogt, C.; Cohen, S.; Kainer, K.U.; Willumeit, R.; Feyerabend, F. Degradable biomaterials based on magnesium corrosion. *Curr. Opin. Solid State Mater. Sci.* **2008**, *12*, 63–72. [CrossRef]
98. Hanzi, A.C.; Gerber, I.; Schinhammer, M.; Loffler, J.F.; Uggowitzer, P.J. On the in vitro and in vivo degradation performance and biological response of new biodegradable Mg-Y-Zn alloys. *Acta Biomater.* **2010**, *6*, 1824–1833. [CrossRef] [PubMed]
99. KubÁSek, J.; VojtĚCh, D. Structural and corrosion characterization of biodegradable Mg–Re (Re = Gd, Y, Nd) alloys. *Trans. Nonferrous Met. Soc. China* **2013**, *23*, 1215–1225. [CrossRef]
100. Chen, L.; Bin, Y.; Zou, W.; Wang, X.; Li, W. The influence of Sr on the microstructure, degradation and stress corrosion cracking of the Mg alloys—ZK40xSr. *J. Mech. Behav. Biomed. Mater.* **2017**, *66*, 187–200. [CrossRef] [PubMed]
101. Li, Z.; Gu, X.; Lou, S.; Zheng, Y. The development of binary Mg-Ca alloys for use as biodegradable materials within bone. *Biomaterials* **2008**, *29*, 1329–1344. [CrossRef] [PubMed]
102. Gu, X.; Zheng, Y.; Cheng, Y.; Zhong, S.; Xi, T. In vitro corrosion and biocompatibility of binary magnesium alloys. *Biomaterials* **2009**, *30*, 484–498. [CrossRef] [PubMed]
103. Lietaert, K.; Weber, L.; Van Humbeeck, J.; Mortensen, A.; Luyten, J.; Schrooten, J. Open cellular magnesium alloys for biodegradable orthopaedic implants. *J. Magn. Alloys* **2013**, *1*, 303–311. [CrossRef]
104. Mutlu, I.; Oktay, E. Influence of fluoride content of artificial saliva on metal release from 17-4 PH stainless steel foam for dental implant applications. *J. Mater. Sci. Technol.* **2013**, *29*, 582–588. [CrossRef]
105. Cui, W.; Beniash, E.; Gawalt, E.; Xu, Z.; Sfeir, C. Biomimetic coating of magnesium alloy for enhanced corrosion resistance and calcium phosphate deposition. *Acta Biomater.* **2013**, *9*, 8650–8659. [CrossRef] [PubMed]
106. Erdmann, N.; Angrisani, N.; Reifenrath, J.; Lucas, A.; Thorey, F.; Bormann, D.; Meyer-Lindenberg, A. Biomechanical testing and degradation analysis of MgCa$_{0.8}$ alloy screws: A comparative in vivo study in rabbits. *Acta Biomater.* **2011**, *7*, 1421–1428. [CrossRef] [PubMed]
107. Kramer, M.; Schilling, M.; Eifler, R.; Hering, B.; Reifenrath, J.; Besdo, S.; Windhagen, H.; Willbold, E.; Weizbauer, A. Corrosion behavior, biocompatibility and biomechanical stability of a prototype magnesium-based biodegradable intramedullary nailing system. *Mater. Sci. Eng. C Mater. Biol. Appl.* **2016**, *59*, 129–135. [CrossRef] [PubMed]
108. Chaya, A.; Yoshizawa, S.; Verdelis, K.; Myers, N.; Costello, B.J.; Chou, D.T.; Pal, S.; Maiti, S.; Kumta, P.N.; Sfeir, C. In vivo study of magnesium plate and screw degradation and bone fracture healing. *Acta Biomater.* **2015**, *18*, 262–269. [CrossRef] [PubMed]
109. Mc, B.E. Absorbable metal in bone surgery: A further report on the use of magnesium alloys. *J. Am. Med. Assoc.* **1938**, *111*, 2464–2467.
110. Windhagen, H.; Radtke, K.; Weizbauer, A.; Diekmann, J.; Noll, Y.; Kreimeyer, U.; Schavan, R.; Stukenborg-Colsman, C.; Waizy, H. Biodegradable magnesium-based screw clinically equivalent to titanium screw in hallux valgus surgery: Short term results of the first prospective, randomized, controlled clinical pilot study. *Biomed. Eng. Online* **2013**, *12*, 62. [CrossRef] [PubMed]
111. Ezechieli, M.; Ettinger, M.; Konig, C.; Weizbauer, A.; Helmecke, P.; Schavan, R.; Lucas, A.; Windhagen, H.; Becher, C. Biomechanical characteristics of bioabsorbable magnesium-based (mgyrezr-alloy) interference screws with different threads. *Knee Surg. Sports Traumatol. Arthrosc.* **2016**, *24*, 3976–3981. [CrossRef] [PubMed]
112. Tie, D.; Feyerabend, F.; Muller, W.D.; Schade, R.; Liefeith, K.; Kainer, K.U.; Willumeit, R. Antibacterial biodegradable Mg-Ag alloys. *Eur. Cells Mater.* **2013**, *25*, 284–298. [CrossRef]
113. Jahn, K.; Saito, H.; Taipaleenmaki, H.; Gasser, A.; Hort, N.; Feyerabend, F.; Schluter, H.; Rueger, J.M.; Lehmann, W.; Willumeit-Romer, R.; et al. Intramedullary Mg$_2$Ag nails augment callus formation during fracture healing in mice. *Acta Biomater.* **2016**, *36*, 350–360. [CrossRef] [PubMed]
114. Bian, D.; Zhou, W.; Deng, J.; Liu, Y.; Li, W.; Chu, X.; Xiu, P.; Cai, H.; Kou, Y.; Jiang, B.; et al. Development of magnesium-based biodegradable metals with dietary trace element germanium as orthopaedic implant applications. *Acta Biomater.* **2017**, *64*, 421–436. [CrossRef] [PubMed]
115. Koo, Y.; Lee, H.B.; Dong, Z.; Kotoka, R.; Sankar, J.; Huang, N.; Yun, Y. The effects of static and dynamic loading on biodegradable magnesium pins in vitro and in vivo. *Sci. Rep.* **2017**, *7*, 14710. [CrossRef] [PubMed]
116. Liu, B.; Zheng, Y.F. Effects of alloying elements (Mn, Co, Al, W, Sn, B, C and S) on biodegradability and in vitro biocompatibility of pure iron. *Acta Biomater.* **2011**, *7*, 1407–1420. [CrossRef] [PubMed]

117. Yang, K.; Zhou, C.; Fan, H.; Fan, Y.; Jiang, Q.; Song, P.; Fan, H.; Chen, Y.; Zhang, X. Bio-functional design, application and trends in metallic biomaterials. *Int. J. Mol. Sci.* **2017**, *19*, 24. [CrossRef] [PubMed]

118. Liu, B.; Zheng, Y.F.; Ruan, L. In vitro investigation of Fe30Mn6Si shape memory alloy as potential biodegradable metallic material. *Mater. Lett.* **2011**, *65*, 540–543. [CrossRef]

119. Hermawan, H.; Dube, D.; Mantovani, D. Developments in metallic biodegradable stents. *Acta Biomater.* **2010**, *6*, 1693–1697. [CrossRef] [PubMed]

120. Capek, J.; Msallamova, S.; Jablonska, E.; Lipov, J.; Vojtech, D. A novel high-strength and highly corrosive biodegradable Fe-Pd alloy: Structural, mechanical and in vitro corrosion and cytotoxicity study. *Mater. Sci. Eng. C Mater. Biol. Appl.* **2017**, *79*, 550–562. [CrossRef] [PubMed]

121. Hufenbach, J.; Wendrock, H.; Kochta, F.; Kühn, U.; Gebert, A. Novel biodegradable Fe-Mn-C-S alloy with superior mechanical and corrosion properties. *Mater. Lett.* **2017**, *186*, 330–333. [CrossRef]

122. Wang, H.; Zheng, Y.; Liu, J.; Jiang, C.; Li, Y. In vitro corrosion properties and cytocompatibility of Fe-Ga alloys as potential biodegradable metallic materials. *Mater. Sci. Eng. C* **2017**, *71*, 60–66. [CrossRef] [PubMed]

123. Babcock, W.W. Metallic sutures and ligatures. *Surg. Clin. N. Am.* **1947**, *27*, 1435–1460. [CrossRef]

124. Chellamani, K.; Veerasubramanian, D.; Balaji, R. Surgical Sutures: An Overview. *J. Acad. Ind. Res.* **2013**, *1*, 778–782.

125. Seitz, J.-M.; Wulf, E.; Freytag, P.; Bormann, D.; Bach, F.-W. The manufacture of resorbable suture material from magnesium. *Adv. Eng. Mater.* **2010**, *12*, 1099–1105. [CrossRef]

126. Seitz, J.-M.; Utermöhlen, D.; Wulf, E.; Klose, C.; Bach, F.-W. The manufacture of resorbable suture material from magnesium—Drawing and stranding of thin wires. *Adv. Eng. Mater.* **2011**, *13*, 1087–1095. [CrossRef]

127. Bai, J.; Yin, L.; Lu, Y.; Gan, Y.; Xue, F.; Chu, C.; Yan, J.; Yan, K.; Wan, X.; Tang, Z. Preparation, microstructure and degradation performance of biomedical magnesium alloy fine wires. *Prog. Natl. Sci. Mater. Int.* **2014**, *24*, 523–530. [CrossRef]

128. Zhang, S.; Liu, J.; Yan, J.; Chen, Y.; Zhao, C.; Zhang, Y.; Jiang, M.; Xu, H.; Ni, J.; Zhang, X. In vivo degradation and biocompatibility of linear cutter staples made of high purity magnesium. *Eur. Cells Mater.* **2014**, *28*, 76.

129. Wu, H.; Zhao, C.; Ni, J.; Zhang, S.; Liu, J.; Yan, J.; Chen, Y.; Zhang, X. Research of a novel biodegradable surgical staple made of high purity magnesium. *Bioact. Mater.* **2016**, *1*, 122–126. [CrossRef]

130. Cao, J.; Jiang, K.W.; Yang, X.D.; Shen, Z.L.; Guo, P.; Yan, Y.C.; Cui, Y.C.; Han, L.; Lv, Y.; Ye, Y.J.; et al. [animal experimental study of biodegradable magnesium alloy stapler for gastrointestinal anastomosis]. *Zhonghua Wei Chang Wai Ke Za Zhi Chin. J. Gastrointest. Surg.* **2013**, *16*, 772–776.

131. Moore, W.R.; Graves, S.E.; Bain, G.I. Synthetic bone graft substitutes. *ANZ J. Surg.* **2001**, *71*, 354–361. [CrossRef] [PubMed]

132. Zhang, J.; Liu, W.; Schnitzler, V.; Tancret, F.; Bouler, J.-M. Calcium phosphate cements for bone substitution: Chemistry, handling and mechanical properties. *Acta Biomater.* **2014**, *10*, 1035–1049. [CrossRef] [PubMed]

133. Zimmermann, G.; Moghaddam, A. Allograft bone matrix versus synthetic bone graft substitutes. *Injury* **2011**, *42*, S16–S21. [CrossRef] [PubMed]

134. Claes, L.; Hoellen, I.; Ignatius, A. Biodegradable bone cements. *Der Orthop.* **1997**, *26*, 459–462. [CrossRef] [PubMed]

135. Geffers, M.; Groll, J.; Gbureck, U. Reinforcement strategies for load-bearing calcium phosphate biocements. *Materials* **2015**, *8*, 2700–2717. [CrossRef]

136. Vaishya, R.; Chauhan, M.; Vaish, A. Bone cement. *J. Clin. Orthop. Trauma* **2013**, *4*, 157–163. [CrossRef] [PubMed]

137. Bohner, M. Design of ceramic-based cements and putties for bone graft substitution. *Eur. Cells Mater.* **2010**, *20*, 3–10. [CrossRef]

138. Dos Santos, L.A.; Carrodéguas, R.G.; Boschi, A.O.; Fonseca de Arruda, A.C. Fiber-enriched double-setting calcium phosphate bone cement. *J. Biomed. Mater. Res. Part A* **2003**, *65*, 244–250. [CrossRef] [PubMed]

139. Von Gonten, A.; Kelly, J.; Antonucci, J.M. Load-bearing behavior of a simulated craniofacial structure fabricated from a hydroxyapatite cement and bioresorbable fiber-mesh. *J. Mater. Sci. Mater. Med.* **2000**, *11*, 95–100. [CrossRef] [PubMed]

140. Xu, H.H.K.; Quinn, J.B. Calcium phosphate cement containing resorbable fibers for short-term reinforcement and macroporosity. *Biomaterials* **2002**, *23*, 193–202. [CrossRef]

141. Zhang, Y.; Xu, H.H. Effects of synergistic reinforcement and absorbable fiber strength on hydroxyapatite bone cement. *J. Biomed. Mater. Res. Part A* **2005**, *75*, 832–840. [CrossRef] [PubMed]

142. Pan, Z.; Jiang, P.; Fan, Q.; Ma, B.; Cai, H. Mechanical and biocompatible influences of chitosan fiber and gelatin on calcium phosphate cement. *J. Biomed. Mater. Res. Part B Appl. Biomater.* **2007**, *82*, 246–252. [CrossRef] [PubMed]

143. Krüger, R.; Groll, J. Fiber reinforced calcium phosphate cements–on the way to degradable load bearing bone substitutes? *Biomaterials* **2012**, *33*, 5887–5900. [CrossRef] [PubMed]

144. Krüger, R.; Seitz, J.-M.; Ewald, A.; Bach, F.-W.; Groll, J. Strong and tough magnesium wire reinforced phosphate cement composites for load-bearing bone replacement. *J. Mech. Behav. Biomed. Mater.* **2013**, *20*, 36–44. [CrossRef] [PubMed]

145. Li, T.; He, Y.; Zhou, J.; Tang, S.; Yang, Y.; Wang, X. Effects of scandium addition on biocompatibility of biodegradable Mg–1.5Zn–0.6Zr alloy. *Mater. Lett.* **2018**, *215*, 200–202. [CrossRef]

146. Salleh, E.M.; Ramakrishnan, S.; Hussain, Z. Synthesis of biodegradable Mg-Zn alloy by mechanical alloying: Effect of milling time. *Procedia Chem.* **2016**, *19*, 525–530. [CrossRef]

147. Gui, Z.; Kang, Z.; Li, Y. Mechanical and corrosion properties of Mg-Gd-Zn-Zr-Mn biodegradable alloy by hot extrusion. *J. Alloys Compd.* **2016**, *685*, 222–230. [CrossRef]

148. Dai, J.; Zhang, X.; Yin, Q.; Ni, S.; Ba, Z.; Wang, Z. Friction and wear behaviors of biodegradable Mg-6Gd-0.5Zn-0.4Zr alloy under simulated body fluid condition. *J. Magn. Alloys* **2017**, *5*, 448–453. [CrossRef]

149. Kubásek, J.; Vojtěch, D.; Jablonská, E.; Pospíšilová, I.; Lipov, J.; Ruml, T. Structure, mechanical characteristics and in vitro degradation, cytotoxicity, genotoxicity and mutagenicity of novel biodegradable Zn–Mg alloys. *Mater. Sci. Eng. C* **2016**, *58*, 24–35. [CrossRef] [PubMed]

150. Shen, C.; Liu, X.; Fan, B.; Lan, P.; Zhou, F.; Li, X.; Wang, H.; Xiao, X.; Li, L.; Zhao, S. Mechanical properties, in vitro degradation behavior, hemocompatibility and cytotoxicity evaluation of Zn–1.2Mg alloy for biodegradable implants. *RSC Adv.* **2016**, *6*, 86410–86419. [CrossRef]

151. Dorozhkin, S.V. Calcium orthophosphate coatings on magnesium and its biodegradable alloys. *Acta Biomater.* **2014**, *10*, 2919–2934. [CrossRef] [PubMed]

152. Song, Y.W.; Shan, D.Y.; Han, E.H. Electrodeposition of hydroxyapatite coating on AZ91D magnesium alloy for biomaterial application. *Mater. Lett.* **2008**, *62*, 3276–3279. [CrossRef]

153. Rojaee, R.; Fathi, M.; Raeissi, K. Controlling the degradation rate of AZ91 magnesium alloy via sol–gel derived nanostructured hydroxyapatite coating. *Mater. Sci. Eng. C* **2013**, *33*, 3817–3825. [CrossRef] [PubMed]

154. Su, Y.; Li, G.; Lian, J. A chemical conversion hydroxyapatite coating on AZ60 magnesium alloy and its electrochemical corrosion behaviour. *Int. J. Electrochem. Sci.* **2012**, *7*, 11497–11511.

155. Zhang, Y.; Ma, Y.; Chen, M.; Wei, J. Effects of anodizing biodegradable Mg–Zn–Zr alloy on the deposition of Ca–P coating. *Surf. Coat. Technol.* **2013**, *228*, S111–S115. [CrossRef]

156. Tsubakino, H.; Yamamoto, A.; Fukumoto, S.; Watanabe, A.; Sugahara, K.; Inoue, H. High-purity magnesium coating on magnesium alloys by vapor deposition technique for improving corrosion resistance. *Mater. Trans.* **2003**, *44*, 504–510. [CrossRef]

157. Johnson, I.; Akari, K.; Liu, H. Nanostructured hydroxyapatite/poly (lactic-*co*-glycolic acid) composite coating for controlling magnesium degradation in simulated body fluid. *Nanotechnology* **2013**, *24*, 375103. [CrossRef] [PubMed]

158. Butev, E.; Esen, Z.; Bor, S. In vitro bioactivity investigation of alkali treated Ti6Al7Nb alloy foams. *Appl. Surf. Sci.* **2015**, *327*, 437–443. [CrossRef]

159. Tang, H.; Wu, T.; Xu, F.; Tao, W.; Jian, X. Fabrication and characterization of Mg(OH)$_2$ films on AZ31 magnesium alloy by alkali treatment. *Int. J. Electrochem. Sci.* **2017**, *12*, 1377–1388. [CrossRef]

160. Feng, J.; Chen, Y.; Liu, X.; Liu, T.; Zou, L.; Wang, Y.; Ren, Y.; Fan, Z.; Lv, Y.; Zhang, M. In-situ hydrothermal crystallization Mg(Oh)$_2$ films on magnesium alloy AZ91 and their corrosion resistance properties. *Mater. Chem. Phys.* **2013**, *143*, 322–329. [CrossRef]

161. Shadanbaz, S.; Dias, G.J. Calcium phosphate coatings on magnesium alloys for biomedical applications: A review. *Acta Biomater.* **2012**, *8*, 20–30. [CrossRef] [PubMed]

162. Harun, W.S.W.; Asri, R.I.M.; Alias, J.; Zulkifli, F.H.; Kadirgama, K.; Ghani, S.A.C.; Shariffuddin, J.H.M. A comprehensive review of hydroxyapatite-based coatings adhesion on metallic biomaterials. *Ceram. Int.* **2018**, *44*, 1250–1268. [CrossRef]

163. Zhao, N.; Zhu, D. Collagen self-assembly on orthopedic magnesium biomaterials surface and subsequent bone cell attachment. *PLoS ONE* **2014**, *9*, e110420. [CrossRef] [PubMed]

Metals **2018**, *8*, 212

164. Wang, Z.-L.; Yan, Y.-H.; Wan, T.; Yang, H. Poly (L-lactic acid)/hydroxyapatite/collagen composite coatings on AZ31 magnesium alloy for biomedical application. *Proc. Inst. Mech. Eng. Part H J. Eng. Med.* **2013**, *227*, 1094–1103. [CrossRef] [PubMed]

165. Chen, C.-Z.; Shi, X.-H.; Zhang, P.-C.; Bai, B.; Leng, Y.-X.; Huang, N. The microstructure and properties of commercial pure iron modified by plasma nitriding. *Solid State Ion.* **2008**, *179*, 971–974. [CrossRef]

166. Zhu, S.; Huang, N.; Xu, L.; Zhang, Y.; Liu, H.; Lei, Y.; Sun, H.; Yao, Y. Biocompatibility of Fe–O films synthesized by plasma immersion ion implantation and deposition. *Surf. Coat. Technol.* **2009**, *203*, 1523–1529. [CrossRef]

167. Zhou, J.; Yang, Y.; Alonso Frank, M.; Detsch, R.; Boccaccini, A.R.; Virtanen, S. Accelerated degradation behavior and cytocompatibility of pure iron treated with sandblasting. *ACS Appl. Mater. Interfaces* **2016**, *8*, 26482–26492. [CrossRef] [PubMed]

168. Dobatkin, S.V.; Lukyanova, E.A.; Martynenko, N.S.; Anisimova, N.Y.; Kiselevskiy, M.V.; Gorshenkov, M.V.; Yurchenko, N.Y.; Raab, G.I.; Yusupov, V.S.; Birbilis, N.; et al. Strength, corrosion resistance, and biocompatibility of ultrafine-grained mg alloys after different modes of severe plastic deformation. *IOP Conf. Ser. Mater. Sci. Eng.* **2017**, *194*, 012004. [CrossRef]

metals

MDPI

Review

The Prospects of Zinc as a Structural Material for Biodegradable Implants—A Review Paper

Galit Katarivas Levy [1,*], Jeremy Goldman [2] and Eli Aghion [1]

[1] Department of Materials Engineering, Ben-Gurion University of the Negev, P.O. Box 652, 8410501 Beer-Sheva, Israel; egyon@bgu.ac.il

[2] Biomedical Engineering Department, Michigan Technological University, Houghton, MI 49931, USA; jgoldman@mtu.edu

* Correspondence: levyga@post.bgu.ac.il; Tel.: +972-50-6944074

Received: 31 August 2017; Accepted: 22 September 2017; Published: 1 October 2017

Abstract: In the last decade, iron and magnesium, both pure and alloyed, have been extensively studied as potential biodegradable metals for medical applications. However, broad experience with these material systems has uncovered critical limitations in terms of their suitability for clinical applications. Recently, zinc and zinc-based alloys have been proposed as new additions to the list of degradable metals and as promising alternatives to magnesium and iron. The main byproduct of zinc metal corrosion, Zn^{2+}, is highly regulated within physiological systems and plays a critical role in numerous fundamental cellular processes. Zn^{2+} released from an implant may suppress harmful smooth muscle cells and restenosis in arteries, while stimulating beneficial osteogenesis in bone. An important limitation of pure zinc as a potential biodegradable structural support, however, lies in its low strength (σ_{UTS} ~30 MPa) and plasticity ($\varepsilon < 0.25\%$) that are insufficient for most medical device applications. Developing high strength and ductility zinc with sufficient hardness, while retaining its biocompatibility, is one of the main goals of metallurgical engineering. This paper will review and compare the biocompatibility, corrosion behavior and mechanical properties of pure zinc, as well as currently researched zinc alloys.

Keywords: zinc; zinc alloys; biodegradable; biocompatible; corrosion degradation; mechanical properties

1. Introduction

Over the last 4 decades, innovations in biomaterials and medical technology have attracted remarkable attention for their potential to improve human life, by replacing and repairing soft and hard tissues, such as bone, cartilage, blood vessels, or even entire organs [1,2]. During this time, metals have become widely used as orthopedic implants, cardiovascular interventional devices, and tissue engineering scaffolds, due to their high strength and toughness compared with polymers and ceramic materials [3,4]. Traditional metallic biomaterials with high corrosion resistance, such as titanium alloys, stainless steels, cobalt–chromium alloys and tantalum are generally used as permanent implants in patients [3–5]. In most applications, however, the function of the implant is temporary and no longer needed after full recovery of the treated site. Furthermore, the permanent presence of the implant can lead to chronic deleterious effects. For instance, metal ions can be released from implanted devices due to defects in the surface oxide film, eventually resulting in implant fracture. In other cases, a chronic inflammatory response against the implant may undermine the therapeutic function of the device. In such circumstances, a second operation may be necessary to extract the implant, resulting in additional injury and expense [1,3,6,7].

Biodegradable metals represent an alternative approach to the traditional paradigm of corrosion resistant metals [1,8]. Biodegradable metals are expected to corrode gradually and harmlessly in vivo, maintain mechanical integrity during the critical tissue healing phase, and then dissolve completely upon fulfilling their mission [8]. In the last decade, iron and magnesium, both pure and alloyed, have been extensively studied as potential biodegradable metals for medical applications [8–15]. However, broad experience with these material systems has uncovered critical limitations in terms of their suitability for clinical applications [5,16–22]. For instance, the corrosion rates for Fe and Fe-based alloys are generally substantially below clinical needs, producing similar problems as found with permanent implants [8,16,23,24]. Their corrosion products do not appear to be excreted or metabolized at a satisfactory rate, but rather accumulate and repel neighboring cells and biological matrices, rather than allowing cells to integrate around and within the original footprint of the degrading implant [25]. On the other hand, pure and alloyed Mg exhibits insufficient mechanical strength as well as excessive corrosion rates, accompanied by hydrogen gas evolution, pH increases, and premature loss of mechanical integrity [7,26–29].

Recently, zinc and zinc-based alloys were proposed as new additions to the list of degradable metals and as promising alternatives to magnesium and iron [30–34]. The following are advantageous characteristics of zinc and its alloys for use in medical applications:

- Similar to magnesium and iron, zinc is an essential trace element in the human body. It is a component of more than 300 enzymes and an even greater number of other proteins, highlighting its indispensable role in human health. Optimal nucleic acid and protein metabolism, as well as cell growth, division, and function, require sufficient availability of zinc [30,35]. From this perspective, zinc ions released from the implant during the degradation phase could integrate into the normal metabolic activity of the host without producing systemic toxic side effects [36].
- Zinc exhibits high chemical activity, with an electrode potential (-0.762 V) falling between that of magnesium (-2.372 V) and iron (-0.444 V) [36–38]. Pure zinc metal, therefore, exhibits moderate degradation rates (faster than the slowly degrading Fe and its alloys, but slower than the rapidly degrading Mg and its alloys) due to passive layers of moderate stability, formed by corrosion products [19,39–41].
- Zinc and zinc-based alloys are easier to cast and process due to their low melting points, low chemical reactivity and good machinability [37,42,43]. For instance, unlike Mg based alloys, the melting of zinc alloys can more conveniently be performed in air [37].

An important limitation of pure zinc as a potential biodegradable structural support, lies in its low strength (σ_{UTS} ~30 MPa) and plasticity ($\varepsilon < 0.25\%$) characteristics that are insufficient for most medical applications [42,44,45]. Developing zinc with high strength and sufficient hardness is one of the main goals of metallurgical engineering, to broaden its utility as a biomedical implant. One of the most powerful tools to improve a metal's mechanical performance is the addition of alloying elements to the pure metal matrix [36,44]. Researchers have developed Zn-based alloys with new and more attractive features, modifying their chemical composition and microstructure, in order to improve mechanical properties, by principally using solid solution and second-phase strengthening [45]. Furthermore, improvements in the mechanical properties of pure zinc can be produced by the thermomechanical refinement of grain size, by extrusion, rolling, etc. [9]. This paper will review and compare the biocompatibility, corrosion behavior and mechanical properties of pure zinc as well as currently researched zinc alloys. The various types of zinc alloys, in terms of compositions and phase constituents, are summarized in Table 1.

Table 1. Common biomedical zinc alloys.

Family	Representive Alloys and Alloying Elements (wt %)	Main Phases	References
Zn–Mg	Zn–0.15Mg	α-Zn, Mg_2Zn_{11}	[9]
	Zn–0.5Mg	α-Zn, Mg_2Zn_{11}	[9]
	Zn–1Mg, ZnMg1	α-Zn, Mg_2Zn_{11}	[9,22,42,46,47]
	Zn–1.2Mg	α-Zn, Mg_2Zn_{11}	[3]
	Zn–1.5Mg, ZnMg1.5	α-Zn, Mg_2Zn_{11}	[36,42]
	Zn–3Mg, ZnMg3	α-Zn, Mg_2Zn_{11}	[9,42,48]
	Zn–1.5Mg–0.1Ca	α-Zn, Mg_2Zn_{11}, $CaZn_{13}$	[36]
	Zn–1Mg–0.5Ca	α-Zn, Mg_2Zn_{11}, $CaZn_{13}$	[47]
	Zn–1Mg–1Ca	α-Zn, Mg_2Zn_{11}, $CaZn_{13}$	[44]
	Zn–1Mg–0.1Sr	Zn, $MgZn_2$, $SrZn_{13}$	[49]
	Zn–1Mg–0.5Sr	Zn, $MgZn_2$, $SrZn_{13}$	[49]
	Zn–1.5Mg–0.1Sr	α-Zn, Mg_2Zn_{11}, $SrZn_{13}$	[36]
	Zn–1Mg–1Sr	α-Zn, Mg_2Zn_{11}, $SrZn_{13}$	[44]
	Zn–1Mg–0.1Mn	Zn, $MgZn_2$	[50]
	Zn–1.5Mg–0.1Mn	Zn, $MgZn_2$	[50]
Zn–Ca	Zn–1Ca	α-Zn, $CaZn_{13}$	[22]
	Zn–1Ca–1Sr	α-Zn, $CaZn_{13}$, $SrZn_{13}$	[44]
Zn–Sr	Zn–1Sr	α-Zn, $SrZn_{13}$	[22]
Zn–Al	Zn–0.5Al	Zn, Al	[9,51]
	Zn–1Al	Zn, Al	[9,52]
	Zn–3Al	Zn, Al	[52]
	Zn–5Al	Zn, Al	[52]
	ZnAl4Cu1	Zn, Al	[42]
	ZA0.1Mg	α-Zn, $Mg_2(Zn,Al)_{11}$	[51]
	ZA0.3Mg	α-Zn, $Mg_2(Zn,Al)_{11}$	[51]
	ZA0.5Mg/Zn–0.5Al–0.5Mg	α-Zn, $Mg_2(Zn,Al)_{11}$	[51,53]
	Zn–0.5Al–0.5Mg–0.1Bi	Zn, $Mg_2(Zn,Al)_{11}$, Mg_3Bi_2	[53]
	Zn–0.5Al–0.5Mg–0.3Bi	Zn, $Mg_2(Zn,Al)_{11}$, Mg_3Bi_2	[53]
	Zn–0.5Al–0.5Mg–0.5Bi	Zn, $Mg_2(Zn,Al)_{11}$, Mg_3Bi_2	[53]
	3.5–.5Al, 0.75–1.25Cu, 0.03–0.08Mg	Zn, Al	[31]
	3.5–4.3Al, 2.5–3.2Cu, 0.03–0.06Mg	Zn, Al	[31]
	5.6–6Al, 1.2–1.6Cu	Zn, Al	[31]
Zn–Cu	Zn–1Cu	η-Zn, ε-$CuZn_5$	[54]
	Zn–2Cu	η-Zn, ε-$CuZn_5$	[54]
	Zn–3Cu	η-Zn, ε-$CuZn_5$	[54,55]
	Zn–4Cu	η-Zn, ε-$CuZn_5$	[37,54]
	Zn–3Cu–0.1Mg	Zn, $CuZn_5$, Mg_2Zn_{11}	[55]
	Zn–3Cu–0.5Mg	Zn, $CuZn_5$, Mg_2Zn_{11}	[55]
	Zn–3Cu–1Mg	Zn, $CuZn_5$, Mg_2Zn_{11}	[55]
Zn–Li	Zn–2Li	Zn, α-$LiZn_4$	[25]
	Zn–4Li	Zn, α-$LiZn_4$	[25]
	Zn–6Li	Zn, α-$LiZn_4$	[25]
	Zn–Li	Zn, α-$LiZn_4$	[56]
Zn–Ag	Zn–2.5Ag	η-Zn, ε-$AgZn_3$	[45]
	Zn–5Ag	η-Zn, ε-$AgZn_3$	[45]
	Zn–7Ag	η-Zn, ε-$AgZn_3$	[45]

2. Zinc in the Human Body

The essential role of zinc in the human body was discovered in 1961 when Iranian farmers subsisting on a zinc deficiency diet (unrefined flat bread, potatoes, and milk) were found to suffer from a group of syndromes, consisting of anemia, hypogonadism, and dwarfism. Since this discovery, interest in the biochemical and clinical aspects of zinc has increased markedly [57]. Presently, it is well known that zinc is one of the most abundant nutritionally essential elements in the human body [44]. Zinc is present in all organs, tissues, fluids and body secretions, with 86% of its mass residing in skeletal muscle and bone, 6% in the skin, 5% in the liver, 1.5% in the brain and the remaining distributed amongst the other tissues [35,58]. At the cellular level, 30–40% is located in the nucleus,

50% in the cytoplasm, organelles and specialized vesicles (for digestive enzymes or hormone storage) and the remaining portion in the cell membrane [35,44]. The human zinc requirement is estimated at 15 mg/day [8,10,11,37] and due to its high importance, the body has developed sophisticated mechanisms to remove zinc from dietary constituents and transport it to desired locations. The human body is able to absorb zinc from the environment, regulate its concentration in body fluids, transport it safely to all tissues of the body and to the sites where its presence is required, and safely excrete excess amounts from the body through the kidneys [35].

As a vital element, zinc plays an important role in numerous physiological systems, including immune, sexual, neurosensory (cognition and vision), and cell development and growth [19,30]. Zinc participates in numerous fundamental biological functions, such as nucleic acid metabolism, signal transduction, apoptosis regulation, and gene expression [22]. Zinc is a critical component of enzymes involved in protein synthesis and energy production. More than 1200 proteins are predicted to contain, bind, or transport Zn^{2+}, for example, zinc-finger proteins [44]. At the cellular level, zinc maintains the structural integrity of biomembranes and is essential for cell proliferation, differentiation and signaling [19,59]. Zinc plays an important role in bone formation, mineralization, and preservation of bone mass and can be found in the bone extracellular matrix, where it is co-deposited with calcium hydroxyapatite [19,22,60]. Indeed, a zinc decrease in bone matrix correlates with aging and skeletal disease [22,61].

Table 2 summarizes the systemic symptoms resulting from zinc deficiency or excess. Zinc deficiency contributes to retarded growth, impaired parturition (dystocia), neuropathy, decreased food intake, diarrhea, dermatitis, hair loss, bleeding tendencies, hypotension, and hypothermia [44]. Zinc deficiency is generally due to insufficient dietary intake. However, it may also be a consequence of malabsorption and chronic illnesses, such as diabetes, malignancy, liver disease, and sickle cell disease [35,57]. On the other hand, excessive amounts of Zn^{2+} in the body may be detrimental to vital organs, such as the kidney, liver, spleen, brain, and heart [19]. In addition, a prolonged overdose of zinc results in copper deficiency, provokes hypocupremia, anemia, leucopenia, and neutropenia, and impairs the Cu–Zn–superoxide dismutase antioxidant enzyme [30,62].

Table 2. The systemic symptoms resulting from zinc deficiency and excess [35,63,64].

Organ or System	Zinc Deficiency	Zinc Excess
Brain	Decreased nerve conduction, neuropsychiatric and neurosensory disorders, mental lethargy	Lethargy, focal neuronal deficits.
Respiratory tract	-	Respiratory disorder after inhalation of zinc smoke, metal fume fever.
Immune system	Impaired immune system function, increased susceptibility to pathogens	Altered lymphocyte function.
Thymus	Thymic atrophy	-
Skin	Skin lesions, decreased wound healing, acrodermatitis	-
Gastrointestinal tract	-	Nausea/vomiting, epigastric pain, diarrhea.
Reproductive system	Infertility, retarded genital development, hypogonadism	-
Prostate	-	Elevated risk of prostate cancer.

3. Corrosion Behavior of Zinc in the Physiological Environment

The degradation model, proposed by Zheng et al. [8], for biodegradable metals in a neutral physiological environment, generally occurs via cathodic and anodic reactions. Metallic corrosion produces hydroxides, oxides, and hydrogen gas by-products [65]. In the specific case of zinc, as presented in Figure 1, the reactions include a number of intermediate species, according to the following:

$$\text{Anodic reaction}: 2Zn \rightarrow 2Zn^{2+} + \overline{4}e^- \tag{1}$$

$$\text{Cathodic reaction}: O_2 + 2H_2O + \overline{4}e^- \rightarrow 4OH^- \tag{2}$$

$$\text{Zn(OH)}_2 \text{ formation}: 2Zn^{2+} + 4OH^- \rightarrow 2Zn(OH)_2 \tag{3}$$

$$\text{ZnO formation}: Zn(OH)_2 \rightarrow ZnO + H_2O \tag{4}$$

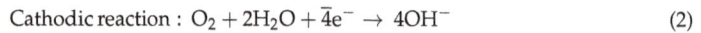

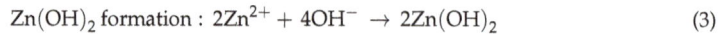

When exposed to body fluid, zinc is oxidized into metal cations following the anodic reaction in Equation (1). The generated electrons are consumed by a cathodic reaction, corresponding to the dissolved oxygen reduction in Equation (2). $Zn(OH)_2$ and ZnO corrosion products are likely to form on the metal surface, without gas evolution, according to Equations (3) and (4). Hence, gas release is not expected as a result of zinc corrosion, in contrast to the highly problematic hydrogen gas that is released as a byproduct of magnesium corrosion [8,30,36,59].

It should be noted that the physiological environment is highly aggressive, particularly due to the high concentration of chloride ions. These ions destabilize the equilibrium between dissolution and formation of the corrosion product layer, given that chloride ions are able to convert the surface into soluble chloride salts as follows:

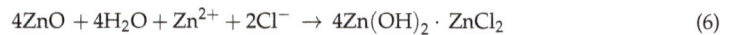

$$6Zn(OH)_2 + Zn^{2+} + 2Cl^- \rightarrow 6Zn(OH)_2 \cdot ZnCl_2 \tag{5}$$

$$4ZnO + 4H_2O + Zn^{2+} + 2Cl^- \rightarrow 4Zn(OH)_2 \cdot ZnCl_2 \tag{6}$$

The dissolution of the $Zn(OH)_2$ and ZnO surface film components promotes further dissolution of the exposed metal. Cycles of cathodic and anodic reactions expose the fresh metal substrate to the physiological solution, form corrosion products, and convert the product into soluble salts. With progressive exposure, an irregular particle may be separated from the zinc matrix and enter the surrounding medium [8,36].

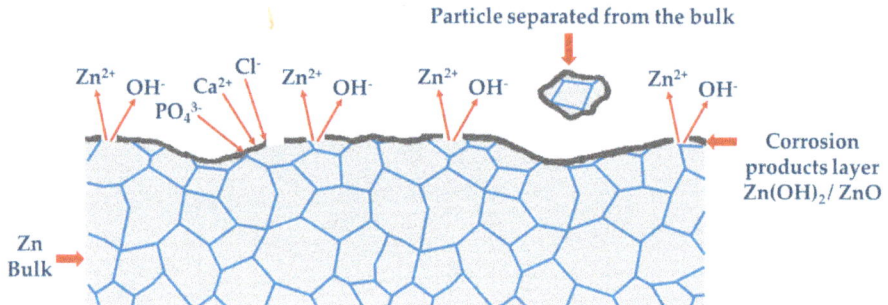

Figure 1. Model of the zinc degradation process in physiological fluids.

According to Bowen et al. [59], thin layers of zinc oxide were the only product observed during early stages (1.5 and 3 months) on the surface of zinc wires (99.99% purity) implanted into rat aorta. However, as corrosion progressed to 4.5 and 6 months, the corrosion layer thickened and contained three different phases/layers: calcium phosphate, zinc oxide, and zinc carbonate. The calcium phosphate layer appeared on the exterior surface without forming a true bulk product. Hence, the calcium phosphate layer was not thought to play a significant role in zinc biocorrosion. The compact corrosion layer included the two other phases, zinc oxide, and zinc carbonate, with ZnO appearing in formations isolated from one another by the zinc carbonate phase. This complex corrosion layer suggests that corrosion products of zinc in body fluid might be similar to those reported for magnesium. These ceramic degradation products can accumulate as a function of the local tissue's physiological mass transfer rate and therefore may impact tissue healing and remodeling [30,59].

Another important factor that needs to be considered in the corrosion of zinc is the pH value of the solution. According to Pourbaix diagram, shown in Figure 2, zinc is present as hydrated $Zn^{2+}_{(aq)}$, over the entire physiological range of pH values and biological standard reduction potentials (~820 mV to ~−670 mV). The purple arrow in Figure 2 indicates the range of biologically-important standard potentials determined at pH 7.4 [66]. As reported by Thomas et al. [67] in their investigation of zinc corrosion as a function of pH, in the pH range 7 to 10, the lowered cathodic reaction rates reduce overall zinc corrosion rates, and the surface oxides thermodynamically predicted to form in this pH range do not form an effective corrosion protection barrier. Hence, zinc metals in physiological environments with a pH of 7.4 will be dissolved over time, as is required for biodegradable medical implants.

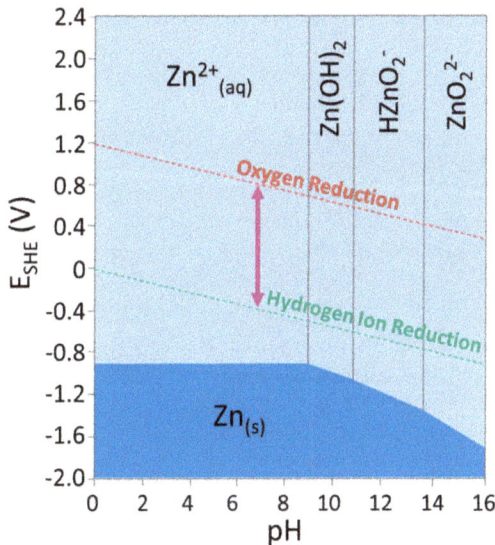

Figure 2. Pourbaix diagram of zinc. The purple arrow shows the range of biological standard reduction potentials at pH 7.4: from ~820 mV to ~−670 mV [66].

4. Biocompatibility and Biological Performances of Zinc and Evaluated Zinc Alloys

4.1. Pure Zinc

After implantation, Zn^{2+} will be released from the zinc implant to the surrounding extracellular space and eventually into the bloodstream. Thus, the cellular responses to high extracellular Zn^{2+} will impact the healing process and biocompatibility with zinc-bearing implants [16]. Furthermore, the response of the host's immune system will determine whether the implant becomes biointegrated and continues to function as designed, or encapsulated in a dense fibrous tissue that may compromise the intended function of the implant. The inflammatory cells are sensitive to the implant's corrosion behavior and cellular accessibility of the surface product, with more porous and actively corroding interfaces eliciting more benign inflammatory responses, relative to surfaces that are more resistant to biocorrosion [20,68]. The inflammatory response can therefore potentially be regulated by controlling implant degradation behavior and rates through alloying and processing.

4.1.1. In Vitro Examination

Törne et al. [32] investigated the initial degradation of zinc exposed to simulated (phosphate buffered saline (PBS), Ringer's saline solution) and physiological body fluids (human plasma, whole blood). They found a decrease in the corrosion rate over time when immersed in body fluids and an increase when placed in simulated body fluids. The passivation film that formed in physiological body fluids was uniform and was composed of inorganic corrosion products and organic materials (biomolecules). In contrast, the simulated body fluids promoted localized corrosion with thick porous products, primarily composed of zinc phosphates and carbonates.

The cellular response of human bone marrow mesenchymal stem cells (hMSC) and human vascular cells (HCECs, HASMC, HAEC) to Zn^{2+} was recently investigated [16,26,69,70]. According to Zhu et al. [70], Zn biomaterial can support hMSC adhesion and proliferation and zinc ions can lead to enhanced regulation of genes, cell survival/growth and differentiation, extracellular matrix (ECM) mineralization, and osteogenesis. The stimulation of osteogenesis by ionic zinc has been supported by numerous studies [71–79]. In the case of human vascular cells, Zn^{2+} at low concentrations can enhance cell viability, proliferation, adhesion, spreading, migration, and F-actin and vinculin expression, while decreasing cell adhesion strength. In contrast, high concentrations of Zn^{2+} elicited opposite effects. Gene expression profiles revealed that the most affected functional genes were related to angiogenesis, inflammation, cell adhesion, vessel tone, and platelet aggregation [26]. Since the Zn^{2+} concentration is intrinsically related to the implant degradation rate, a slow corrosion rate with controlled release of Zn^{2+} is desirable to maintain a low concentration profile of Zn^{2+} in the local tissues, in order to benefit cellular functions [16]. A high concentration of Zn^{2+} may overwhelm metal divalent ion-dependent intracellular signaling pathways, stimulate oxidative stress pathways, inducing apoptosis or necrosis, and generally contribute to negative side effects in cell populations adjacent to the implant.

4.1.2. In Vivo Examination

Bowen et al. studied the in vivo performance of zinc wires (99.99%) over 6 months when implanted into the abdominal aorta of adult rats [59,80]. They found uniform corrosion with linearly increasing corrosion rates over the residence time. The corrosion rate after 1.5 months was below the 0.02 mm/year degradable stent benchmark and increased to 0.05 mm/year (~0.4 and ~0.97 mg/day, respectively) after 6 months, both of which are far below the daily allowance of zinc (15 mg/day) [59]. In addition, a histological examination indicated excellent biocompatibility with the arterial tissue as well as tissue regeneration within the original footprint of the degrading implant. Intriguingly, observations of low cellular density and a distinct lack of smooth muscle cells adjacent to the implant interface indicates that the Zn^{2+} ions released from a zinc implant may suppress restenosis pathways [80]. Yang et al. [81] investigated the degradation of pure zinc stents over one year in a rabbit abdominal aorta model. They reported an excellent biocompatibility, without severe inflammation, platelet aggregation, thrombosis formation or obvious intimal hyperplasia. The degradation rates were matched to the artery healing process. Moreover, the pure zinc stent retained its mechanical integrity for 6 months and degraded to $41.75 \pm 29.72\%$ of its original volume after 12 months of implantation.

4.2. Zinc Alloys

The biocompatibility of absorbable biomaterial components must be considered, given that all elements of the metal alloy will eventually pass through the human body [10]. Table 3 summarizes the pathophysiology and toxicology of zinc and the alloying elements used in current zinc alloys. Table 4 lists the corrosion performance of different zinc alloys, including their in vitro and in vivo corrosion rates. In Table 4, the Zn–0.5Al–0.5Mg–0.5Bi alloy exhibits the highest corrosion rate (0.28 mm/year), which for standard implants is still far below the daily allowance of zinc (15 mg/day). The tabulated data supports the notion that zinc release during the stent degradation process should be considered safe to human systems, although toxicity at the local level will need to be examined on a case-by-case basis.

In the ideal design of degradable biomaterials, elements with potential toxicological effects should be avoided, and their use should be minimized if they cannot be excluded [10]. Since Mg, Ca, Cu, Mn and Sr are essential for humans, these elements should be the first choices as alloying elements for biomedical zinc alloys. Li et al. studied the in vitro and in vivo performance of Zn-1X binary alloys with the nutrient alloying elements, Mg, Ca and Sr [22]. The in vitro results for human morphologies on the Zn–1X alloy surfaces, and generated pseudopods and extrace umbilical vein endothelial cells ECV304 and Human osteosarcoma MG63 demonstrated healthy cell llular matrix secretions, compared to an unhealthy morphology for cells cultured on pure Zn. When experimental Zn–1X pins were implanted into mouse femurs, new bone formation in the absence of inflammation was observed around the implantation site, in particular for the Zn–1Sr alloy. The corrosion rates of the Zn–1X pins were 0.17, 0.19 and 0.22 mm/year for Zn–1Mg, Zn–1Ca and Zn–1Sr pins, respectively, which is far below the daily allowance of zinc [22]. Even when adding higher amounts of Mg (3 wt %) to pure zinc, to a concentration of 0.75 mg/mL (Zn: 0.49 ppm and Mg: 10.75 ppm), the Zn–3Mg alloy extract exhibited acceptable cytotoxic effects on human osteoblasts. Results of the three main parameters of cell–material interaction, which includes cell health, cell functionality, and inflammatory responses, demonstrated acceptable cellular toxicity [19]. The addition of Mn to the Zn–Mg system increased the susceptibility for galvanic micro-cell corrosion due to the higher electrode potential of Mn [50]. The Zn–xCu system exhibited a slightly increased corrosion rate with increasing Cu concentrations, compared to pure Zn, but the increase was not significantly different among the Zn–xCu alloys. In vitro testing demonstrated the cytocompatibilty of Zn–xCu alloys with human endothelial cells as well as an antibacterial property when the Cu concentration was above 2 wt % [54]. The addition of 0.1–1% Mg to the Zn–3Cu system increased the corrosion rate, as Mg was added [55].

The second choice of alloying element should be those that have been found to improve the properties of magnesium alloys, including Li and Al. However, given the potential toxicity of Li and Al, the addition of these two elements should be limited to low weight percent alloys. Moreover, the degradation rate of zinc alloys containing these elements should match the tolerance or effective dose range of these elements in the human body. Zhao et al. [25,56] investigated the effect of Li as a zinc alloying element. They reported the overall quantities of lithium released from a zinc–lithium implant (with 0.7 wt % of Li) at two orders of magnitude below the daily bodily consumption allowances. The in vivo implantation of Zn–1Li demonstrated positive biocompatibility, with no significant differences in serum zinc observed before, and 1–3 months after, implantation. According to Bowen et al. [52], biodegradation of a Zn–Al stent, with a weight of ~50 mg and a 5 wt % Al content—assuming bioabsorption of the entire stent in 2 years and a stable corrosion rate—would result in a daily intake of ~0.003 mg of Al, substantially below the ~10 mg daily intake for an average person. In their in vivo studies [20,52], Zn–Al (1, 3 and 5% Al) strips were implanted in the wall of the abdominal aorta of adult Sprague–Dawley rats. The Zn–Al systems exhibited acceptable compatibility with surrounding arterial tissue, as the histopathological analysis failed to identify necrotic tissue in the samples examined, although indications of chronic and acute inflammation were both identified. Moreover, the pattern of corrosion was modified through Al additions, with intergranular corrosion observed in all Zn–Al alloys. Intergranular corrosion accelerates the oxidation of zinc to zinc oxide, whose volume expansion produces implant cracking and fragmentation. The addition of 0.1–0.5 wt % Mg to the Zn–Al system decreased the corrosion rate with increasing Mg content, after 720 h of immersion in simulated body fluid (SBF) solution. The cytotoxicity results demonstrated acceptable biocompatibility of the Zn–0.5Al–0.5Mg alloy, with superior antibacterial activity relative to the other alloys [51].

Presently, the pathophysiology and toxicology of Ag and Bi remain unclear. Silver has been used clinically for decades to treat burns and assist with wound healing, due to its antibacterial properties, and has diverse medical applications [45,82,83]. Bi is generally considered less toxic than other heavy metal elements, such as antimony, and purified bismuth metal has been used to prepare a number of pharmaceutical products [53,83]. In two studies, the addition of Ag (2.5–7%) to pure zinc matrix and the addition of Bi (0.1–0.5%) to ternary Zn–Al–Mg alloy slightly increased the degradation rate. This is likely due to the increased galvanic coupling between the Zn and AgZn$_3$/Mg$_3$Bi$_2$ secondary phases [45,53]. In vitro cytotoxicity tests indicated a more toxic effect on MC3T3-E1 cells for Zn–Al–Mg–Bi alloys, compared to the Zn–Al–Mg alloy [53].

Table 3. Summary of the pathophysiology and toxicology of zinc and select alloying elements [8,10,11,83] and their effect on zinc alloys.

Element	Blood Serum Level	Daily Allowance	Pathophysiology	Toxicology	Effect on Zinc Alloys
			Essential Elements		
Mg	17.7–25.8 mg/L	700 mg	Activator of many enzymes; co-regulator of protein synthesis and muscle contraction; stabilizer of DNA and RNA	Excessive Mg leads to nausea	Mg: ↑mechanical properties & ↑corrosion rates [22]
Ca	36.8–39.8 mg/L	800 mg	More than 99% have structural functions in the skeleton; the solution Ca has signaling functions, including muscle contraction, blood clotting, cell function, etc.	Inhibit the intestinal absorption of other essential minerals	Ca: ↑mechanical properties & ↑corrosion rates [22]
Fe	5000–17,600 mg/L	10–20 mg	Component of several metalloproteins; crucial in vital biochemical activities, i.e., oxygen sensing and transport	Iron toxicity gives rise to lesions in the gastrointestinal tract, shock and liver damage	Fe: ↑corrosion rates by galvanic corrosion mechanism [15]
			Essential Trace Elements		
Zn	0.8–1.14 mg/L	15 mg	Trace element; appears in all enzyme classes; most Zn appears in muscle	Neurotoxic and hinders bone development at higher concentrations	-
Cu	4.51–8.32 mg/L	1–3 mg	Cu plays a vital role in the immune system; has beneficial effects on endothelial cell proliferation and has been reported to enhance antibacterial properties [55]	Excessive Cu (>1 mg/day) can cause neurodegenerative diseases, including Alzheimer's, Menkes and Wilson's diseases [55]	↑Cu (1–4%): ↑mechanical properties & ↓corrosion rates [37,54]
Mn	<0.0008 mg/L	4 mg	Activator of enzymes, Mn deficiency is related to osteoporosis, diabetes mellitus, and atherosclerosis	Excessive Mn results in neurotoxicity	Mn improves the casting process. Mn ↑ susceptibility of galvanic micro-cell corrosion [50]
			Other Elements		
Sr	0.17 mg [a]	2 mg	99% is located in bone; shows dose dependent metabolic effects on bone; low doses stimulate new bone formation	High doses induce skeletal abnormalities	Sr: ↑mechanical properties & ↑corrosion rates [22]
Li	0.002–0.004 mg/L	0.2–0.6 mg	Used in the treatment of manic depressive psychoses	Plasma concentrations of 2 mM are associated with reduced kidney function and neurotoxicity, 4 mM may be fatal	Li: ↑ultimate tensile strength, ↓ductility & ↓corrosion rate [56]
Al	0.0021–0.0048 mg/L	-	-	Primarily accumulates in the bone and nervous systems; implicated in the pathogenesis of Alzheimer's disease; can cause muscle fiber damage; decreases osteoblast viability	Al: ↑mechanical properties & ↑corrosion rates

[a] Sr concentration in total blood [8].

Table 4. In vitro and in vivo corrosion rates of different zinc alloys.

Alloy	In Vitro Corrosion Rate		In Vivo Corrosion Rate (mm/year)	References
	Electrochemical (μA/cm²)	Immersion (mm/year)		
Zn	1.8–9.2 (Hank's) 0.05 (plasma) 0.04 (whole blood) 0.035 (PBS)	0.027–0.13 (Hank's)	0.02–0.05 (1.5-6 months)	[9,20,22,23,31,32,45,53,59]
Zn *	8.98	-	-	[9]
Zn-Mg				
Zn-0.15Mg c	11.52 (Hank's)	0.17 (Hank's)	-	[9]
Zn-0.15Mg *,c	10.98 (Hank's)	-	-	[9]
Zn-0.5Mg c	11.73 (Hank's)	0.175 (Hank's)	-	[9]
Zn-0.5Mg *,c	11.01 (Hank's)	-	-	[9]
Zn-1Mg f /ZnMg1 b	9.9–11.9 (Hank's) 0.28–1.2 (SBF) 0.74 (PBS)	0.085–0.18 (Hank's) 0.06–0.28 (SBF) 0.027 (PBS)	0.17	[9,22,42,46,47]
Zn-1Mg *	11.32 (Hank's)	0.12 (SBF)	-	[9,46]
Zn-1.2Mg	7.7 (Hank's)	0.08 (Hank's)	-	[3]
Zn-1.2Mg *	12.4 (Hank's)	0.11 (Hank's)	-	[3]
Zn-1.5Mg d /ZnMg1.5 b	8.8 (SBF)	0.063 (Hank's) 0.05 (SBF)	-	[4,36]
Zn-3Mg/ZnMg3 b	9.01 (Hank's) 7.4 (SBF)	0.13 (Hank's) 0.06–0.21 (SBF)	-	[9,42,48]
Zn-3Mg *	8.6 (Hank's)	0.13 (SBF)	-	[9]
Zn-3Mg ***	-	0.12 (Hank's)	-	[48]
Zn-1.5Mg-0.1Ca d	4.3 (PBS)	0.37 (PBS)	-	[36]
Zn-1Mg-0.5Ca	0.17 (Hank's)	0.09 (Hank's)	-	[47]
Zn-1Mg-1Ca e	7.85 (Hank's)	-	-	[44]
Zn-1Mg-0.15Sr	7.83 (Hank's)	-	-	[49]
Zn-1Mg-0.5Sr	-	-	-	[49]
Zn-1.5Mg-0.15Sr d	-	0.1 (Hank's)	-	[36]
Zn-1Mg-1Sr e	0.175 (Hank's)	0.095 (Hank's)	-	[44]
Zn-1Mg-0.1Mn	17.21 (Hank's)	0.12 (Hank's)	-	[50]
Zn-1.5Mg-0.1Mn	9.34 (Hank's)	0.09 (Hank's)	-	[50]
Zn-Ca				
Zn-1Ca f	10.75 (Hank's)	0.09 (Hank's)	0.19	[22]
Zn-1Ca-1Sr e	0.185 (Hank's)	0.11 (Hank's)	-	[44]
Zn-Sr				
Zn-1Sr f	11.76 (Hank's)	0.095 (Hank's)	0.22	[22]
Zn-Al				
Zn-0.5Al	11.08 (Hank's) 20 (SBF)	0.14 (Hank's) 15 (SBF)	-	[9,51]
Zn-0.5Al *,c	9.6 (Hank's)	-	-	[9]
Zn-1Al	11.11 (Hank's)	0.16 (Hank's)	-	[9]

Table 4. *Cont.*

Alloy	In Vitro Corrosion Rate		In Vivo Corrosion Rate (mm/year)	References
	Electrochemical (μA/cm²)	Immersion (mm/year)		
Zn-1Al *	9.7 (Hank's)	-	-	[9]
ZnAl4Cu1 b	5.2 (SBF)	0.07 (SBF)	-	[42]
ZA0.1Mg	17 (SBF)	0.13 (SBF)	-	[51]
ZA0.3Mg	11.2 (SBF)	0.11 (SBF)	-	[51]
ZA0.5Mg/Zn-0.5Al-0.5Mg	9.5 (SBF)	0.11–0.15 (SBF)	-	[51,53]
Zn-0.5Al-0.5Mg-0.1Bi	12 (SBF)	0.17 (SBF)	-	[53]
Zn-0.5Al-0.5Mg-0.3Bi	16 (SBF)	0.2 (SBF)	-	[53]
Zn-0.5Al-0.5Mg-0.5Bi	23 (SBF)	0.28 (SBF)	-	[53]
ZA4-1	2.986 (Hank's)	-	-	[31]
ZA4-3	7.209 (Hank's)	-	-	[31]
ZA6-1	5.331 (Hank's)	-	-	[31]
Zn–Cu				
Zn-1Cu *,a	-	0.033 (c-SBF)	-	[54]
Zn-2Cu *,a	-	0.027 (c-SBF)	-	[54]
Zn-3Cu *,a	0.372 (Hank's)	0.012 (Hank's) / 0.03 (c-SBF)	-	[54,55]
Zn-4Cu *,a	4.1 (Hank's)	0.009 (Hank's) / 0.025 (c-SBF)	-	[37,54]
Zn-3Cu-0.1Mg *	1.18 (Hank's)	0.023 (Hank's)	-	[55]
Zn-3Cu-0.5Mg *	1.56 (Hank's)	0.03 (Hank's)	-	[55]
Zn-3Cu-1Mg *	12.4 (Hank's)	0.0432 (Hank's)	-	[55]
Zn–Ag				
Zn-2.5Ag	9.2 (Hank's)	0.079 (Hank's)	-	[45]
Zn-5Ag	9.7 (Hank's)	0.081 (Hank's)	-	[45]
Zn-7Ag	9.9 (Hank's)	0.084 (Hank's)	-	[45]
Zn–Li				
Zn-2Li	0.011 (SBF)	-	-	[25]
Zn-4Li	0.004 (SBF)	-	-	[25]
Zn-6Li	0.0038 (SBF)	-	-	[25]
Zn-1Li	-	-	0.02–0.05	[56]

SBF: simulated body fluid; PBS: phosphate buffered saline; * Hot extrusion; *** Homogenisation; a Data gathered from figure in literature [54] (Figure 7); b Data gathered from figure in literature [42] (Figure 5b); c Data gathered from figure in literature [9] (Figure 12); d Data gathered from figure in literature [36] (Figure 5); e Data gathered from figure in literature [44] (Figure 7); f Data gathered from figure in literature [22] (Figure 3b).

5. Mechanical Properties of Zinc Alloys

Implanted biomaterials, such as bone plates and stents, should be compatible with the mechanical properties of the substituted tissue [10]. Table 5 compares the mechanical properties of zinc and current zinc-based alloys to bone and arterial tissues. It is apparent that the mechanical properties of pure zinc are insufficient for bone and arterial medical device applications, in particular compared to cortical bone. The most effective way to address this issue is by adding alloying elements and/or through the refinement of grain size by thermomechanical processing [9].

Zinc alloys exhibit a wide range of ultimate tensile strengths and elongations, from 87 to 399 MPa and from 0.9% to ~170%, respectively. It can be seen from Table 5 that even minor alloying can significantly improve mechanical properties; for example, adding 0.15% Mg to pure zinc improves its ultimate tensile strength from 18 MPa to 250 MPa, and the elongation fraction from 0.32% to 22% [9,44]. Moreover, hot rolling and hot extrusion contribute to the strength and ductility of zinc alloys [3,9,44,46]. For example: the yield strength (YS), ultimate tensile strength (UTS) and elongation of as-cast Zn–1Mg–1Ca are 80 MPa, 130 MPa, and 1%, respectively. Meanwhile, the YS, UTS and elongation properties of Zn-based ternary alloy samples are improved to 138 MPa, 197 MPa, and 8.5% after hot rolling and 205 MPa, 250 MPa, and 5.2% after hot extrusion, respectively [44]. Therefore, it is feasible to satisfy strength requirements for zinc alloys through conventional metallurgical approaches.

Table 5. Mechanical properties of bone and arterial tissues, compared with current biomedical zinc alloys.

Tissue/Alloy	Mechanical Properties				References
	Yield Strength (YS) (MPa)	Ultimate Tensile Strength (UTS) (MPa)	Elongation (%)	Hardness (HV)	
Cortical bone	104.9–114.3	35–283	5–23	-	[10,11]
Cancellous bone	-	1.5–38	-	-	[10,11]
Arterial wall	-	0.5–1.72	-	-	[10,11]
Zn [c]	10	18	0.32	38	[44]
Zn *,c	35	60	3.5	-	[44]
Zn **,c,d	30–110	50–140	5.8–36	39	[44,52]
Zn–Mg					
Zn—0.15Mg *	114	250	22	52	[9]
Zn–0.5Mg *	159	297	13	65	[9]
Zn–1Mg/ZnMg1	180	340	6	75–86	[9,22,42,47]
Zn–1Mg *,f	175	250	12	-	[46]
Zn–1.2Mg	116	130	1.4	93	[3]
Zn–1.2Mg *	220	362	21	96	[3]
Zn–1.5Mg	112	150	1.3	155	[36]
Zn–3Mg/ZnMg3	-	104	2.3	201	[48]
Zn–3Mg *	291	399	1	117	[9]
Zn–3Mg ***	-	88	8.8	175	[48]
Zn–1.5Mg–0.1Ca	173	241	1.72	150	[36]
Zn–1Mg–0.5Ca	-	150	1.34	116	[47]
Zn–1Mg–1Ca [c]	80	130	1	90	[44]
Zn–1Mg–1Ca *,c	205	250	5.2	-	[44]
Zn–1Mg–1Ca **,c	138	197	8.5	105	[44]
Zn–1Mg–0.1Sr	109	133	1.4	94	[49]
Zn–1Mg–0.5Sr	129	144	1.1	109	[49]
Zn–1.5Mg–0.1Sr	130	209	2.0	145	[36]
Zn–1Mg–1Sr [c]	85	135	1.2	85	[44]
Zn–1Mg–1Sr *,c	200	250	7.3	-	[44]
Zn–1Mg–1Sr **,c	140	200	9.7	90	[44]
Zn–1Mg–0.1Mn	114	132	98	1.11	[50]
Zn–1.5Mg–0.1Mn	115	122	149	0.77	[50]

Table 5. *Cont.*

Tissue/Alloy	Yield Strength (YS) (MPa)	Ultimate Tensile Strength (UTS) (MPa)	Elongation (%)	Hardness (HV)	References
		Zn–Ca			
Zn–1Ca	119	165	2	73	[22]
Zn–1Ca–1Sr [c]	83	140	1.1	90	[44]
Zn–1Ca–1Sr *,c	210	260	6.8	-	[44]
Zn–1Ca–1Sr **,c	145	203	8.6	85	[44]
		Zn–Sr			
Zn–1Sr	120	171	2	61	[22]
		Zn–Al			
Zn–0.5Al *	119	203	33	59	[9]
Zn–1Al *	134	223	24	73	[9]
Zn–1Al **,d	190	220	24	-	[52]
Zn–3Al **,d	200	240	30	-	[52]
Zn–5Al **,d	240	300	16	-	[52]
ZnAl4Cu1	171	210	1	80	[42]
ZA0.1Mg	-	87	1.6	79	[51]
ZA0.3Mg	-	93	1.7	89	[51]
ZA0.5Mg/Zn–0.5Al–0.5Mg	-	92–102	1.73–2.1	94	[51,53]
Zn–0.5Al–0.5Mg–0.1Bi	-	102	2.4	102	[53]
Zn–0.5Al–0.5Mg–0.3Bi	-	108	2.7	109	[53]
Zn–0.5Al–0.5Mg–0.5Bi	-	98	1.97	99	[53]
ZA4–1	75	180	~112	50	[31]
ZA4–3	110	200	~130	55	[31]
ZA6–1	175	275	~170	65	[31]
		Zn–Cu			
Zn–1Cu *	149	186	21	-	[54]
Zn–2Cu *	199	240	46	-	[54]
Zn–3Cu *	213	257	47	-	[54,55]
Zn–4Cu *	227–250	270	51	-	[37,54]
Zn–3Cu–0.1Mg *,b	340	355	5	-	[55]
Zn–3Cu–0.5Mg *,b	390	400	2	-	[55]
Zn–3Cu–1Mg *,b	427	441	0.9	-	[55]
		Zn–Ag			
Zn–2.5Ag *,a	174	200	35	-	[45]
Zn–5Ag *,a	236	250	36	-	[45]
Zn–7Ag *,a	258	287	32	-	[45]
		Zn–Li			
Zn–2Li **,e	240	360	14.2	98	[25]
Zn–4Li **,e	420	440	13.7	115	[25]
Zn–6Li **,e	470	560	2.2	136	[25]
Zn–1Li	238	274	17	97	[56]

* Hot extrusion; ** Hot rolling; *** Homogenisation; [a] Data gathered from figure in literature [45]; (Figure 8a); [b] Data gathered from figure in literature [55] (Figure 5b); [c] Data gathered from figure in literature [44] (Figures 3 and 4); [d] Data gathered from figure in literature [52] (Figure 7); [e] Data gathered from figure in literature [25] (Figure 7); [f] Data gathered from figure in literature [46] (Figure 8).

6. Concluding Remarks and Perspectives

The previous several years have seen rapid growth in the research and development of zinc-based medical devices, due to their biological and biodegradability properties that match the human body. Zn is one of the most abundant essential elements in the human body, playing essential roles in human health. Moreover, zinc overcomes the limitations inherent to iron and magnesium, both pure and alloyed. This includes more suitable corrosion rates as well as easier casting and processing. In this review, current research progress for zinc and zinc-based alloys has been presented and discussed.

In the case of pure zinc, in vivo studies have demonstrated a great potential for use as biodegradable stents. The degradation of pure zinc stents proceeded with excellent biocompatibility to local cells and tissue and was aligned with the time course of arterial healing. Although the pure zinc stent poorly retained its mechanical integrity during the healing phase, this can be overcome by using zinc alloys with superior strength.

Despite the advantages, the use of pure Zn as a biodegradable metal is limited due to its insufficient strength, plasticity and hardness for most medical applications. Adding alloying elements and refining grain sizes via thermomechanical processing are commonly applied to modify the mechanical properties of metallic materials. A number of zinc alloys have been developed with nontoxic and biocompatible alloying elements that have achieved suitable mechanical properties to serve as structure support for arteries or bone, with promising preliminary results in cell culture and small animal models. For instance, Zn–Mg and Zn–Al at alloying concentrations of less than one percent, have enhanced mechanical properties and achieved adequate strength and ductility, without using excessive quantities of potentially toxic alloying elements. The mechanical properties of these materials are generally enhanced considerably from the as-cast state, following extrusion to break up embrittling intermetallics and to refine the grain size. Because each biomedical device (e.g., stent, bone plate or screw, etc.) requires unique processing conditions that can dramatically change material properties, it is crucial to evaluate candidate materials through these processing steps.

Looking ahead, researchers are required to translate the wealth of biomedical research data into an understanding of how these materials will behave within test subjects. Hence, it will be important to clarify the molecular mechanisms that stimulate beneficial cellular remodeling activities in response to Zn^{2+} and the alloying element ions. Next generation zinc implants will need to have tailored corrosion rates in order to optimize favorable cellular responses and minimize toxicity and negative inflammatory reactions, specific to the host tissue.

Author Contributions: All the authors wrote the paper.

Conflicts of Interest: The authors declare no conflict of interest.

References

1. Katarivas Levy, G.; Aghion, E. Influence of heat treatment temperature on corrosion characteristics of biodegradable EW10X04 Mg alloy coated with Nd. *Adv. Eng. Mater.* **2016**, *18*, 269–276. [CrossRef]
2. Holzapfel, B.M.; Reichert, J.C.; Schantz, J.-T.; Gbureck, U.; Rackwitz, L.; Nöth, U.; Jakob, F.; Rudert, M.; Groll, J.; Hutmacher, D.W. How smart do biomaterials need to be? A translational science and clinical point of view. *Adv. Drug Deliv. Rev.* **2013**, *65*, 581–603. [CrossRef] [PubMed]
3. Shen, C.; Liu, X.; Fan, B.; Lan, P.; Zhou, F.; Li, X.; Wang, H.; Xiao, X.; Li, L.; Zhao, S. Mechanical properties, in vitro degradation behavior, hemocompatibility and cytotoxicity evaluation of Zn–1.2 Mg alloy for biodegradable implants. *RSC Adv.* **2016**, *6*, 86410–86419. [CrossRef]
4. Niinomi, M. Metallic biomaterials. *J. Artif. Organs* **2008**, *11*, 105–110. [CrossRef] [PubMed]
5. Katarivas Levy, G.; Ventura, Y.; Goldman, J.; Vago, R.; Aghion, E. Cytotoxic characteristics of biodegradable EW10X04 Mg alloy after nd coating and subsequent heat treatment. *Mater. Sci. Eng. C* **2016**, *62*, 752–761. [CrossRef] [PubMed]
6. Pospíšilová, I.; Soukupová, V.; Vojtěch, D. Influence of calcium on the structure and mechanical properties of biodegradable zinc alloys. In *Materials Science Forum*; Trans Tech Publication Ltd.: Stafa-Zurich, Switzerland, 2017; pp. 400–403.
7. Aghion, E.; Levy, G.; Ovadia, S. In Vivo behavior of biodegradable Mg–Nd–Y–Zr–Ca alloy. *J. Mater. Sci.* **2012**, *23*, 805–812. [CrossRef] [PubMed]
8. Zheng, Y.; Gu, X.; Witte, F. Biodegradable metals. *Mater. Sci. Eng. C* **2014**, *77*, 1–34. [CrossRef]
9. Mostaed, E.; Sikora-Jasinska, M.; Mostaed, A.; Loffredo, S.; Demir, A.; Previtali, B.; Mantovani, D.; Beanland, R.; Vedani, M. Novel zn-based alloys for biodegradable stent applications: Design, development and in vitro degradation. *J. Mech. Behav. Biomed. Mater.* **2016**, *60*, 581–602. [CrossRef] [PubMed]

10. Gu, X.-N.; Zheng, Y.-F. A review on magnesium alloys as biodegradable materials. *Front. Mater. Sci. China* **2010**, *4*, 111–115. [CrossRef]

11. Witte, F.; Hort, N.; Vogt, C.; Cohen, S.; Kainer, K.U.; Willumeit, R.; Feyerabend, F. Degradable biomaterials based on magnesium corrosion. *Curr. Opin. Solid State Mater. Sci.* **2008**, *12*, 63–72. [CrossRef]

12. Jablonská, E.; Vojtěch, D.; Fousová, M.; Kubásek, J.; Lipov, J.; Fojt, J.; Ruml, T. Influence of surface pre-treatment on the cytocompatibility of a novel biodegradable ZnMg alloy. *Mater. Sci. Eng. C* **2016**, *68*, 198–204. [CrossRef] [PubMed]

13. Demir, A.G.; Monguzzi, L.; Previtali, B. Selective laser melting of pure Zn with high density for biodegradable implant manufacturing. *Addit. Manuf.* **2017**, *15*, 20–28. [CrossRef]

14. Wang, C.; Yu, Z.; Cui, Y.; Zhang, Y.; Yu, S.; Qu, G.; Gong, H. Processing of a novel Zn alloy micro-tube for biodegradable vascular stent application. *J. Mater. Sci. Technol.* **2016**, *32*, 925–929. [CrossRef]

15. Luo, M.; Shen, W.; Wang, Y.; Allen, M. In Vitro Degradation of Biodegradable Metal Zn And Zn/Fe-Couples and Their Application as Conductors in Biodegradable Sensors. In Proceedings of the 2015 Transducers-2015 18th International Conference on Solid-State Sensors, Actuators and Microsystems (TRANSDUCERS), Anchorage, AK, USA, 21–25 June 2015; pp. 1370–1373.

16. Ma, J.; Zhao, N.; Zhu, D. Endothelial cellular responses to biodegradable metal zinc. *ACS Biomater. Sci. Eng.* **2015**, *1*, 1174–1182. [CrossRef] [PubMed]

17. Aghion, E.; Levy, G. The effect of Ca on the in vitro corrosion performance of biodegradable Mg-Nd-Y-Zr alloy. *J. Mater. Sci.* **2010**, *45*, 3096–3101. [CrossRef]

18. Levy, G.; Aghion, E. Effect of diffusion coating of Nd on the corrosion resistance of biodegradable Mg implants in simulated physiological electrolyte. *Acta Biomater.* **2013**, *9*, 8624–8630. [CrossRef] [PubMed]

19. Murni, N.; Dambatta, M.; Yeap, S.; Froemming, G.; Hermawan, H. Cytotoxicity evaluation of biodegradable Zn–3Mg alloy toward normal human osteoblast cells. *Mater. Sci. Eng. C* **2015**, *49*, 560–566. [CrossRef] [PubMed]

20. Guillory, R.J.; Bowen, P.K.; Hopkins, S.P.; Shearier, E.R.; Earley, E.J.; Gillette, A.A.; Aghion, E.; Bocks, M.; Drelich, J.W.; Goldman, J. Corrosion characteristics dictate the long-term inflammatory profile of degradable zinc arterial implants. *ACS Biomater. Sci. Eng.* **2016**, *2*, 2355–2364. [CrossRef]

21. Hakimi, O.; Ventura, Y.; Goldman, J.; Vago, R.; Aghion, E. Porous biodegradable EW62 medical implants resist tumor cell growth. *Mater. Sci. Eng. C* **2016**, *61*, 516–525. [CrossRef] [PubMed]

22. Li, H.; Xie, X.; Zheng, Y.; Cong, Y.; Zhou, F.; Qiu, K.; Wang, X.; Chen, S.; Huang, L.; Tian, L. Development of biodegradable Zn-1X binary alloys with nutrient alloying elements Mg, Ca and Sr. *Sci. Rep.* **2015**, *5*, 10719. [CrossRef] [PubMed]

23. Liu, X.; Sun, J.; Yang, Y.; Pu, Z.; Zheng, Y. In vitro investigation of ultra-pure Zn and its mini-tube as potential bioabsorbable stent material. *Mater. Lett.* **2015**, *161*, 53–56. [CrossRef]

24. Kubásek, J.; Vojtěch, D.; Jablonská, E.; Pospíšilová, I.; Lipov, J.; Ruml, T. Structure, mechanical characteristics and in vitro degradation, cytotoxicity, genotoxicity and mutagenicity of novel biodegradable Zn–Mg alloys. *Mater. Sci. Eng. C* **2016**, *58*, 24–35. [CrossRef] [PubMed]

25. Zhao, S.; McNamara, C.T.; Bowen, P.K.; Verhun, N.; Braykovich, J.P.; Goldman, J.; Drelich, J.W. Structural characteristics and in vitro biodegradation of a novel Zn-Li alloy prepared by induction melting and hot rolling. *Metall. Mater. Trans. A* **2017**, *3*, 1204–1215. [CrossRef]

26. Ma, J.; Zhao, N.; Zhu, D. Bioabsorbable zinc ion induced biphasic cellular responses in vascular smooth muscle cells. *Sci. Rep.* **2016**, *6*, 26661. [CrossRef] [PubMed]

27. Kubasek, J.; Vojtěch, D. Zn-based alloys as an alternative biodegradable materials. *Proc Met.* **2012**, *5*, 23–25.

28. Kubasek, J.; Pospisilova, I.; Vojtech, D.; Jablonska, E.; Ruml, T. Structural, mechanical and cytotoxicity characterization of as-cast biodegradable Zn–xMg (x = 0.8%–8.3%) alloys. *Mater. Tehnol.* **2014**, *48*, 623–629.

29. Dunne, C.F.; Katarivas Levy, G.; Hakimi, O.; Aghion, E.; Twomey, B.; Stanton, K.T. Corrosion behaviour of biodegradable magnesium alloys with hydroxyapatite coatings. *Surf. Coat. Technol.* **2016**, *289*, 37–44. [CrossRef]

30. Seitz, J.M.; Durisin, M.; Goldman, J.; Drelich, J.W. Recent advances in biodegradable metals for medical sutures: A critical review. *Adv. Healthc. Mater.* **2015**, *4*, 1915–1936. [CrossRef] [PubMed]

31. Wang, C.; Yang, H.; Li, X.; Zheng, Y. In vitro evaluation of the feasibility of commercial Zn alloys as biodegradable metals. *J. Mater. Sci. Technol.* **2016**, *32*, 909–918. [CrossRef]

32. Törne, K.; Larsson, M.; Norlin, A.; Weissenrieder, J. Degradation of zinc in saline solutions, plasma, and whole blood. *J. Biomed. Mater. Res. Part B* **2016**, *104*, 1141–1151. [CrossRef] [PubMed]

33. Bolz, A.; Popp, T. Implantable, Bioresorbable Vessel Wall Support, in Particular Coronary Stent. U.S. Patent 6,287,332 B1, 11 September 2001.

34. Zhou, G.; Gongqi, Q.; Gong, H. Kind of Absorbable High Strength and Toughness Corrosion-Resistant Zinc Alloy Implant Material for Human Body. U.S. Patent 20170028107 A1, 2 February 2017.

35. Plum, L.M.; Rink, L.; Haase, H. The essential toxin: Impact of zinc on human health. *Int. J. Environ. Res. Public Health* **2010**, *7*, 1342–1365. [CrossRef] [PubMed]

36. Liu, X.; Sun, J.; Qiu, K.; Yang, Y.; Pu, Z.; Li, L.; Zheng, Y. Effects of alloying elements (Ca and Sr) on microstructure, mechanical property and in vitro corrosion behavior of biodegradable Zn–1.5Mg alloy. *J. Alloys Compd.* **2016**, *664*, 444–452. [CrossRef]

37. Niu, J.; Tang, Z.; Huang, H.; Pei, J.; Zhang, H.; Yuan, G.; Ding, W. Research on a Zn-Cu alloy as a biodegradable material for potential vascular stents application. *Mater. Sci. Eng. C* **2016**, *69*, 407–413. [CrossRef] [PubMed]

38. Huang, T.; Zheng, Y.; Han, Y. Accelerating degradation rate of pure iron by zinc ion implantation. *Regen. Biomater.* **2016**, *3*, 205–215. [CrossRef] [PubMed]

39. Vojtech, D.; Pospisilova, I.; Michalcova, A.; Maixner, J. Microstructure and mechanical properties of the micrograined hypoeutectic zn-mg alloy. *Int. J. Miner. Metall. Mater.* **2016**, *23*, 1167–1176.

40. Pospíšilová, I.; Vojtěch, D. Zinc alloys for biodegradable medical implants. In *Materials Science Forum*; Trans Tech Publlication: Stafa-Zurich, Switzerland, 2014; pp. 457–460.

41. Zhao, L.; Zhang, Z.; Song, Y.; Liu, S.; Qi, Y.; Wang, X.; Wang, Q.; Cui, C. Mechanical properties and in vitro biodegradation of newly developed porous Zn scaffolds for biomedical applications. *Mater. Des.* **2016**, *108*, 136–144. [CrossRef]

42. Vojtěch, D.; Kubásek, J.; Šerák, J.; Novák, P. Mechanical and corrosion properties of newly developed biodegradable Zn-based alloys for bone fixation. *Acta Biomater.* **2011**, *7*, 3515–3522. [CrossRef] [PubMed]

43. Guleryuz, L.; Ipek, R.; Arıtman, I.; Karaoglu, S. Microstructure and Mechanical Properties of Zn-Mg Alloys as Implant Materials Manufactured by Powder Metallurgy Method. In *AIP Conference Proceedings*; AIP Publishing: Melville, NY, USA, 2017.

44. Li, H.; Yang, H.; Zheng, Y.; Zhou, F.; Qiu, K.; Wang, X. Design and characterizations of novel biodegradable ternary Zn-based alloys with iia nutrient alloying elements mg, ca and sr. *Mater. Des.* **2015**, *83*, 95–102. [CrossRef]

45. Sikora-Jasinska, M.; Mostaed, E.; Mostaed, A.; Beanland, R.; Mantovani, D.; Vedani, M. Fabrication, mechanical properties and in vitro degradation behavior of newly developed ZnAg alloys for degradable implant applications. *Mater. Sci. Eng. C* **2017**, *77*, 1170–1181. [CrossRef] [PubMed]

46. Gong, H.; Wang, K.; Strich, R.; Zhou, J.G. In vitro biodegradation behavior, mechanical properties, and cytotoxicity of biodegradable Zn–Mg alloy. *J. Biomed. Mater. Res. Part B* **2015**, *103*, 1632–1640. [CrossRef] [PubMed]

47. Katarivas Levy, G.L.; Leon, A.; Kafri, A.; Ventura, Y.; Drelich, J.W.; Goldman, J.; Vago, R.; Aghion, E. Evaluation of biodegradable Zn–1%Mg and Zn–1%Mg–0.5%Ca alloys for biomedical applications. *J. Mater. Sci. Mater. Med.* **2017**. accepted. [CrossRef]

48. Dambatta, M.; Izman, S.; Kurniawan, D.; Farahany, S.; Yahaya, B.; Hermawan, H. Influence of thermal treatment on microstructure, mechanical and degradation properties of Zn–3Mg alloy as potential biodegradable implant material. *Mater. Des.* **2015**, *85*, 431–437. [CrossRef]

49. Liu, X.; Sun, J.; Yang, Y.; Zhou, F.; Pu, Z.; Li, L.; Zheng, Y. Microstructure, mechanical properties, in vitro degradation behavior and hemocompatibility of novel Zn–Mg–Sr alloys as biodegradable metals. *Mater. Lett.* **2016**, *162*, 242–245. [CrossRef]

50. Liu, X.; Sun, J.; Zhou, F.; Yang, Y.; Chang, R.; Qiu, K.; Pu, Z.; Li, L.; Zheng, Y. Micro-alloying with Mn in Zn–Mg alloy for future biodegradable metals application. *Mater. Des.* **2016**, *94*, 95–104. [CrossRef]

51. Bakhsheshi-Rad, H.; Hamzah, E.; Low, H.; Kasiri-Asgarani, M.; Farahany, S.; Akbari, E.; Cho, M. Fabrication of biodegradable Zn-Al-Mg alloy: Mechanical properties, corrosion behavior, cytotoxicity and antibacterial activities. *Mater. Sci. Eng. C* **2017**, *73*, 215–219. [CrossRef] [PubMed]

52. Bowen, P.; Seitz, J.; Guillory, R.; Braykovich, J.; Zhao, F.; Goldman, J.; Drelich, J. Evaluation of wrought Zn-Al alloys (1, 3, and 5 wt % al) through mechanical and in vivo corrosion testing for stent applications. *J. Biomed. Mater. Res. Part B* **2016**. [CrossRef]

53. Bakhsheshi-Rad, H.; Hamzah, E.; Low, H.; Cho, M.; Kasiri-Asgarani, M.; Farahany, S.; Mostafa, A.; Medraj, M. Thermal characteristics, mechanical properties, in vitro degradation and cytotoxicity of novel biodegradable Zn–Al–Mg and Zn–Al–Mg–xBi alloys. *Acta Metall. Sin. (Engl. Lett.)* **2017**, *30*, 201–211. [CrossRef]

54. Tang, Z.; Niu, J.; Huang, H.; Zhang, H.; Pei, J.; Ou, J.; Yuan, G. Potential biodegradable Zn-Cu binary alloys developed for cardiovascular implant applications. *J. Mech. Behav. Biomed. Mater.* **2017**, *72*, 182–191. [CrossRef] [PubMed]

55. Tang, Z.; Huang, H.; Niu, J.; Zhang, L.; Zhang, H.; Pei, J.; Tan, J.; Yuan, G. Design and characterizations of novel biodegradable Zn-Cu-Mg alloys for potential biodegradable implants. *Mater. Des.* **2017**, *117*, 84–94. [CrossRef]

56. Zhao, S.; Seitz, J.-M.; Eifler, R.; Maier, H.J.; Guillory, R.J.; Earley, E.J.; Drelich, A.; Goldman, J.; Drelich, J.W. Zn-Li alloy after extrusion and drawing: Structural, mechanical characterization, and biodegradation in abdominal aorta of rat. *Mater. Sci. Eng. C* **2017**, *76*, 301–312. [CrossRef] [PubMed]

57. Roohani, N.; Hurrell, R.; Kelishadi, R.; Schulin, R. Zinc and its importance for human health: An integrative review. *J. Res. Med. Sci.* **2013**, *18*, 144. [PubMed]

58. Jackson, M. Physiology of zinc: General aspects. In *Zinc in Human Biology*; Springer: Berlin, Germany, 1989; pp. 1–14.

59. Bowen, P.K.; Drelich, J.; Goldman, J. Zinc exhibits ideal physiological corrosion behavior for bioabsorbable stents. *Adv. Mater.* **2013**, *25*, 2577–2582. [CrossRef] [PubMed]

60. Moonga, B.S.; Dempster, D.W. Zinc is a potent inhibitor of osteoclastic bone resorption in vitro. *J. Bone Miner. Res.* **1995**, *10*, 453–457. [CrossRef] [PubMed]

61. Yamaguchi, M. Role of zinc in bone formation and bone resorption. *J. Trace Elem. Exp. Med.* **1998**, *11*, 119–135. [CrossRef]

62. Bowen, P.K.; Shearier, E.R.; Zhao, S.; Guillory, R.J.; Zhao, F.; Goldman, J.; Drelich, J.W. Biodegradable metals for cardiovascular stents: From clinical concerns to recent Zn-alloys. *Adv. Healthc. Mater.* **2016**, *5*, 1121–1140. [CrossRef] [PubMed]

63. Kaltenberg, J.; Plum, L.M.; Ober-Blobaum, J.L.; Honscheid, A.; Rink, L.; Haase, H. Zinc signals promote IL-2-dependent proliferation of T cells. *Eur. J. Immunol.* **2010**, *40*, 1496–1503. [CrossRef] [PubMed]

64. Shankar, A.H.; Prasad, A.S. Zinc and immune function: The biological basis of altered resistance to infection. *Am. J. Clin. Nutr.* **1998**, *68*, 447s–463s. [PubMed]

65. Kaur, G. Biodegradable metals as bioactive materials. In *Bioactive Glasses: Potential Biomaterials for Future Therapy*; Springer International Publishing: Cham, Switzerland, 2017.

66. Krezel, A.; Maret, W. The biological inorganic chemistry of zinc ions. *Arch. Biochem. Biophys.* **2016**, *611*, 3–19. [CrossRef] [PubMed]

67. Thomas, S.; Birbilis, N.; Venkatraman, M.S.; Cole, I.S. Corrosion of zinc as a function of pH. *J. Sci. Eng.* **2012**, *68*. [CrossRef]

68. Drelich, A.; Zhao, S.; Guillory, R.J.; Drelich, J.W.; Goldman, J. Long-term surveillance of zinc implant in murine artery: Surprisingly steady biocorrosion rate. *Acta Biomater.* **2017**, *58*, 539–549. [CrossRef] [PubMed]

69. Shearier, E.R.; Bowen, P.K.; He, W.; Drelich, A.; Drelich, J.; Goldman, J.; Zhao, F. In vitro cytotoxicity, adhesion, and proliferation of human vascular cells exposed to zinc. *ACS Biomater. Sci. Eng.* **2016**, *2*, 634–642. [CrossRef] [PubMed]

70. Zhu, D.H.; Su, Y.C.; Young, M.L.; Ma, J.; Zheng, Y.F.; Tang, L.P. Biological responses and mechanisms of human bone marrow mesenchymal stem cells to Zn and Mg biomaterials. *ACS Appl. Mater. Interfaces* **2017**, *9*, 27453–27461. [CrossRef] [PubMed]

71. Qiao, Y.Q.; Zhang, W.J.; Tian, P.; Meng, F.H.; Zhu, H.Q.; Jiang, X.Q.; Liu, X.Y.; Chu, P.K. Stimulation of bone growth following zinc incorporation into biomaterials. *Biomaterials* **2014**, *35*, 6882–6897. [CrossRef] [PubMed]

72. Al Qaysi, M.; Petrie, A.; Shah, R.; Knowles, J.C. Degradation of zinc containing phosphate-based glass as a material for orthopedic tissue engineering. *J. Mater. Sci.* **2016**. [CrossRef]

73. Zhong, Z.; Ma, J. Fabrication, characterization, and in vitro study of zinc substituted hydroxyapatite/silk fibroin composite coatings on titanium for biomedical applications. *J. Biomater. Appl.* **2017**, *32*, 399–409. [CrossRef] [PubMed]

74. Yu, J.M.; Xu, L.Z.; Li, K.; Xie, N.; Xi, Y.H.; Wang, Y.; Zheng, X.B.; Chen, X.S.; Wang, M.Y.; Ye, X.J. Zinc-modified calcium silicate coatings promote osteogenic differentiation through TGF-β/Smad pathway and osseointegration in osteopenic rabbits. *Sci. Rep.* **2017**, *7*, 3440. [CrossRef] [PubMed]

75. Cama, G.; Nkhwa, S.; Gharibi, B.; Lagazzo, A.; Cabella, R.; Carbone, C.; Dubruel, P.; Haugen, H.; Di Silvio, L.; Deb, S. The role of new zinc incorporated monetite cements on osteogenic differentiation of human mesenchymal stem cells. *Mater. Sci. Eng. C* **2017**, *78*, 485–494. [CrossRef] [PubMed]

76. Ferreira, E.C.S.; Bortolin, R.H.; Freire-Neto, F.P.; Souza, K.S.C.; Bezerra, J.F.; Ururahy, M.A.G.; Ramos, A.M.O.; Himelfarb, S.T.; Abreu, B.J.; Didone, T.V.N.; et al. Zinc supplementation reduces RANKL/OPG ratio and prevents bone architecture alterations in ovariectomized and type 1 diabetic rats. *Nutr. Res.* **2017**, *40*, 48–56. [CrossRef] [PubMed]

77. Yu, W.L.; Sun, T.W.; Qi, C.; Ding, Z.Y.; Zhao, H.K.; Zhao, S.C.; Shi, Z.M.; Zhu, Y.J.; Chen, D.Y.; He, Y.H. Evaluation of zinc-doped mesoporous hydroxyapatite microspheres for the construction of a novel biomimetic scaffold optimized for bone augmentation. *Int. J. Nanomed.* **2017**, *12*, 2293–2306. [CrossRef] [PubMed]

78. Yusa, K.; Yamamoto, O.; Takano, H.; Fukuda, M.; Iino, M. Zinc-modified titanium surface enhances osteoblast differentiation of dental pulp stem cells in vitro. *Sci. Rep.* **2016**, *6*. [CrossRef] [PubMed]

79. An, S.F.; Gong, Q.M.; Huang, Y.H. Promotive effect of zinc ions on the vitality, migration, and osteogenic differentiation of human dental pulp cells. *Biol. Trace Elem. Res.* **2017**, *175*, 112–121. [CrossRef] [PubMed]

80. Bowen, P.K.; Guillory, R.J.; Shearier, E.R.; Seitz, J.-M.; Drelich, J.; Bocks, M.; Zhao, F.; Goldman, J. Metallic zinc exhibits optimal biocompatibility for bioabsorbable endovascular stents. *Mater. Sci. Eng. C* **2015**, *56*, 467–472. [CrossRef] [PubMed]

81. Yang, H.; Wang, C.; Liu, C.; Chen, H.; Wu, Y.; Han, J.; Jia, Z.; Lin, W.; Zhang, D.; Li, W.; et al. Evolution of the degradation mechanism of pure zinc stent in the one-year study of rabbit abdominal aorta model. *Biomaterials* **2017**, *145*, 92–105. [CrossRef] [PubMed]

82. Lansdown, A.B. Silver in health care: Antimicrobial effects and safety in use. In *Biofunctional Textiles and the Skin*; Karger: Basel, Switzerland, 2006.

83. Mertz, W. *Trace Elements in Human and Animal Nutrition*; Academic Press, INC.: Orlando, FL, USA, 1986.

metals

MDPI

Article

The Suitability of Zn–1.3%Fe Alloy as a Biodegradable Implant Material

Alon Kafri [1], Shira Ovadia [2], Jeremy Goldman [3,*], Jaroslaw Drelich [4] and Eli Aghion [1]

[1] Department of Materials Engineering, Ben-Gurion University of the Negev, Beer-Sheva 8410501, Israel; kafri09@gmail.com (A.K.); egyon@bgu.ac.il (E.A.)

[2] Faculty of Health Science, Ben-Gurion University of the Negev, P.O. Box 653, Beer-Sheva 84105, Israel; shiraov@bgu.ac.il

[3] Department of Biomedical Engineering, Michigan Technological University, Houghton, MI 49931, USA

[4] Department of Materials Science and Engineering, Michigan Technological University, Houghton, MI 49931, USA; jwdrelic@mtu.edu

* Correspondence: jgoldman@mtu.edu

Received: 5 February 2018; Accepted: 26 February 2018; Published: 28 February 2018

Abstract: Efforts to develop metallic zinc for biodegradable implants have significantly advanced following an earlier focus on magnesium (Mg) and iron (Fe). Mg and Fe base alloys experience an accelerated corrosion rate and harmful corrosion products, respectively. The corrosion rate of pure Zn, however, may need to be modified from its reported ~20 μm/year penetration rate, depending upon the intended application. The present study aimed at evaluating the possibility of using Fe as a relatively cathodic biocompatible alloying element in zinc that can tune the implant degradation rate via microgalvanic effects. The selected Zn–1.3wt %Fe alloy composition produced by gravity casting was examined in vitro and in vivo. The in vitro examination included immersion tests, potentiodynamic polarization and impedance spectroscopy, all in a simulated physiological environment (phosphate-buffered saline, PBS) at 37 °C. For the in vivo study, two cylindrical disks (seven millimeters diameter and two millimeters height) were implanted into the back midline of male Wister rats. The rats were examined post implantation in terms of weight gain and hematological characteristics, including red blood cell (RBC), hemoglobin (HGB) and white blood cell (WBC) levels. Following retrieval, specimens were examined for corrosion rate measurements and histological analysis of subcutaneous tissue in the implant vicinity. In vivo analysis demonstrated that the Zn–1.3%Fe implant avoided harmful systemic effects. The in vivo and in vitro results indicate that the Zn–1.3%Fe alloy corrosion rate is significantly increased compared to pure zinc. The relatively increased degradation of Zn–1.3%Fe was mainly related to microgalvanic effects produced by a secondary $Zn_{11}Fe$ phase.

Keywords: zinc; bioabsorbable; biodegradable; implants; encapsulation; in vivo

---

1. Introduction

The development of bioabsorbable metal implants has drawn major attention over the last two decades. Early studies mainly focused on pure Fe and Mg and their alloys. Bioabsorbable implants based on Fe exhibit good biocompatibility and excellent mechanical properties [1–5]. However, they suffer from a harmful mode of corrosion that produces a voluminous iron oxide layer [6]. The steady accumulation of the iron oxide reduces the lumen cross section, compromises the integrity of the arterial wall by repelling neighboring tissue and cells [7], and stimulates inflammation [8]. Bioabsorbable implants based on Mg are attractive mainly due to their excellent biocompatibility [8–20]. Unfortunately, Mg and its alloys exhibit an excessive corrosion rate associated with a potentially harmful release of hydrogen gas [15,18]. The challenges faced with both Fe and Mg systems have motivated a search for new metallic materials, among them Zn and Zn-based alloys.

Early in vivo studies with pure zinc arterial implants have reported a promising biocompatibility [21,22]. The biodegradation rate is near the projected ideal value for vascular stents, and between that of Mg and Fe. In addition, Zn has important physiological roles, exerting strong anti-atherogenic properties [23], participating in nucleic acid metabolism, signal transduction, apoptosis regulation, and gene expression [24,25]. Zinc is not toxic to the human body at the low levels expected to be released from a typical stent [26]. Our extensive preliminary observations have confirmed the lack of overt toxicity at the interface between a zinc implant and biological tissue, despite even 20 months residence time in the body [21,27–29]. However, at high levels zinc can cause toxicity, manifested in poor growth and anemia [30,31]. To avoid negative responses, the corrosion rate for zinc-based implants may need to be optimized for biodegradable application. Our previous studies demonstrated that the corrosion rate of a Zn implant can be reduced in the first several weeks after deployment into an arterial environment by manipulating the oxide film characteristics [32] or adding a polymeric layer [33]. In addition, earlier research carried out by the authors [27] suggested that the low degradation rate of arterial implants made from pure zinc can provoke a long-term process of inflammation and fibrous encapsulation. Encapsulation can be considered as a major factor contributing to implant failure, as it isolates the implant from the surrounding tissue and can potentially cause a discontinuation of the biodegradation process [34,35]. This was supported by Yue et al. [36] which indicated that too low degradation rate is a challenge for future application of Zn base alloys as adequate biodegradable implant.

The need to improve the strength of Zn and Zn-based materials, manipulate their corrosion rate and improve their biocompatibility motivated the search for new Zn-based alloys [37–51]. In order to obtain a higher corrosion rate without compromising biocompatibility, the alloying element must be biocompatible and have the potential to increase the corrosion rate of the Zn-based alloy. Since Fe is known as a biocompatible element and the intermetallic phases formed between Zn and Fe all have higher electrochemical potential [52,53], it was considered a suitable alloying element. The present study aimed at exploring the prospects of Zn–1.3wt %Fe (denoted as Zn–1.3%Fe) alloy as a biodegradable implant material in terms of in vitro and in vivo behavior.

2. Experimental Procedure

Ingots of pure Zn and Zn–1.3%Fe alloy were produced by gravity casting in a rectangular steel die having the following dimensions: 6 × 25 × 4.5 cm. Prior to the casting process, pure zinc bars (99.99%) and pure iron (99%) powder (−325 mesh) in the desired amount were placed in a graphite crucible and heated to 750 °C for 3 h in a furnace, with stirring every 30 min. All the test samples produced from the cast ingot were machined from the central part of the ingot, producing surface roughness type N5. The chemical compositions of the obtained ingots were determined using an Inductively Coupled Plasma Optical Emission Spectrometer (ICP-SPECTRO, ARCOS FHS-12, Kelve, Germany) method, with results shown in Table 1.

Table 1. Chemical composition of pure Zn and Zn–1.3%Fe alloy.

Material System	Fe (wt %)	Pb (wt %)	Al (wt %)	Cu (wt %)	Cd (wt %)
Pure Zn	0.0054	0.0011	0.0026	0.0007	0.0017
Zn–1.3%Fe	1.31	0.0010	0.0079	0.0006	0.0017

Microstructure examination was carried out using scanning electron microscopy (SEM) with a JEOL JSM-5600 (JEOL, Tokyo, Japan) equipped with an Energy-dispersive X-ray spectroscopy (EDS) detector (Thermo Fisher Scientific, Waltham, MA, USA) for spot elemental analysis. Identification of internal phases was conducted using an X-ray diffractometer RIGAKU-2100H (RIGAKU, Tokyo,

Japan) with Cu-Kα. The diffraction parameter was 40 kV/30 mA and the scanning rate was 2°/min. Indication of mechanical properties was obtained by Vickers hardness measurements performed in a Zwick/Roell Indentec of Quantarad Technologies, with an applied load of 3 kg. The tensile tests were performed at room temperature using a universal material test machine (Hounsfield H25KT, Horsham, PA, USA) at a rate of 0.5mm/min.

The in vitro corrosion behavior was examined by an immersion test for up to 20 days, according to the ASTM G31-12a standard as well as by electrochemical characterization and stress corrosion analysis. All the in vitro corrosion tests were carried out in phosphate-buffered saline (PBS) solution at 37 °C with pH levels of ~7.4. The selected corrosion medium and environmental conditions aimed at simulating the natural physiological environment [54–56]. Monitoring of pH levels was conducted by daily replacement of the PBS solution. The electrochemical characterization included potentiodynamic polarization analysis and impedance spectroscopy (EIS) using a Bio-Logic SP-200 potentiostat equipped with EC-Lab software V11.02. The three-electrode cell used for the electrochemical analysis included a saturated calomel reference electrode (SCE), a platinum counter electrode, and the tested sample as the working electrode, with an exposure area of 1cm^2. The scanning rate of the potentiodynamic polarization analysis was 1 mV/s; the EIS measurements were carried out between 10 kHz and 100 mHz at 10 mV amplitude over the open circuit potential. Prior to electrochemical testing, the samples were cleaned in an ultrasonic bath for 5 min, washed with alcohol, and dried in hot air.

The metals used for the in vivo assessment included pure zinc as the biodegradable reference material, the Zn–1.3%Fe allow, and titanium alloy Ti–6Al–4V as a biostable reference material. The implant geometry of cylindrical disks of 7 mm diameter and 2 mm height was obtained by regular machining from cast Zn base ingots and wrought billet in the case of the titanium ally. Prior to implantation, the disks were ultrasonically cleaned in ethanol for 5 min and then in acetone for 3 min before final air drying.

All the animal experiments were approved by the Ben-Gurion University of the Negev (BGU) Committee for the Ethical Care and Use of Laboratory Animals (BGU-IACUC). The experiments were performed according to the Israel Animal Welfare law (1994) and the NRC Guide for the Care and Use of Laboratory Animals (2011). BGU's animal care and use program is approved by the Association for the Assessment and Accreditation of Laboratory Animal Care International (AAALAC). The in vivo experiments were carried out in the BGU rodent facility. Nine 250 g male Wistar rats (Envigo, Jerusalem, Israel) were selected for the in vivo assessment. The rats were divided into 3 groups. The first group (n = 3) was implanted with pure Zn disks, the second group (n = 3) with the Zn–1.3%Fe alloy, and the third group (n = 3) with Ti–6Al–4V.

For implantation, the rats were anesthetized with an inhalation anesthesia machine using 3% isoflurane (Terrel TM Piramal Critical Care, Inc., Bethlehem, PA, USA) in 500 mL/min 100% oxygen. Two cylindrical disks were implanted aseptically subcutaneously in the midline of the back of the rats, one between scapulas and another in mid-lumbar area. Post alloy implantation, the rats were placed on a heating pad until they recovered form anesthesia and were able to ambulate. For postoperative analgesia, rats received 100 mg/kg Dipyrone (Vitamed Pharmaceutical Industries LTD, Binyamina, Israel) in the drinking water for 3 days. Rats were monitored daily for surgical wound appearance, locomotion in the cage, grooming activity and general wellbeing for the first week. This was followed by a weekly evaluation of body weight, attitude and incision appearance. Blood was collected from the retro-orbital sinus under isoflurane anesthesia as described above. Whole blood (1 mL) was collected in EDTA for a complete blood count for red blood cells (RBC), hemoglobin (HGB) and white blood cells (WBC). Serum (1 mL) was collected for determination of Zn levels. Blood work was analyzed on the day of implantation and at 4 and 8 weeks post-implantation. The blood biochemical parameters in terms of RBC, HGB and WBC can indicate abnormal situations, such as a significant increase in Zn content or infection. At 14 weeks the rats were euthanized with intraperitoneal 150 mg/kg Pentobarbital (CTS Chemical Industries Ltd., Hod Hasharon, Israel) for alloy and tissue harvest. Tissue from each alloy location was placed in 10% formaldehyde for histology. Removal of corrosion products to calculate the corrosion rate was preformed according to ASTM G1 using 10%NH$_4$Cl solution, which allowed for a calculation of the corrosion rate.

The statistical analysis was implemented using a one-way ANOVA to determine the significance of paired comparisons. A *p* value < 0.05 was selected for statistical difference between means.

3. Results

The typical microstructure of pure zinc and Zn–1.3%Fe alloy are shown in cross-sectional view (Figure 1), along with a spot chemical analysis at various points, shown in Table 2. A regular structure without any second phase is present in pure zinc. The Zn–1.3%Fe alloy contained Zn matrix and an Fe-rich phase that was dispersed homogenously across the bulk of the alloy. The Fe content within the Fe-rich phase, as obtained by EDS analysis, was between 7 and 10 wt % corresponding with Fe content in the Zn-delta phase composition [52].

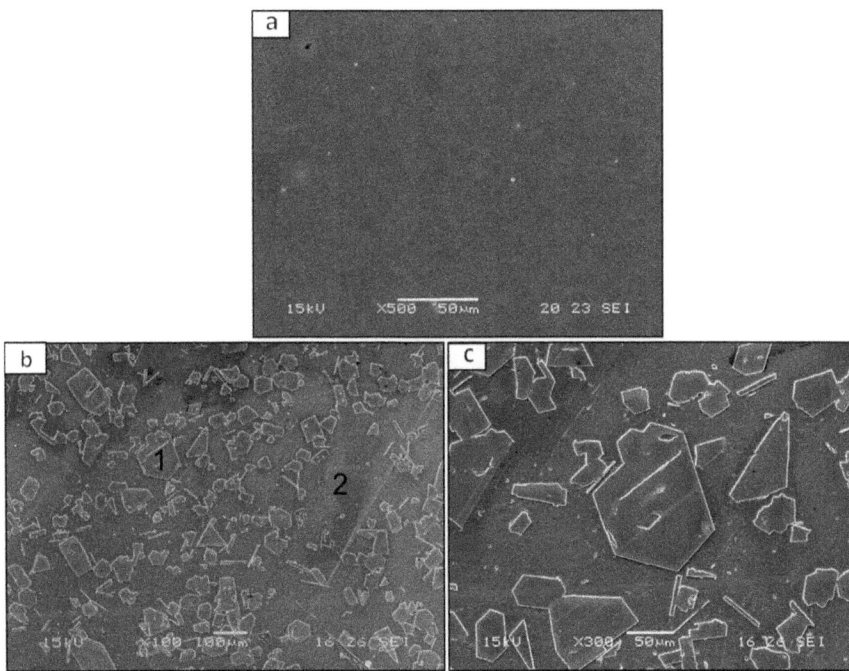

Figure 1. Microstructure at cross-section of (a) Pure Zn. (b,c) Zn–1.3%Fe alloy, macro and close-up view respectively.

Table 2. Chemical composition of Zn–1.3%Fe at different points shown in Figure 1b.

El Element (wt %) Tested Area	Zn	Fe
Point 1	90 ± 1	9.8 ± 0.3
Point 2	100 ± 1	0.13 ± 0.08

The X-ray diffraction analysis (Figure 2) revealed the presence of the two major phases: a pure Zn and Fe rich phase. Additional precipitating phases in the Zn–1.3%Fe alloy appear to be Delta Zn (ICDD 045-1184), with a possible contribution of Zeta Zn. This result is in accordance with the EDS analysis of the Fe-rich phase.

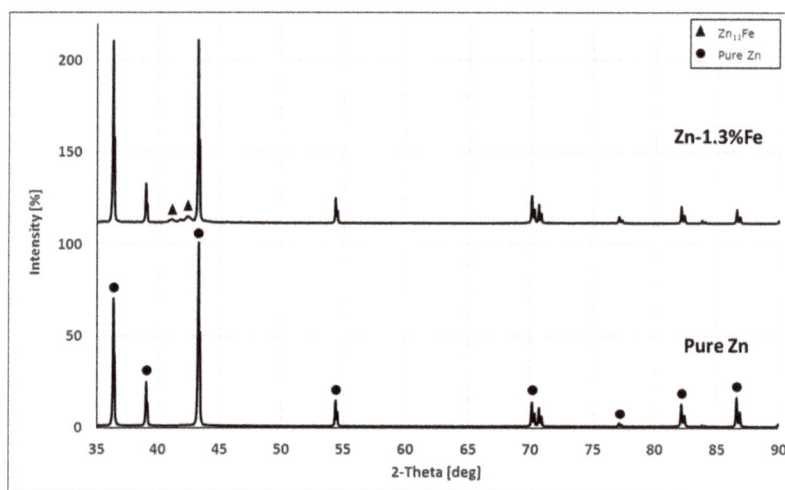

Figure 2. X-ray diffraction analysis of pure Zn and Zn–1.3%Fe alloy.

Hardness tests were conducted to begin evaluating the mechanical properties. The hardness of pure Zn and Zn–1.3%Fe alloy were 40 ± 3 and 56 ± 2 HV (hardness Vickers), respectively. The increased Zn–1.3%Fe alloy hardness is attributed to the additional Delta Zn phase.

The mechanical properties of pure Zn and Zn–1.3%Fe alloy in as-cast conditions are shown in Figure 3. While the yield strength (YS) and ultimate tensile strength (UTS) of Zn–1.3%Fe alloy was improved compared to pure Zn, the ductility was relatively reduced.

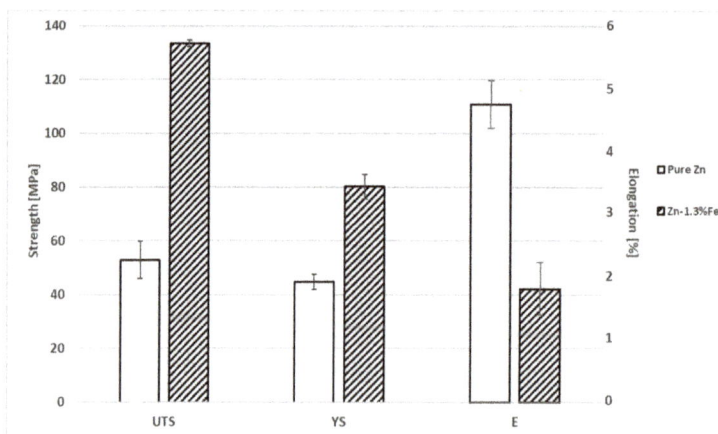

Figure 3. Mechanical properties of pure Zn and Zn–1.3%Fe.

Close-up views of the external surfaces of pure Zn and Zn–1.3%Fe alloy after immersion tests of 10 and 20 days are shown in Figure 4. While a general corrosion attack was observed in both samples, it was significantly more intense in the Zn–1.3%Fe alloy for both exposure times. The corrosion attack against pure Zn was relatively mild and did not cover the entire sample surface, the attack against Zn–1.3%Fe alloy was more severe and covered the entire sample surface.

The calculated corrosion rates for pure Zn and Zn–1.3%Fe alloy, as obtained by the immersion tests, are shown in Figure 5. The corrosion rate of the Zn–1.3%Fe alloy was nearly twice that of pure Zn after 10 and 20 days of exposure.

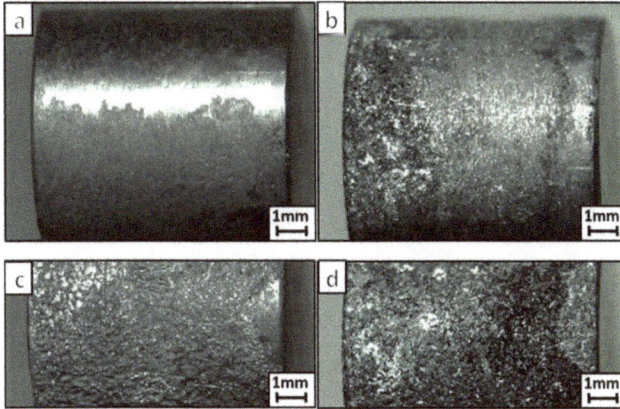

Figure 4. Optical microscopy showing corrosion attack at the surface of pure Zn and Zn–1.3%Fe alloy after immersion tests in phosphate-buffered saline (PBS) solution at 37 °C. (**a,b**) pure Zn and Zn–1.3%Fe alloy after 10 days of exposure. (**c,d**) pure Zn and Zn–1.3%Fe alloy after 20 days of exposure.

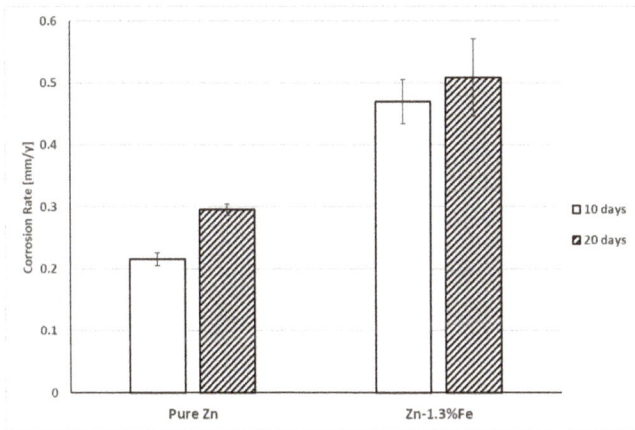

Figure 5. Corrosion rate measurements of pure Zn and Zn–1.3%Fe alloy after immersion tests in PBS solution at 37 °C.

In order to further investigate the corrosion attack mechanism in Zn–1.3%Fe alloy, close-up cross sectional views of the corroded area were inspected, as shown in Figure 6. The images revealed a uniform corrosion attack at the external surface (Figure 6a). This phenomenon is in line with the basic degradation requirements of biodegradable implants in terms of preserving their mechanical integrity during the critical period after implantation and undergoing degradation without risk of premature implant fracture during the mechanical scaffolding phase. At higher magnification (Figure 6b), intact $Zn_{11}Fe$ particles are seen in the vicinity of corroded matrix. In accordance with our hypothesis, the cathodic $Zn_{11}Fe$ particles (-0.87 Vsce vs. -1.03 Vsce for pure Zn [53]) acted as a cathodic phase to the Zn matrix, causing accelerated degradation by micro-galvanic effect. This mechanism can explain the higher corrosion rate of the Zn–1.3%Fe alloy compared to pure Zn as quantified by the immersion test.

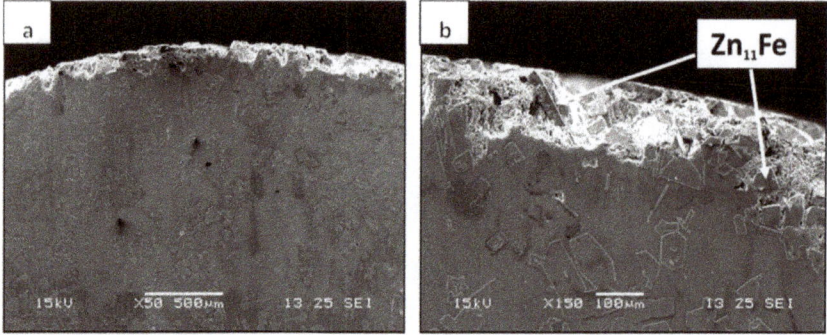

Figure 6. Surface cross section showing corrosion attack of Zn–1.3%Fe alloy after 10 days of exposure: (**a**) Macro view. (**b**) Close-up view.

The electrochemical analysis by potentiodynamic polarization is shown in Figure 7, and the corresponding Tafel extrapolation measurements are listed in Table 3. The polarization curve of the Zn–1.3%Fe alloy shifted to higher current densities compared to pure Zn, indicating a reduced corrosion resistance. Tafel extrapolation obtained from the polarization curves support this observation, identifying a corrosion rate of 0.013 millimeter per year (mm/y) for Zn–1.3%Fe compared to 0.010 mm/y for the pure Zn sample.

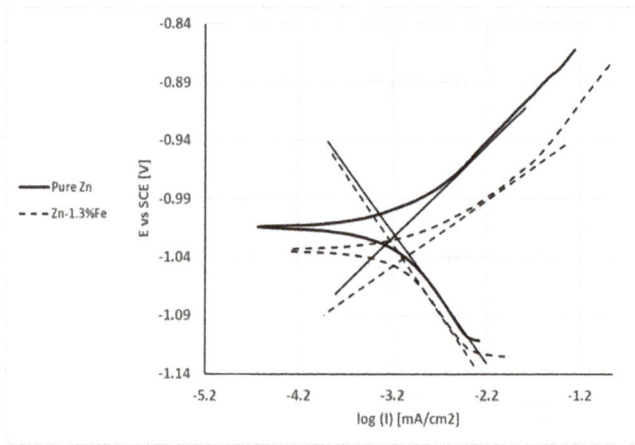

Figure 7. Potentiodynamic polarization curves obtained in PBS solution.

Table 3. Corrosion potential (E_{corr}), corrosion current (I_{corr}) and corrosion rate (C.R.) obtained by Tafel extrapolation from polarization curves shown in Figure 7.

C.R (mm/y)	Icorr ($\mu A/cm^2$)	Ecorr (V)	Parameter Specimen	C.R (mm/y)
0.010	0.67	−1.02	Pure Zn	0.010
0.013	0.89	−1.04	Zn–1.3%Fe	0.013

The impedance spectroscopy analysis in terms of impedance modifications after immersion times of 1 and 48 h in PBS solution are shown by Nyquist plots in Figure 8 and Bode plots in Figure 9. The relatively larger radius of curvature in the Nyquist diagram of the pure Zn sample indicates that its corrosion resistance was significantly increased compared to the Zn–1.3%Fe sample. Bode diagrams reveal that after one hour of exposure to the PBS solution, the corrosion resistance of both samples are nearly the same (around 10,000 Ohm), while after 48 h of exposure the corrosion resistance of the Zn–1.3%Fe sample decreased from about 8130 Ohm to 4365 Ohm. The results obtained by the potentiodynamic polarization and impedance spectroscopy analysis are in accordance with the results from the immersion tests.

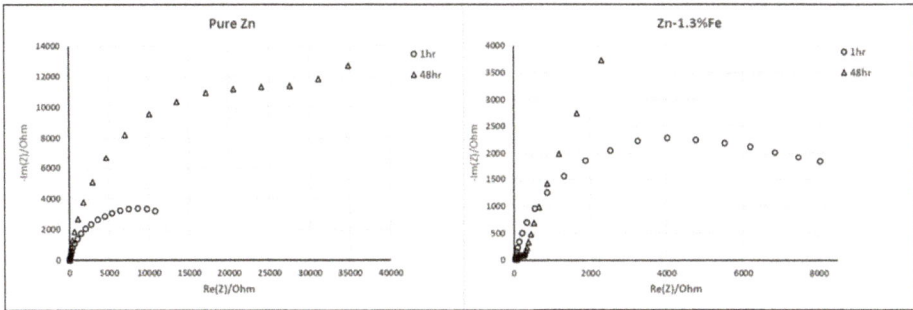

Figure 8. Nyquist plots obtained after immersion in PBS solution for up to 48 h.

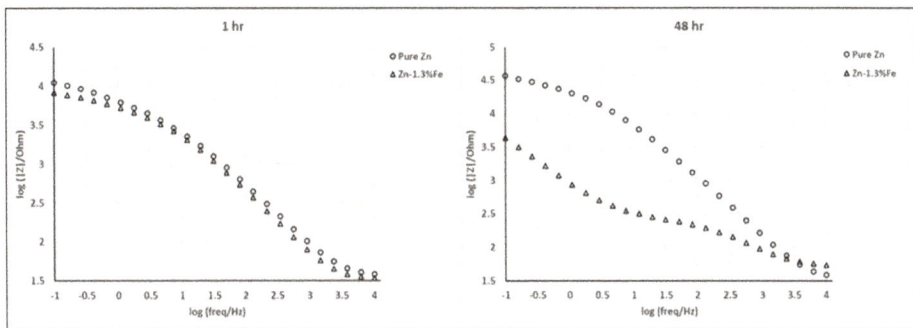

Figure 9. Bode plots obtained after immersion in PBS solution for up to 48 h.

The in vivo evaluation demonstrated a normal weight gain for all the rats irrespective of implant type Zn, Figure 10. Post-implantation behavior, cage locomotion and surgical wound appearance were all normal, with no signs of compromised health.

Zinc serum levels and blood biochemical parameters (RBC, HGB and WBC) examined before and after implantation are shown in Figures 11–14. The Zn content was slightly increased in the case of pure Zn and Zn–1.3%Fe implantations compared to titanium, as expected, but in all cases was within the normal range of 168–190 [µg/dL] [57]. Furthermore, none of the implanted samples exerted a negative effect on the hematological profiles of the rats, as all of the measurements were within the normal range [58]: RBC 7.62–9.99 [106/µL], HGB 13.6–17.4 [g/dL], WBC 1.98–11.06 [103/µL].

The corrosion rate of pure Zn and Zn–1.3%Fe implants at 14 weeks post implantation are shown in Figure 15. The corrosion rate of Zn–1.3%Fe was significantly increased relative to pure Zn.

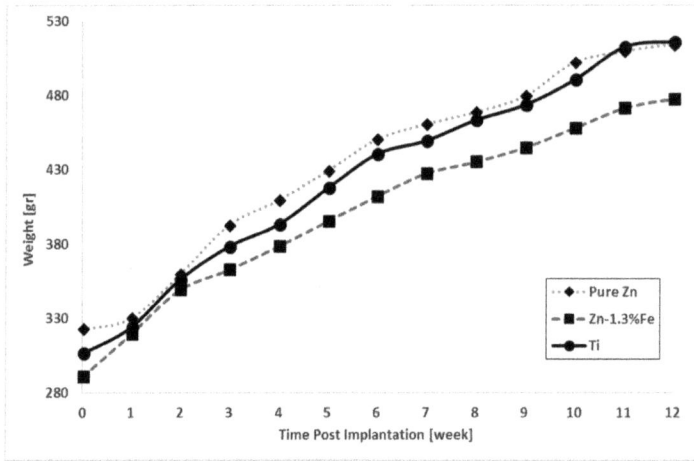

Figure 10. Rat body weights pre- and post-implantation.

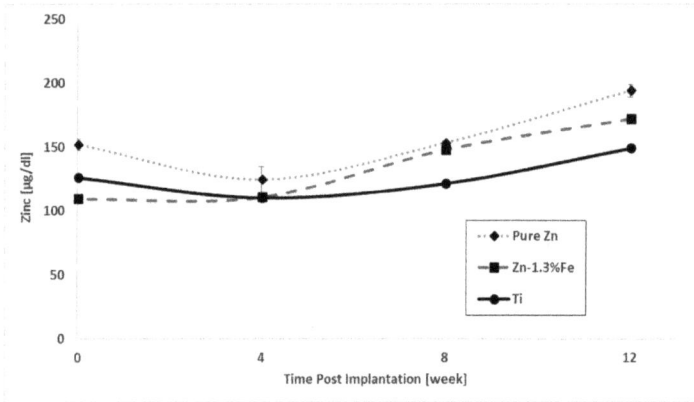

Figure 11. Serum Zinc level pre- and post-implantation.

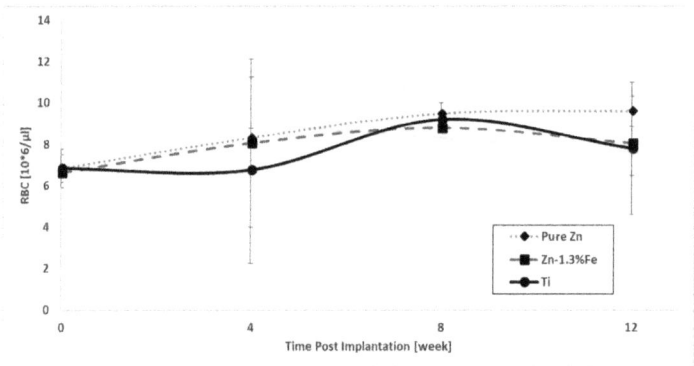

Figure 12. Red blood cells (RBC) level pre- and post-implantation.

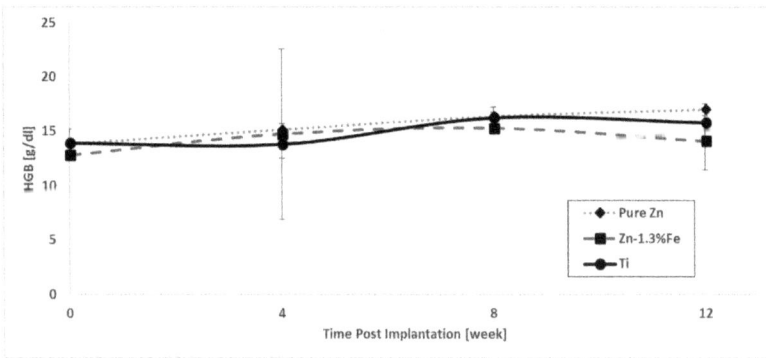

Figure 13. Hemoglobin (HGB) level pre- and post-implantation.

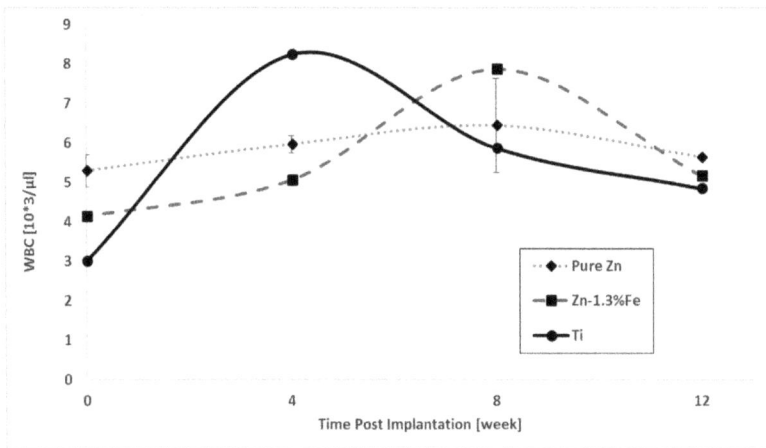

Figure 14. White blood cells (WBC) level pre- and post-implantation.

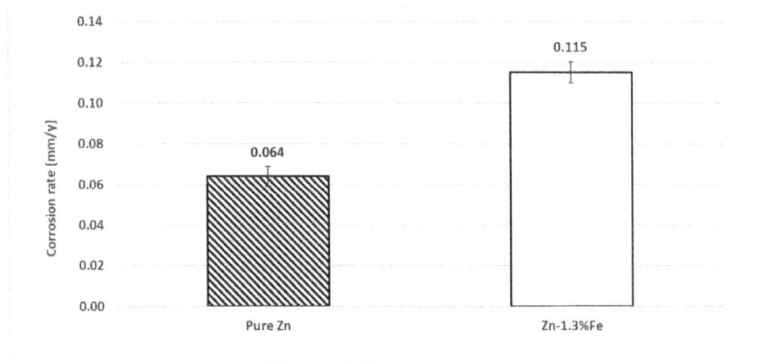

Figure 15. Corrosion rate of pure Zn and Zn–1.3%Fe implants after 14 weeks of implantation.

Histology analyses of the subcutaneous tissue around the tested implants (pure Zn, Zn–1.3%Fe and Ti–6Al–4V as reference alloy) at 14 weeks post-implantation are shown in Figure 16. None of the zinc-bearing implants promoted a negative effect on the live tissue or provoked an inflammatory response compared to the reference Ti alloy.

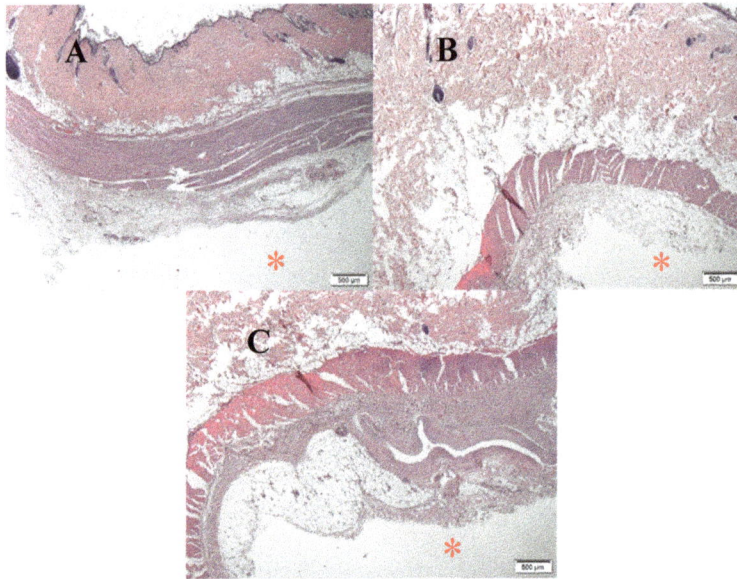

Figure 16. Histological analysis of subcutaneous tissue in the vicinity of the implants. (**A**) Pure Zn; (**B**) Zn–1.3%Fe; (**C**) Ti–6Al–4V. * marks the implants location.

4. Discussion

Biodegradable implants that strongly resist corrosion in physiological environments may elicit negative inflammatory responses, as macrophages increase their activity to metabolize foreign material [27]. Indeed, earlier in vivo work with zinc vascular implants demonstrated that the corrosion rate may be a critical regulator of inflammation [27]. In this work, we were surprised to find that the highest purity zinc implant (~0.01 wt % impurities) produced the strongest negative inflammatory response, and introductions of mainly iron impurities of ~0.3 wt % into the implant substantially ameliorated the inflammatory response. This discovery raised the intriguing prospect that inflammatory responses to zinc implants could be controlled by the implant's corrosion behavior. In the present proof of concept work, we intended to develop a zinc alloy system that imparts tunable corrosion rates (by adjusting alloying element concentration) in order to regulate inflammatory responses. Herein, we have developed such an implant material as well as characterized the effects of alloying on material properties. The ultimate goal, not pursued in the present study, is to identify an optimal corrosion rate for avoiding harmful inflammation when deploying zinc based medical implants.

Several methods can be used to increase the corrosion rate of pure zinc. These methods include surface film modification, coatings, microstructure manipulation via thermal and mechanical processing, and alloying. The intent of the present study was to use the biocompatible alloying element Fe to produce micro-galvanic effects to increase the corrosion rate of pure zinc. This approach was also adopted by other researchers, for instance in Zn–Cu base systems such as Zn–Cu, by Niu et al. [40], Zn–Cu–Mg by Tang et al. [44] and Zn–3Cu by Yue et al. [36]. The use of Cu as an alloying element to produce a micro-galvanic effect was justified by the higher standard potential of Cu (0.337 V compared to –0.763 V for Zinc) as well as the micro-galvanic effects generated by a secondary phase CuZn$_5$. The potential to achieve a micro-galvanic effect in the present study is based on the elevated electrochemical potential of all Zn–Fe precipitate phases relative to pure zinc. The potentials of Zn–Fe phases as identified by Lee et al. [53] are as follows (vs. SCE): Delta phase −0.822 V, Zeta phase −0.940 V, Gamma phase −0.772 V, while pure zinc is −1.03 V.

In-vitro examination of the gravity cast Zn–1.3%Fe alloy demonstrated a nearly 100% increase in corrosion rate compared to pure zinc. This was supported by the electrochemical analysis in terms of potentiodynamic polarization and impedance examination as well as the in vivo corrosion rate measurements at 14 weeks post implantation. The mechanical properties of pure zinc in as cast conditions obtained by this study are in line with the reported results of Gong et al. [37], and Li et al. [39]. The increased strength of Zn–1.3%Fe with concomitant reduction in ductility is attributed to the effect of Fe as an alloying element. Nevertheless, the strength and ductility of Zn–1.3%Fe alloy in the as cast condition are still relatively low. In order to improve these properties, additional metal forming processing, such as hot extrusion [59], are required to obtain an adequate structural material for practical degradable implant applications.

The in vivo study demonstrated that the pure Zn and Zn–1.3%Fe alloy avoided negative systemic effects on the wellbeing behavior and normal growth of the rats during the 12 weeks of implantation. Furthermore, the implantation of these two materials did not increase the amount of zinc in blood beyond the acceptable level. This was supported by the absence of anemia in terms of RBC and HGB levels. In addition, there were no signs of infection following implantation of either Zn base system, as there was no increase in the amount of white blood cells (WBC). This was confirmed by the histology analyses of subcutaneous tissue in the vicinity of the tested Zn base implants. Although not performed directly at the implant-tissue interface, the histology shows a general absence of harmful effects at 14 weeks post implantation from the implant or corrosion activity. The corrosion rate of pure Zn and Zn–1.3%Fe alloy under in vivo conditions was relatively reduced compared to their corrosion rate under in vitro conditions. This result is in accordance with the observations of Torne et al. [56], and Witte et al. [60] who have reported that the aggressiveness of the in vivo physiological environment towards the biodegradable metal implants is relatively lower compared to the in vitro submersion of metals under PBS solution. Further contributing to the disparity, the two cylindrical disks implanted in the midline of the back of the rats were exposed to a relatively low blood flow that practically generates a relatively reduced corrosion attack compared to the in vitro conditions.

Together, the results of the present study indicate that Zn–1.3%Fe alloy can significantly increase the corrosion rate of pure Zn. Consequently, this alloy can overcome the inherent low degradation rate of pure Zn and can be considered as a potential structural material candidate for biodegradable Zn base implants.

5. Conclusions

Alloying of Zn with Fe at a concentration of 1.3 wt % nearly doubled the corrosion rate in in vitro and in vivo conditions. The relatively higher corrosion of Zn–1.3%Fe is attributed to the microgalvanic effect generated mainly by the formation of Delta phase (Zn$_{11}$Fe). The absence of undesirable systemic effects, including weight growth, wellbeing activities and hematological characteristics (RBC, HGB and WBC) of the rats during 14 weeks of implantation, as well as obtaining adequate histology results in subcutaneous tissue close to the tested implants suggests that the Zn–1.3%Fe alloy can be considered as a potential material for biodegradable implants.

Acknowledgments: The authors would like to thank Avi Leon for his assistance in the experimental work.

Author Contributions: Eli Aghion and Alon Kafri conceived and designed the experiments; Alon kafri and Shira Ovadia performed the experiments; Eli Aghion, Alon Kafri, Jeremy Goldman and Jaroslaw Drelich analyzed the data; Eli Aghion and Alon Kafri wrote the paper.

Conflicts of Interest: The authors declare no conflict of interest.

References

1. Schinhammer, M.; Hänzi, A.C.; Löffler, J.F.; Uggowitzer, P.J. Design strategy for biodegradable Fe-based alloys for medical applications. *Acta Biomater.* **2010**, *6*, 1705–1713. [CrossRef] [PubMed]
2. Peuster, M.; Wohlsein, P.; Brugmann, M.; Ehlerding, M.; Seidler, K.; Fink, C.; Brauer, H.; Fischer, A.; Hausdorf, G. A novel approach to temporary stenting: Degradable cardiovascular stents produced from corrodible metal-results 6–18 months after implantation into New Zealand white rabbits. *Heart* **2001**, *86*, 563–569. [CrossRef] [PubMed]
3. Feng, Q.; Zhang, D.; Xin, C.; Liu, X.; Lin, W.; Zhang, W.; Chen, S.; Sun, K. Characterization and in vivo evaluation of a bio-corrodible nitrided iron stent. *J. Mater. Sci. Mater. Med.* **2013**, *24*, 713–724. [CrossRef] [PubMed]
4. Mueller, P.P.; Arnold, S.; Badar, M.; Bormann, D.; Bach, F.W.; Drynda, A.; Meyer-Lindenberg, A.; Hauser, H.; Peuster, M. Histological and molecular evaluation of iron as degradable medical implant material in a murine animal model. *J. Biomed. Mater. Res. Part A* **2012**, *100A*, 2881–2889. [CrossRef] [PubMed]
5. Liu, B.; Zheng, Y.F. Effects of alloying elements (Mn, Co, Al, W, Sn, B, C and S) on biodegradability and in vitro biocompatibility of pure iron. *Acta Biomater.* **2011**, *7*, 1407–1420. [CrossRef] [PubMed]
6. Pierson, D.; Edick, J.; Tauscher, A.; Pokorney, E.; Bowen, P.; Gelbaugh, J.; Stinson, J.; Getty, H.; Lee, C.H.; Drelich, J.; et al. A simplified in vivo approach for evaluating the bioabsorbable behavior of candidate stent materials. *J. Biomed. Mater. Res. Part B Appl. Biomater.* **2012**, *100B*, 58–67. [CrossRef] [PubMed]
7. Bowen, P.K.; Drelich, J.; Buxbaum, R.E.; Rajachar, R.M.; Goldman, J. New approaches in evaluating metallic candidates for bioabsorbable stents. *Emerg. Mater. Res.* **2012**, *1*, 237–255. [CrossRef]
8. Schümann, K.; Ettle, T.; Szegner, B.; Elsenhans, B.; Solomons, N.W. On risks and benefits of iron supplementation recommendations for iron intake revisited. *J. Trace Elem. Med. Biol.* **2007**, *21*, 147–168. [CrossRef] [PubMed]
9. Staiger, M.P.; Pietak, A.M.; Huadmai, J.; Dias, G. Magnesium and its alloys as orthopedic biomaterials: A review. *Biomaterials* **2006**, *27*, 1728–1734. [CrossRef] [PubMed]
10. Chen, D.; He, Y.; Tao, H.; Zhang, Y.; Jiang, Y.; Zhang, X.; Zhang, S. Biocompatibility of magnesium-zinc alloy in biodegradable orthopedic implants. *Int. J. Mol. Med.* **2011**, *28*, 343–348. [PubMed]
11. Dunne, C.F.; Levy, G.K.; Hakimi, O.; Aghion, E.; Twomey, B.; Stanton, K.T. Corrosion behaviour of biodegradable magnesium alloys with hydroxyapatite coatings. *Surf. Coat. Technol.* **2016**, *289*, 37–44. [CrossRef]
12. Hakimi, O.; Aghion, E. Corrosion performance of biodegradable Mg-6%Nd-2%Y-0.5%Zr produced by melt spinning technology. *Adv. Eng. Mater.* **2014**, *16*, 364–370. [CrossRef]
13. Zhang, S.; Zhang, X.; Zhao, C.; Li, J.; Song, Y.; Xie, C.; Tao, H.; Zhang, Y.; He, Y.; Jiang, Y. Research on an Mg-Zn alloy as a degradable biomaterial. *Acta Biomater.* **2010**, *6*, 626–640. [CrossRef] [PubMed]
14. Persaud-Sharma, D.; McGoron, A. Biodegradable Magnesium Alloys: A Review of Material Development and Applications. *J. Biomim. Biomater. Tissue Eng.* **2012**, *12*, 25–39. [CrossRef] [PubMed]
15. Song, G. Control of biodegradation of biocompatable magnesium alloys. *Corros. Sci.* **2007**, *49*, 1696–1701. [CrossRef]
16. Ghali, E. Testing of Aluminum, Magnesium, and Their Alloys. In *Uhlig's Corrosion Handbook*, 3rd ed.; John Wiley & Sons, Inc.: Hoboken, NJ, USA, 2011; pp. 1103–1106.
17. Hakimi, O.; Aghion, E.; Goldman, J. Improved stress corrosion cracking resistance of a novel biodegradable EW62 magnesium alloy by rapid solidification, in simulated electrolytes. *Mater. Sci. Eng. C* **2015**, *51*, 226–232. [CrossRef] [PubMed]
18. Aghion, E.; Levy, G. The effect of Ca on the in vitro corrosion performance of biodegradable Mg-Nd-Y-Zr alloy. *J. Mater. Sci.* **2010**, *45*, 3096–3101. [CrossRef]
19. Levy, G.; Aghion, E. Effect of diffusion coating of Nd on the corrosion resistance of biodegradable Mg implants in simulated physiological electrolyte. *Acta Biomater.* **2013**, *9*, 8624–8630. [CrossRef] [PubMed]
20. Cheng, J.; Liu, B.; Wu, Y.H.; Zheng, Y.F. Comparative invitro study on pure metals (Fe, Mn, Mg, Zn and W) as biodegradable metals. *J. Mater. Sci. Technol.* **2013**, *29*, 619–627. [CrossRef]
21. Bowen, P.K.; Shearier, E.R.; Zhao, S.; Guillory, R.J., II; Zhao, F.; Goldman, J.; Drelich, J.W. Biodegradable Metals for Cardiovascular Stents: From Clinical Concerns to Recent Zn-Alloys. *Adv. Healthc. Mater.* **2016**, *5*, 1121–1140. [CrossRef] [PubMed]

22. Bowen, P.K.; Drelich, J.; Goldman, J. Zinc exhibits ideal physiological corrosion behavior for bioabsorbable stents. *Adv. Mater.* **2013**, *25*, 2577–2582. [CrossRef] [PubMed]

23. Hennig, B.; Toborek, M.; McClain, C.J. Antiatherogenic properties of zinc: Implications in endothelial cell metabolism. *Nutrition* **1996**, *12*, 711–717. [CrossRef]

24. Al-Maroof, R.A.; Al-Sharbatti, S.S. Serum zinc levels in diabetic patients and effect of zinc supplementation on glycemic control of type 2 diabetics. *Saudi Med. J.* **2006**, *27*, 344–350. [PubMed]

25. Hambidge, K.M.; Krebs, N.F. Zinc Deficiency: A Special Challenge. *J. Nutr.* **2007**, *137*, 1101–1105. [CrossRef] [PubMed]

26. Plum, L.M.; Rink, L.; Hajo, H. The essential toxin: Impact of zinc on human health. *Int. J. Environ. Res. Public Health* **2010**, *7*, 1342–1365. [CrossRef] [PubMed]

27. Guillory, R.J.; Bowen, P.K.; Hopkins, S.P.; Shearier, E.R.; Earley, E.J.; Gillette, A.A.; Aghion, E.; Bocks, M.; Drelich, J.W.; Goldman, J. Corrosion Characteristics Dictate the Long-Term Inflammatory Profile of Degradable Zinc Arterial Implants. *ACS Biomater. Sci. Eng.* **2016**, *2*, 2355–2364. [CrossRef]

28. Drelich, A.J.; Zhao, S.; Guillory, R.J.; Drelich, J.W.; Goldman, J. Long-term surveillance of zinc implant in murine artery: Surprisingly steady biocorrosion rate. *Acta Biomater.* **2017**, *58*, 539–549. [CrossRef] [PubMed]

29. Bowen, P.K.; Guillory, R.J., II; Shearier, E.R.; Seitz, J.-M.; Drelich, J.; Bocks, M.; Zhao, F.; Goldman, J. Metallic zinc exhibits optimal biocompatibility for bioabsorbable endovascular stents. *Mater. Sci. Eng. C* **2015**, *56*, 467–472. [CrossRef] [PubMed]

30. Smith, S.E.; Lakson, E.J. Zinc toxicity in rats; antagonistic effects of copper and liver. *J. Biol. Chem.* **1946**, *163*, 29–38. [PubMed]

31. Piao, F.; Yokoyama, K.; Ma, N.; Yamauchi, T. Subacute toxic effects of zinc on various tissues and organs of rats. *Toxicol. Lett.* **2003**, *145*, 28–35. [CrossRef]

32. Drelich, A.J.; Bowen, P.K.; LaLonde, L.; Goldman, J.; Drelich, J.W. Importance of oxide film in endovascular biodegradable zinc stents. *Surf. Innov.* **2016**, *4*, 133–140. [CrossRef]

33. Shomali, A.A.; Guillory, R.J.; Seguin, D.; Goldman, J.; Drelich, J.W. Effect of PLLA coating on corrosion and biocompatibility of zinc in vascular environment. *Surf. Innov.* **2017**, *5*, 211–220. [CrossRef]

34. Tjellström, A.; Rosenhall, U.; Lindström, J.; Hallén, O.; Albrektsson, T.; Brånemark, P.I. Five-Year Experience with Skin-Penetrating Bone-Anchored Implants in the Temporal Bone. *Acta Otolaryngol.* **1983**, *95*, 568–575. [CrossRef] [PubMed]

35. Geetha, M.; Singh, A.K.; Asokamani, R.; Gogia, A.K. Ti based biomaterials, the ultimate choice for orthopaedic implants—A review. *Prog. Mater. Sci.* **2009**, *54*, 397–425. [CrossRef]

36. Yue, R.; Huang, H.; Ke, G.; Zhang, H.; Pei, J.; Xue, G.; Yuan, G. Microstructure, mechanical properties and in vitro degradation behavior of novel Zn-Cu-Fe alloys. *Mater. Charact.* **2017**, *134*, 114–122. [CrossRef]

37. Gong, H.; Wang, K.; Strich, R.; Zhou, J.G. In vitro biodegradation behavior, mechanical properties, and cytotoxicity of biodegradable Zn-Mg alloy. *J. Biomed. Mater. Res. Part B Appl. Biomater.* **2015**, *103*, 1632–1640. [CrossRef] [PubMed]

38. Vojtěch, D.; Kubásek, J.; Šerák, J.; Novák, P. Mechanical and corrosion properties of newly developed biodegradable Zn-based alloys for bone fixation. *Acta Biomater.* **2011**, *7*, 3515–3522. [CrossRef] [PubMed]

39. Li, H.F.; Xie, X.H.; Zheng, Y.F.; Cong, Y.; Zhou, F.Y.; Qiu, K.J.; Wang, X.; Chen, S.H.; Huang, L.; Tian, L.; et al. Development of biodegradable Zn-1X binary alloys with nutrient alloying elements Mg, Ca and Sr. *Sci. Rep.* **2015**, *5*, 10719. [CrossRef] [PubMed]

40. Zhao, S.; McNamara, C.T.; Bowen, P.K.; Verhun, N.; Braykovich, J.P.; Goldman, J.; Drelich, J.W. Structural Characteristics and In Vitro Biodegradation of a Novel Zn-Li Alloy Prepared by Induction Melting and Hot Rolling. *Metall. Mater. Trans. A Phys. Metall. Mater. Sci.* **2017**, *48*, 1204–1215. [CrossRef]

41. Bowen, P.K.; Seitz, J.M.; Guillory, R.J., II; Braykovich, J.P.; Zhao, S.; Goldman, J.; Drelich, J.W. Evaluation of wrought Zn-Al alloys (1, 3, and 5 wt % Al) through mechanical and in vivo testing for stent applications. *J. Biomed. Mater. Res. Part B Appl. Biomater.* **2018**, *106*, 245–258. [CrossRef] [PubMed]

42. Zhao, S.; Seitz, J.M.; Eifler, R.; Maier, H.J.; Guillory, R.J., II; Earley, E.J.; Drelich, A.; Goldman, J.; Drelich, J.W. Zn-Li alloy after extrusion and drawing: Structural, mechanical characterization, and biodegradation in abdominal aorta of rat. *Mater. Sci. Eng. C* **2017**, *76*, 301–312. [CrossRef] [PubMed]

43. Niu, J.; Tang, Z.; Huang, H.; Pei, J.; Zhang, H.; Yuan, G.; Ding, W. Research on a Zn-Cu alloy as a biodegradable material for potential vascular stents application. *Mater. Sci. Eng. C* **2016**, *69*, 407–413. [CrossRef] [PubMed]

44. Tang, Z.; Huang, H.; Niu, J.; Zhang, L.; Zhang, H.; Pei, J.; Tan, J.; Yuan, G. Design and characterizations of novel biodegradable Zn-Cu-Mg alloys for potential biodegradable implants. *Mater. Des.* **2017**, *117*, 84–94. [CrossRef]

45. Bakhsheshi-Rad, H.R.; Hamzah, E.; Low, H.T.; Kasiri-Asgarani, M.; Farahany, S.; Akbari, E.; Cho, M.H. Fabrication of biodegradable Zn-Al-Mg alloy: Mechanical properties, corrosion behavior, cytotoxicity and antibacterial activities. *Mater. Sci. Eng. C* **2017**, *73*, 215–219. [CrossRef] [PubMed]

46. Jin, H.; Zhao, S.; Guillory, R.; Bowen, P.K.; Yin, Z.; Griebel, A.; Schaffer, J.; Earley, E.J.; Goldman, J.; Drelich, J.W. Novel high-strength, low-alloys Zn-Mg (<0.1 wt % Mg) and their arterial biodegradation. *Mater. Sci. Eng. C* **2017**, *84*, 67–79.

47. Dambatta, M.S.; Izman, S.; Kurniawan, D.; Hermawan, H. Processing of Zn-3Mg alloy by equal channel angular pressing for biodegradable metal implants. *J. King Saud Univ. Sci.* **2017**, *29*, 455–461. [CrossRef]

48. Mostaed, E.; Sikora-Jasinska, M.; Mostaed, A.; Loffredo, S.; Demir, A.G.; Previtali, B.; Mantovani, D.; Beanland, R.; Vedani, M. Novel Zn-based alloys for biodegradable stent applications: Design, development and in vitro degradation. *J. Mech. Behav. Biomed. Mater.* **2016**, *60*, 581–602. [CrossRef] [PubMed]

49. Bagha, P.S.; Khaleghpanah, S.; Sheibani, S.; Khakbiz, M.; Zakeri, A. Characterization of nanostructured biodegradable Zn-Mn alloy synthesized by mechanical alloying. *J. Alloys Compd.* **2018**, *735*, 1319–1327. [CrossRef]

50. Xiao, C.; Wang, L.; Ren, Y.; Sun, S.; Zhang, E.; Yan, C.; Liu, Q.; Sun, X.; Shou, F.; Duan, J.; et al. Indirectly extruded biodegradable Zn-0.05wt%Mg alloy with improved strength and ductility: In vitro and in vivo studies. *J. Mater. Sci. Technol.* **2018**. [CrossRef]

51. Zhang, Y.; Li, J.; Li, J. Microstructure, mechanical properties, corrosion behavior and film formation mechanism of Mg-Zn-Mn-xNd in Kokubo's solution. *J. Alloys Compd.* **2018**, *730*, 458–470. [CrossRef]

52. Fransen, M.; Nazikkol, C. Zinc/Iron Phase Transformation Studies on Galvannealed Steel Coatings By X-ray Diffraction. *Advances* **2003**, *46*, 291–296.

53. Lee, H.H.; Hiam, D. Corrosion resistance of galvannealed steel. *Corrosion* **1989**, *45*, 852–856. [CrossRef]

54. Yun, Y.; Dong, Z.; Yang, D.; Schulz, M.J.; Shanov, V.N.; Yarmolenko, S.; Xu, Z.; Kumta, P.; Sfeir, C. Biodegradable Mg corrosion and osteoblast cell culture studies. *Mater. Sci. Eng. C* **2009**, *29*, 1814–1821. [CrossRef]

55. Jaiswal, S.; Kumar, R.M.; Gupta, P.; Kumaraswamy, M.; Roy, P.; Lahiri, D. Mechanical, corrosion and biocompatibility behaviour of Mg-3Zn-HA biodegradable composites for orthopaedic fixture accessories. *J. Mech. Behav. Biomed. Mater.* **2018**, *78*, 442–454. [CrossRef] [PubMed]

56. Törne, K.; Larsson, M.; Norlin, A.; Weissenrieder, J. Degradation of zinc in saline solutions, plasma, and whole blood. *J. Biomed. Mater. Res. Part B Appl. Biomater.* **2016**, *104*, 1141–1151. [CrossRef] [PubMed]

57. Yur, F.; Bildik, A.; Belge, F.; Kilicalp, D. Serum Plasma and Erythrocyte Zinc levels in various animal species. *VAN Vet. J.* **2002**, *13*, 82–83.

58. Quesenberry, K.E.; Carpenter, J.W. Small Rodents. In *Ferrets, Rabbits and Rodents: Clinical Medicine and Surgery*; Elsevier: St. Louis, MO, USA, 2003; p. 348.

59. Sikora-Jasinska, M.; Mostaed, E.; Mostaed, A.; Beanland, R.; Mantovani, D.; Vedani, M. Fabrication, mechanical properties and in vitro degradation behavior of newly developed Zn Ag alloys for degradable implant applications. *Mater. Sci. Eng. C* **2017**, *77*, 1170–1181. [CrossRef] [PubMed]

60. Witte, F.; Fischer, J.; Nellesen, J.; Crostack, H.A.; Kaese, V.; Pisch, A.; Beckmann, F.; Windhagen, H. In vitro and in vivo corrosion measurements of magnesium alloys. *Biomaterials* **2006**, *27*, 1013–1018. [CrossRef] [PubMed]

metals

MDPI

Article

Preparation and Characterization of Zinc Materials Prepared by Powder Metallurgy

Michaela Krystýnová [1,*], Pavel Doležal [1,2], Stanislava Fintová [1,3], Matěj Březina [1], Josef Zapletal [2] and Jaromír Wasserbauer [1]

[1] Materials Research Centre, Faculty of Chemistry, Brno University of Technology, Purkyňova 464/118, 612 00 Brno, Czech Republic; dolezal@fme.vutbr.cz (P.D.); fintova@ipm.cz (S.F.); xcbrezinam@fch.vut.cz (M.B.); wasserbauer@fch.vut.cz (J.W.)

[2] Institute of Materials Science and Engineering, Faculty of Mechanical Engineering, Brno University of Technology, Technická 2896/2, 616 69 Brno, Czech Republic; zapletal@fme.vutbr.cz

[3] Institute of Physics of Materials Academy of Sciences of the Czech Republic, Žižkova 22, 616 62 Brno, Czech Republic

* Correspondence: xckrystynovam@fch.vut.cz; Tel.: +420-541-149-469

Received: 30 August 2017; Accepted: 22 September 2017; Published: 27 September 2017

Abstract: The use of zinc-based materials as biodegradable materials for medical purposes is offered as a possible alternative to corrosion-less resistant magnesium-based materials. Zinc powders with two different particle sizes (7.5 µm and 150 µm) were processed by the methods of powder metallurgy: cold pressing, cold pressing followed by sintering and hot pressing. The microstructure of prepared materials was evaluated in terms of light optical microscopy, and the mechanical properties were analyzed with Vickers microhardness testing and three-point bend testing. Fractographic analysis of broken samples was performed with scanning electron microscopy. Particle size was shown to have a significant effect on compacts mechanical properties. The deformability of 7.5 µm particle size powder was improved by increased temperature during the processing, while in the case of larger powder, no significant influence of temperature was observed. Bending properties of prepared materials were positively influenced by elevated temperature during processing and correspond to the increasing compacting pressures. Better properties were achieved for pure zinc prepared from 150 µm particle size powder compared to materials prepared from 7.5 µm particle size powder.

Keywords: powder metallurgy; zinc; cold pressing; sintering; hot pressing; mechanical properties

1. Introduction

In the field of biodegradable metal materials, magnesium and its alloys are the most studied [1–6]. Due to their suitable mechanical properties such as high specific strength, stiffness and damping ability, they are suitable materials for the preparation of bone implants [1–7]. Because of its low corrosion resistance and therefore difficult degradation control, magnesium is not used in its pure state, but alloyed. Commercial magnesium alloys are not primarily designed for medical applications, but some studies on the biomedical purpose of these alloys were done before, showing good corrosion behavior and biocompatibility [8,9]. For medical purposes, magnesium alloys with calcium, zinc, rare earth elements or manganese are being considered [10–13]. Due to the low density of calcium, these alloys have a similar density as a natural bone. In addition, Ca^{2+} ions are beneficial for human bones, and Mg^{2+} ions support the function of Ca^{2+} ions and generally the treatment of injury [11,14,15].

Zinc has been considered, in the field of biomedical applications, a suitable alloying element for magnesium alloys in terms of improving corrosion resistance and enhancement of mechanical properties [10,16,17]. Considering zinc is a nobler metal than magnesium and zinc is also biocompatible, it can be a suitable biodegradable material [16–18]. Zinc is a nutritionally-essential element in the human body; approximately 85% of Zn in the human body can be found in bone and muscle. It is an integral part of the structure of macromolecules and enzymes, and it participates in a large number of enzymatic reactions. The human zinc requirement for adult males is 10–15 mg/day (upper limit 40 mg/day) [19,20]. The potential for systemic toxicity of metallic zinc should be nonexistent due to the rapid transport of ionic zinc in living tissue. Moreover, higher consumption (up to 100 mg/day) of zinc is considered non-toxic. However, more detailed cytotoxicity tests must be carried out in the future [15,21–23].

Based on its positive effect on the human body, zinc can be considered a suitable base material for the preparation of biodegradable materials. There are only a few works dealing with the use of zinc alloys for biomedical applications. In the case of pure zinc, there is just one article [22] dealing with the applicability of zinc as a bioabsorbable cardiac stent material, according to the author's knowledge. The authors in [22] demonstrated the biocorrosion behavior of zinc by a series of four wire samples implanted in the abdominal aorta of Sprague-Dawley rat for 1.5, 3, 4.5 and 6 months. Results show that zinc corrosion is slower than the corrosion of magnesium and its alloys [24,25]. Zinc implants stay almost intact for approximately four months, and then, corrosion subsequently accelerates. Corrosion products of zinc are formed from zinc oxide and zinc carbonate. Due to the slow corrosion of zinc, there is sufficient time to exclude hydrogen resulting from the corrosion process. That is both the main difference and an advantage over magnesium [21]. However, no mechanical tests results were provided in this work.

Generally, the mechanical properties of zinc are relatively low. The modulus of compressive elasticity of commercially pure zinc is not listed as an exact value because there is no region of strict proportionality in the compressive stress-strain curve. Therefore, this value was determined in the range from 70–140 GPa [25]. The Vickers hardness of this material is 30 HV. The tensile strength of wrought pure zinc is in the range of 120–150 MPa, depending on the direction of rolling [26]. The tensile strength of cast pure zinc is very low with the value approximately 25 MPa [17]. Another article dealing with mechanical properties of zinc showed that applying hot extrusion (300 °C, extrusion ratio 10:1, 2 mm/min) can enhance the ultimate tensile strength to a value of approximately 100 MPa and a Vickers hardness to 44 HV5 [27] and, in [28], to approximately 110 MPa for hot extruded pure zinc material (250 °C, 14:1). Due to the poor mechanical properties, alloying of zinc is a suitable solution. As alloying elements, mostly magnesium [17,27,29], aluminum [30,31] and silver [28] were studied.

Alloying zinc with aluminum provides substantial enhancement of the mechanical properties. That is due to the presence of Al-Zn solid solution and the volume fraction of the lamellar microconstituents from the monotectoid reaction acting as potential barriers for dislocation motion. With an addition of 5.5 wt % of aluminum, the tensile strength of material prepared by hot rolling at 350 °C reaches the value of approximately 308 MPa and a yield strength of approximately 240 MPa [30]. However, the use of Zn-Al alloys in medicine is still limited, because of the uncertainty regarding the toxicity of aluminum. Although, the toxicity of aluminum has never been sufficiently proven [30], its connection to neurological disorders is still discussed in the literature [32]. Furthermore, the corrosion resistance of Zn-Al alloys tended to be lower compared to high purity zinc due to the intergranular corrosion. Moreover, volume expansion associated with the formation of corrosion products led to cracking and fragmentation of implants [30]. Another alloying element with positive influence on mechanical properties and preserving biocompatibility is silver. With the addition of 7.0 wt % of Ag, the ultimate tensile strength of the cast zinc alloy increased up to 287 MPa [28]. Moreover, silver has been used in medicine for healing wounds, and it is used, in the form of nanoparticles, for the prevention of the adherence of bacteria to the surface of implants [33]. However, the presence of secondary phase particles in the Zn-Ag alloys led to micro-galvanic corrosion because they act as anodes. Consequently, Zn-Ag alloys have higher degradation rates in comparison with pure Zn [28].

Probably the most discussed alloying element of zinc-based biodegradable materials is magnesium. Because of the existence of the hard Mg_2Zn_{11} intermetallic phase in the alloy structure, the value of the hardness increases with a higher content of magnesium up to the value of 200 HV (cast alloy, 3 wt % of magnesium). With higher magnesium content (35–45 wt % of magnesium), the hardness of the cast alloys can reach values of approximately 285–300 HV1 for 35 wt % of magnesium (depending on the cast product cooling rate and subsequent microstructure). The hardness increases because of the presence of another strengthening phase in the alloy microstructure ($MgZn_2$). With the higher content of magnesium in the alloy, the content of the $MgZn_2$ phase decreases and the content of the $MgZn$ and Mg_7Zn_3 phases increases. The hardness falls again due to decreasing content of $MgZn_2$ to 255–280 HV1 for the alloy with 45 wt % of magnesium (depending on the cast product cooling rate and subsequent microstructure) [29]. However, the existence of brittle eutectic phases in alloys with the content of magnesium higher than 1 wt % has a negative effect on ultimate tensile strength. Ultimate tensile strength increases up to 150 MPa (1 wt % of Mg) and then decreases to the value of 30 MPa (3 wt % of Mg). That is the same value as for pure zinc prepared by the same method. Furthermore, elongation of Zn-Mg alloys reaches the highest value for the alloy with 1 wt % of magnesium [17]. Another article deals with an enhancement of mechanical properties with hot extrusion processing. Hot extruded alloy with 0.8 wt % of magnesium reaches the ultimate tensile strength of approximately 300 MPa and a Vickers hardness of approximately 80 HV5. With the content of 1.6 wt % of magnesium, hot extruded materials reach the ultimate tensile strength of approximately 360 MPa and a Vickers hardness of approximately 97 HV5 [27].

Most of the available articles deal with the influence of alloying elements and their content on the mechanical properties and corrosion resistance of zinc-based materials prepared by casting or mechanical treatment of cast products. The influence of the preparation method and its parameters on mechanical properties is studied only marginally [17,27–30].

This work deals with three different methods of preparation of zinc materials by powder metallurgy: cold pressing, cold pressing followed by sintering and hot pressing. According to the author's knowledge, there is no available publication related to the preparation of pure zinc by these methods. Besides the influence of the preparation method, this work also focuses on the influence of the particle size of the used powder materials on the resulting microstructure and mechanical properties.

2. Materials and Methods

For the experiment, two different zinc powders (99.8% purity, 7.5 μm and 150 μm mean particle size provided by Goodfellow Cambridge Limited Company (Huntingdon, UK) were used. The smaller, 7.5 μm particle size powder (Zn7.5) was prepared by the electrothermal process, and the larger, 150 μm particle size powder (Zn150) was prepared by air atomization. In order to prevent material oxidation, manipulation with Zn powder was carried out in an inert atmosphere (N_2) in the glove box. The size and shape of metal powder particles were verified and analyzed by a Zeiss Evo LS 10 scanning electron microscope (SEM; Carl Zeiss Ltd., Cambridge, UK).

The particles of Zn powder with the declared particle size of 7.5 μm had spherical shapes, and their size ranged from 1 μm to a maximum value of 20 μm (Figure 1a). The particles tended to clump. The smaller particles attached to the larger ones and created clusters. The particles of zinc powder with a declared particle size of 150 μm were irregularly rod-shaped with a minor amount of round-shaped particles (Figure 1b). The SEM analysis of the powder showed that some particles in their largest dimension reached a size from 40–640 μm.

The purity of powders declared by the supplier was verified by EDS analysis. The larger content of oxygen, approximately 8.0 ± 0.5 wt %, was determined in Zn7.5 powder when compared to Zn150 powder, containing only 2.5 ± 0.5 wt %.

Experimental samples were processed in three ways: (i) By bidirectional cold pressing (CP), (ii) cold pressing followed by sintering (CP-S) and (iii) hot pressing (HP). A hollow cylindrical steel die with the inner diameter of 20 mm was used for zinc powders' compaction. Before it was filled with zinc, the die surface was carburized in order to prevent adhesion of the base powder material to the surface of the die. The pressing of the powders was carried out by the Zwick Z250 Allround-Line universal testing machine (Zwick GmbH & Co.KG, Ulm, Germany) with a velocity of 2 mm/min. Pressures of 100, 200, 300, 400 and 500 MPa were used for prepared powders' compaction.

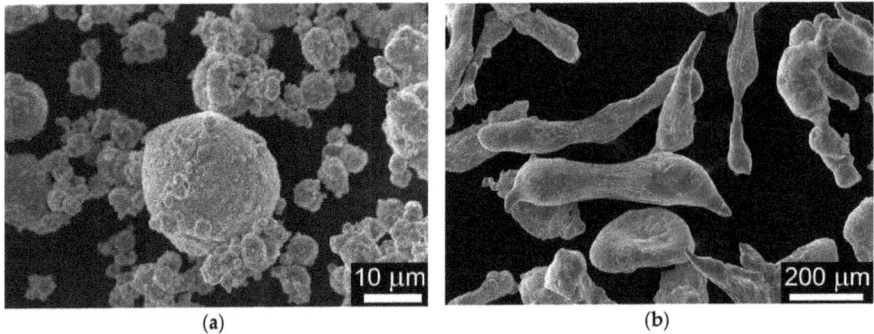

(a)   (b)

Figure 1. Zn particles, SEM (scanning electron microscope): (**a**) Mean particle size 7.5 μm; (**b**) mean particle size 150 μm.

For sintering, cold-pressed samples pressed under 100, 200, 300, 400 and 500 MPa were used. Cold-pressed samples were inserted into glass vials and then filled with argon of purity 4.6 and sealed. The sintering of pressed powder was done for 1 h at a temperature of 400 °C in the laboratory furnace preheated to the required temperature.

Hot pressing was carried out for 1 h at a temperature of 400 °C under 100, 200, 300, 400 and 500 MPa. Prepared bulk materials were in the form of cylindrical tablets with a diameter of 20 mm and a height of 5 mm.

Metallographic evaluation of prepared samples was performed in a conventional manner. Isopropanol was used as a lubricant and a rinse to prevent oxidation of the samples during grinding and polishing. For metallographic evaluation, the Zeiss Axio Z1M (Carl Zeiss AG, Oberkochen, Germany) inverted light optical microscope (OM) and Zeiss Evo LS 10 (Oxford Instruments, Abington, UK) were used. The microhardness of prepared samples was measured with the LM 248 at machine produced by LECO Company (Saint Joseph, MO, USA). The measurement was carried out in accordance with the ISO 6507-1 standard (Vickers method, applied load 25 g) on 10 positions on the sample.

The 3-point bend test was carried out on the Zwick Z020 (Zwick GmbH & Co.KG, Ulm, Germany) universal testing machine according to the ISO 7438 standard. One sample for each preparation condition was used for the 3-point bend test. Samples for testing were detracted from the central part of prepared tablets and ground up to 4 mm × 4 mm × 18 mm proportions with the support span of 16 mm. Fractographic evaluation of the fracture surfaces of broken samples was performed by SEM. Documentation was carried out in the area of applied tensile stress.

3. Results

3.1. Processed Material Analysis

Microstructures of materials prepared from Zn7.5 applying minimal and maximal (100 and 500 MPa) compacting pressures are summarized in Figure 2. The results showed the low deformation ability of the Zn7.5 powder particles and an open porosity of materials prepared under the low compacting pressure of 100 MPa. The porosity of the samples compacted under 100 MPa decreased with applying temperature. While in the case of CP, quite a high level of porosity was observed, in the case of the CP-S sample, the porosity level decreased, and only very limited porosity was observed for the HP Zn7.5 powder sample. Low compacting pressure resulted also in low handling strength of samples, and falling off of individual particles during the metallographic sample preparation was observed (marked by arrows). The HP Zn7.5 powder sample microstructural analysis revealed small particle deformation during the bulk material processing, which was not observed in the case of CP and CP-S samples compacted under 100 MPa.

The microstructure of materials pressed under 500 MPa shows much lower porosity than samples pressed under 100 MPa, considering the same processing conditions (Figure 2). Only small particles' deformation can be seen in the case of the CP sample compacted under 500 MPa (Figure 2b); however, by the addition of temperature (CP-S and HP), the particles' deformation increased (Figure 2e,f).

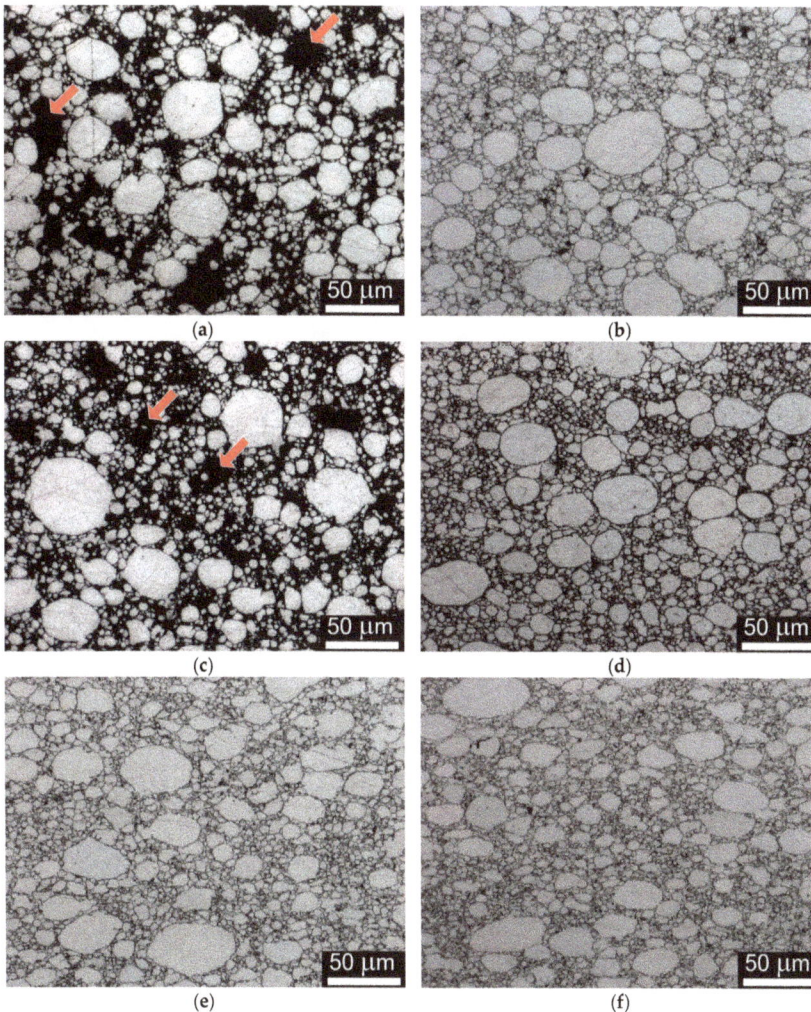

(a)   (b)   (c)   (d)   (e)   (f)

Figure 2. Microstructure of materials prepared from Zn7.5 powder, etched with 5% Nital, optical microscope (OM): (**a**) 100 MPa, CP; (**b**) 500 MPa, CP; (**c**) 100 MPa, CP-S; (**d**) 500 MPa, CP-S; (**e**) 100 MPa, HP; (**f**) 500 MPa, HP (CP: Cold pressing; S: Sintering; HP: Hot pressing).

The microstructure of materials prepared from Zn150 powder under 100 and 500 MPa compacting pressures is shown in Figure 3. In the case of material compacted under a pressure of 100 MPa, closed pores can be seen between the powder particles in the microstructure (Figure 3a). Even larger porosity can be seen in the CP-S sample's microstructure (Figure 3c), while the porosity is not visible in the case of the HP sample (Figure 3e). Compacting pressure of 500 MPa resulted in a microstructure without defects for all used processing methods of Zn150 powder-based materials (Figure 3). Zn150 powder particles' deformation was observed for all the material processing methods and all the used compacting pressures. Particles deformation increased with the addition of temperature, i.e., larger particles' deformation was observed for material processed by CP-S and even larger for material processed by HP when compared to CP material. Larger particle deformation was observed for samples prepared under 500 MPa compared to samples prepared under 100 MPa.

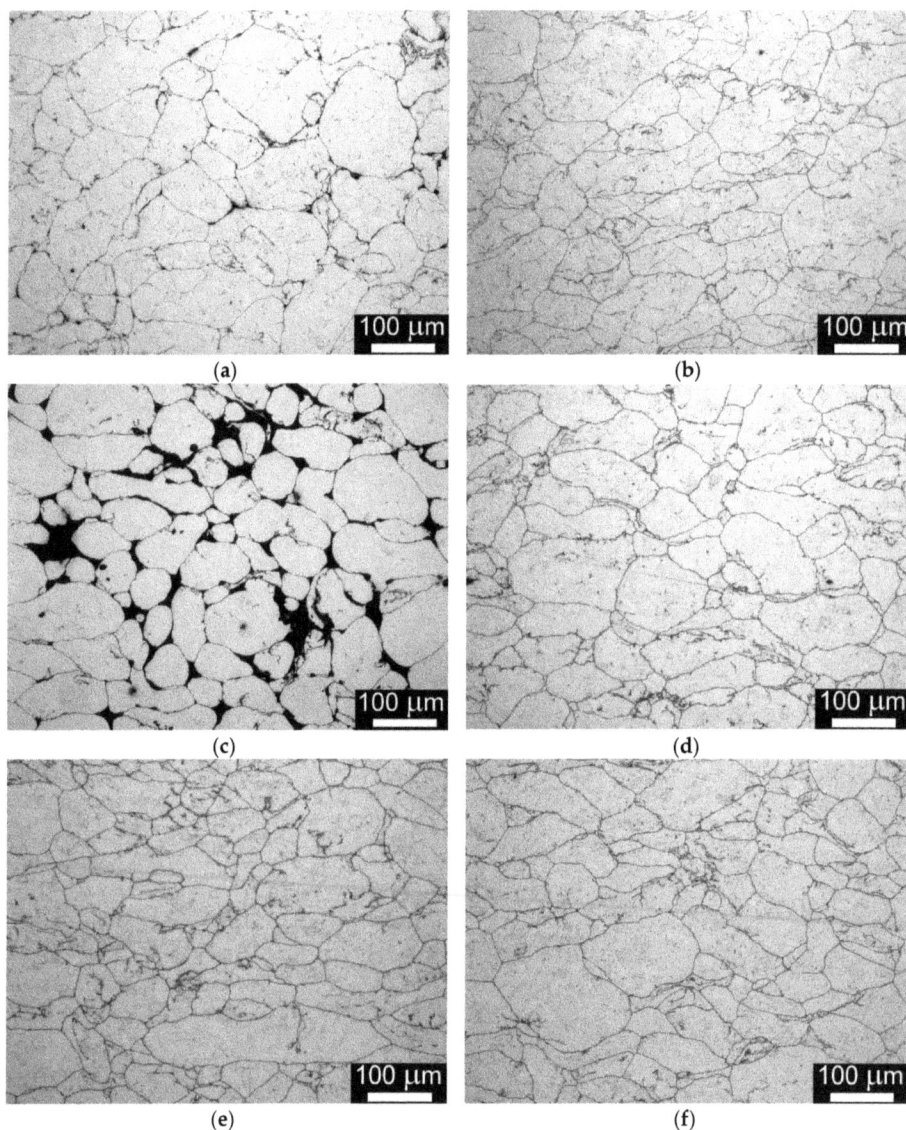

Figure 3. Microstructure of materials prepared from Zn150 powder, etched with 5% Nital, OM: (**a**) 100 MPa, CP; (**b**) 500 MPa, CP; (**c**) 100 MPa, CP-S; (**d**) 500 MPa, CP-S; (**e**) 100 MPa, HP; (**f**) 500 MPa, HP.

The values of microhardness of prepared materials were in the range from 40–49 HV025 for Zn7.5 powder and from 40–45 HV025 for Zn150 powder. No significant dependence on compacting pressure or processing method was found.

3.2. 3-Point Bend Test

The plots representing the evolution of the flexural strength and the maximal displacement before the failure are shown in Figure 4. In some cases, no data are plotted due to the low handling strength of prepared samples or very low obtained values (impossible to determine).

In the case of materials prepared from Zn7.5 powder, there is a clear increasing dependence of flexural strength on the applied compacting pressure. The same tendency of materials prepared from Zn150 powder processed by CP and CP-S can be observed. The different behavior appears only in the case of Zn150 powder HP material where the value of the flexural strength of material processed under 100 and 200 MPa is around 200 MPa, and then, the value decreased to approximately 150 MPa for compacting pressures of 300 and 400 MPa. Material pressed under 500 MPa reaches the value of about 200 MPa again. Generally, CP and CP-S materials prepared from Zn150 powder have higher flexural strength values than materials from Zn7.5 powder prepared with the same method at the same compacting pressures. In the case of materials prepared under compacting pressure of 500 MPa from Zn7.5 powder, the flexural strength increases from 20 MPa for CP to 76 MPa for CP-S material. The flexural strength of materials prepared from Zn150 powder under the same condition increases from 88 MPa for CP to 148 MPa for CP-S material.

The values of the flexural strength of materials prepared by HP are substantially higher when compared to the CP-S and CP materials, especially in the case of materials prepared from Zn7.5 powder. The highest value of flexural strength (322 MPa) is reached with material prepared by HP under compacting pressure of 400 MPa. While in the case of material from Zn7.5 powder, a significant increase of flexural strength is observed for all the applied compacting pressures due to the HP processing (Figure 4a), in the case of Zn150 powder material, no such significant increase of flexural strength is observed, and in the case of compacting pressure of 400 MPa, the values of flexural strength for CP-S and HP material are similar (Figure 4b).

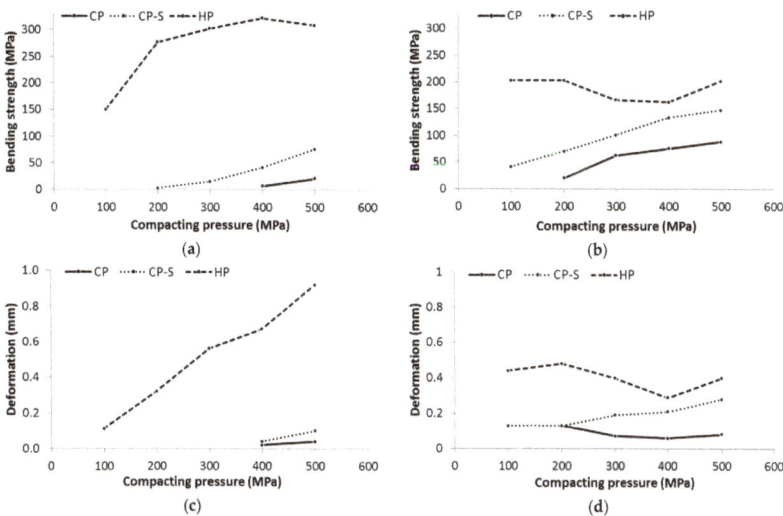

Figure 4. Dependence of flexural strength and displacement on compacting pressure and processing of prepared materials: (**a**) Flexural strength, Zn7.5 powder; (**b**) Flexural strength, Zn150 powder; (**c**) Displacement, Zn7.5 powder; (**d**) Displacement, Zn150 powder.

The displacement before the failure of Zn7.5 powder materials prepared by CP and by CP-S is very low (from 0.04–0.10 mm for 500 MPa compacting pressure) (Figure 4c). However, in the case of HP Zn7.5 powder materials, the displacement before the failure lies in the range from 0.11 mm for the lowest compacting pressure up to 0.92 mm for the pressure of 500 MPa.

The results of maximal displacement before the failure of materials prepared from Zn150 powder (Figure 4d) are relatively steady around the value of 0.2 mm for CP and CP-S materials and around 0.4 mm for HP materials. In the case of HP Zn150 powder material, the values of displacement before failure decrease from 0.49 mm to 0.29 mm for materials compacted under 200 MPa and 400 MPa, respectively (equivalent to the change of flexural strength) and then again increases for material pressed under 500 MPa (Figure 4d).

3.3. Fractographic Analysis

The samples broken during the three-point bend testing were examined with the aim to analyze the fracture mechanism and its correlation with the samples' processing in terms of SEM. Details of fracture surfaces of materials prepared from Zn7.5 powder are shown in Figure 5. Only samples prepared under the lowest and the highest compacting pressures (as possible) with adequate processing were analyzed. The fracture surface of CP material is characterized by crack growth along powder particle boundaries, which is also the case for the material compacted under 200 MPa and sintered at 400 °C (Figure 5a–c). The fracture mechanism can be considered as transgranular. There is visible porosity between individual powder particles in the case of CP materials and CP-S materials compacted under 200 MPa. The fracture surface of material pressed under 500 MPa and then sintered shows lower porosity compared to the previous material states. In addition, crack growth through particle boundaries is accompanied by powder particle cracking (Figure 5d). The combination of transgranular with more pronounced intergranular fracture is characteristic for HP materials. Broken powder particles are characterized by plane facet or by fine non-well developed cleavage facets with a river-like morphology. HP resulted also in the decrease of a number of pores present on the samples' fracture surface, while no visible pores are present on the fracture surface of HP material compacted under 500 MPa (Figure 5f) (showing also the detail of the fracture mechanism).

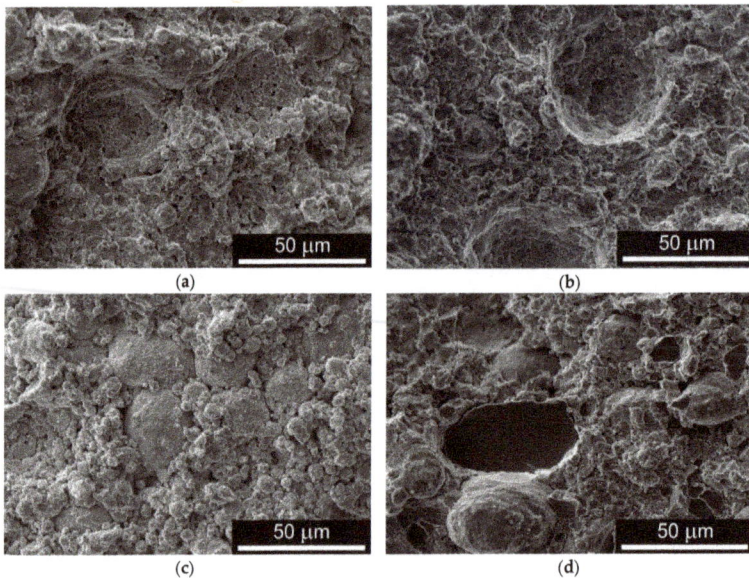

Figure 5. *Cont.*

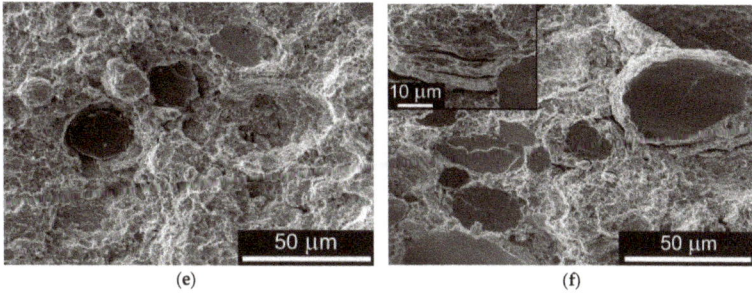

Figure 5. Microstructure of materials prepared from Zn7.5 powder, SEM: (**a**) 400 MPa, CP; (**b**) 500 MPa, CP; (**c**) 200 MPa, CP-S; (**d**) 500 MPa, CP-S; (**e**) 100 MPa, HP; (**f**) 500 MPa, HP.

In Figure 6, a summary of the details of the fracture surfaces of materials prepared from Zn150 powder is shown. Compared with materials prepared from Zn7.5 powder, in the case of Zn150 powder materials, there is a visibly higher influence of compacting pressure on particles' deformation and porosity. In the case of CP materials, the particles are deformed and mechanically bonded (Figure 6a,b). Three-point bend test loading led to particles being pulled out of the structure, while holes remaining on the fracture surface can be seen instead of missing particles in Figure 6a. Similarly to the Zn7.5 powder materials, the transgranular fracture mechanism is characteristic for CP and CP-S samples compacted under 200 MPa (Figure 6a–c). The combination of transgranular with intergranular fracture is characteristic for CP-S compacted under 500 MPa and HP materials (Figure 6d–f). Well-developed cleavage facets with the river-like morphology can be observed on fracture surfaces (broken powder particles) of HP materials (Figure 6e,f), showing also the detail of the fracture mechanism in Figure 6f.

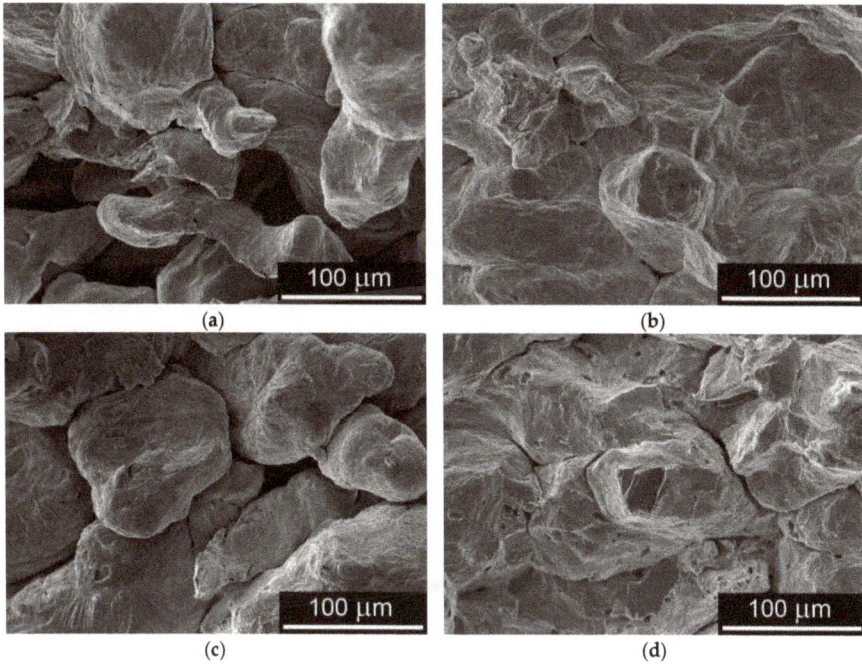

Figure 6. *Cont.*

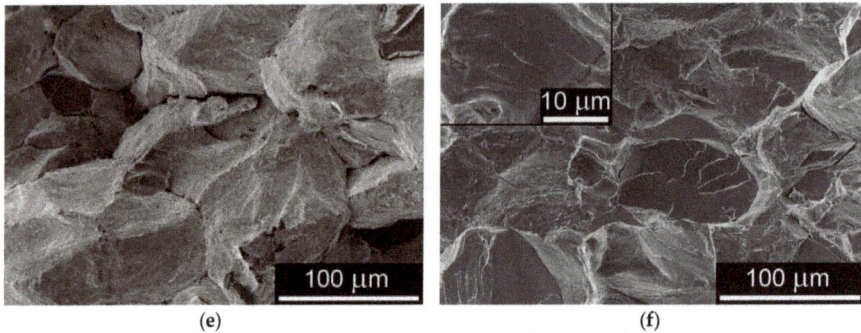

(e)   (f)

Figure 6. Microstructure of materials prepared from Zn150 powder, SEM: (**a**) 200 MPa, CP; (**b**) 500 MPa, CP; (**c**) 100 MPa, CP-S; (**d**) 500 MPa, CP-S; (**e**) 100 MPa, HP; (**f**) 500 MPa, HP.

4. Discussion

Microstructural analysis of materials prepared from Zn7.5 powder (Figure 2) revealed lower deformability of powder particles when compared to Zn150 powder particles (Figure 3). The considerably lower particle size and spherical shape of Zn7.5 powder may be the reason for this behavior.

Even though the mechanical properties of pure zinc powder particles are comparable, the size of the particles plays a similar role as in the case of grain size. The larger size of the particles was connected with their higher deformability during the processing, and the lower particle size resulted in grain boundaries' (particles boundaries) strengthening of material. The strengthening of material was also connected with lower deformability of the material.

The deformability of individual powder particles influenced the material properties and content of porosity. Although the zinc particles were not significantly deformed in some of the cases (mostly Zn7.5, Figure 2), the porosity of all the prepared materials decreased with increasing compacting pressure and even decreased with the application of temperature during the material processing (CP-S and HP). This fact can be attributed to a wide distribution of powder particle size, when the space between large particles was filled up with the smaller ones. However, the porosity of materials from Zn7.5 powder was substantially higher than in the case of material prepared from Zn150 powder, which can be also observed from the fractographic evaluation of broken samples (Figures 5 and 6). This can be explained by the larger range of the powder particle size measured for Zn150 powder (up to 640 µm) compared to Zn7.5 powder (up to 20 µm) and the adequate arrangement of the particles with different size in the compacted sample volume.

CP-S processing of Zn7.5 powder materials did not influence the material microstructure with as much intensity as the increasing compression pressure did. Only the minor effect of the following sintering of CP materials was observed by metallographic analysis (Figure 2). In the case of Zn150 powder materials, the following sintering of CP samples led to the increase of the porosity of materials prepared at low pressures (Figure 3c). Due to the high sintering temperature of 400 °C and high material porosity (of CP samples), the powder particles reduced the surface energy and changed their shape from irregular rod-shaped to spherical during sintering. At the same time, no dimensional changes of prepared samples were observed, which would generally correspond if a solid-state diffusion was the primary sintering mechanism [34]. The shape changes of particles and no shrinkage could lead to pores growing between the particles.

The effect of HP processing is more evident on the Zn7.5 powder-based materials when compared to the Zn150 powder materials. Elevated temperature during the material processing resulted in improved plasticity and enhanced deformability of fine powder particles (Zn7.5 powder). The particles' deformation is observable on materials' microstructures documented in Figure 2e,f. Elevated temperature applied during material processing resulted in activation of more slip systems in the hexagonal close packed (HCP) structure (also characteristic for zinc) [34]. In the case of Zn150 powder materials, the elevated temperature of the processing had no evident influence on material microstructure. In the case of smaller powder particles, the improved deformability of the material due to the activation of more slip systems was much more significant compared to the larger powder particles, which were deformable even at room temperature. Quite good deformability of large powder particles was not significantly improved and remained the same as in the case of CP and CP-S materials from the microstructural point of view, Figure 3.

There is only a small number of articles dealing with evaluation of the mechanical properties of pure zinc [17,25–28]. From the mechanical properties point of view, materials in these studies were characterized mostly by Vickers hardness and tensile tests. The Vickers hardness of experimental materials prepared from Zn7.5 and Zn150 powders in the presented work was determined as 40–49 HV025 and 40–45 HV025, respectively. The values were higher in comparison to the cast pure zinc of 30 HV given in [17]; however, the values were comparable with results of Vickers hardness testing of hot extruded pure zinc of 44 HV5 given in [27]. Due to the obtained microstructure of materials prepared by powder metallurgy, the characteristic mechanical properties should correlate more with the properties of wrought materials than with cast materials. Considering the powder particles' boundaries as grain boundaries, the microstructure of the materials processed from Zn7.5 powder (knowing the real particle size was up to 20 μm) is comparable with the fine recrystallized equi-axed grains with an average grain size of 20 μm characteristic for pure extruded zinc analyzed in [27].

The resistance of prepared powder-based materials against the fracture during the three-point bend test was (besides particles deformability) influenced mainly by the mechanical interlocking of irregularities on the powder particle surfaces, which was promoted by plastic deformation during the pressing. The bonding of particles was responsible for the mechanism of the bulk material fracture. Because of the spherical shape of Zn7.5 powder particles, the flexural strength of compacted samples showed lower values when compared to materials prepared from Zn150 powder. Larger surface area and particles shape of Zn150 powder particles allowed easier contact of particles and interlocking of the surface irregularities, which was even enhanced by the particles' larger deformability compared to the Zn7.5 powders, which was easily observed in the case of CP and CP-S materials, Figure 3.

The surface layer of corrosion products on particles' surface had also a negative influence on powder particles' compaction during sintering and HP (processes performed at elevated temperature, enhancing diffusion processes). Only the diffusion mechanisms contributed to sintering of particles in the case of the absence of an external pressure [35]. The oxide layer on the powder particles could affect the diffusion bonding of particles during sintering, and this can be seen in the case of CP-S Zn150 powder material compacted under 100 MPa (Figure 3c). In this case, the observed porosity was higher than as in the case of CP materials. Due to the surface oxide layer on the powder particles, the sintering process could be affected by the diffusion delay. The role of the surface oxide layer in the sintering of metal powder particles was defined as follows: Shifting the mechanisms of the sintering process from a bulk-transport mechanism to that controlled by surface transport [34]. Even though the content of oxygen was detected to be higher on the Zn7.5 powder particles, the powder particles' low deformability seemed to have a larger influence on material compaction than the oxide layer on the powder particles. Due to the low particle deformability and spherical shape, only limited areas of individual particles were in direct contact, and the conditions for diffusion were worse compared to the larger and more plane particles of Zn150 powder material. However, according to the experimental observations, the influence of the oxide layer on the sinterability of zinc was minimal, and the presence of an oxide layer on powder particles makes essentially no difference in the kinetics of sintering [34].

Even though the influence of processing parameters on Zn powder-based materials' microstructure is minimal, the influence of the processing on the samples' bending properties was observed. CP-S materials reached higher values of flexural strength for both powder particles sizes compared to the CP materials. The influence was, however, only minor compared to the influence of the HP.

In the case of Zn7.5 powder material, the CP samples prepared under 100, 200 and 300 MPa did not reach the required handling strength necessary for three-point bend testing, and only samples prepared under 400 and 500 MPa were tested. However, the values reached for the flexural strength and displacement before fracture were very low (Figure 4a,c), which corresponds to low particle deformation and the subsequent low bonding and high porosity of the material (Figure 2a,b). The transgranular fracture mechanism and porosity present on the fracture surfaces support this theory (Figure 5a,b). Sintering of the Zn7.5 powder material resulted in higher material handling strength; however, the values of flexural strength reached were still low for materials prepared under 100–300 MPa (Figure 4a). The fracture surface of the CP-S sample prepared under 200 MPa corresponded to the microstructural observation and low bending characteristics of the material reached (Figure 5c). Sintering materials prepared under higher pressures resulted in flexural strength values higher than 40 MPa (41 and 76 MPa for 400 and 500 MPa pressure, respectively). However, the displacement before the fracture was still very low and comparable to the one of CP samples prepared under the same pressures (Figure 4c). HP of Zn7.5 powder materials resulted in a significant increase of materials' flexural strength, reaching maximal value of 322 MPa for material prepared under 400 MPa (Figure 4a). The positive influence of compacting pressure of HP-processed Zn7.5 powder material was revealed by three-point bend test up to the compacting pressure of 400 MPa. Samples prepared under 500 MPa reached lower values of flexural strength than the maximum, which can be connected with the limited deformability of the material due to the HCP crystallographic structure of zinc [36]. High compaction of the material during HP under 500 MPa resulted in strong particle bonding, limiting the transgranular fracture mechanism. With increasing compacting pressure, a larger amount of broken powder particles can be observed on the samples' fracture surfaces (Figure 5e,f). The cleavage mechanism playing a role in the material failure was responsible for the measured decrease of the flexural strength (HP material processed under 500 MPa); however, displacement before fracture still increased.

In the case of Zn150 powder materials, only the sample prepared by CP under 100 MPa did not reach the adequate handling strength necessary for the three-point bend test. Due to the higher deformability of larger powder particles compared to the Zn7.5 powder materials, CP and CP-S samples reached higher values of flexural strength and displacement before fracture when compared to the Zn7.5 samples. In the case of CP-S Zn150 powder materials, sintering had a significantly positive influence on material bending properties (improvement of flexural strength approximately twice compared to the CP materials). HP Zn150 powder materials reached even larger values of bending properties than the CP-S materials; however, the improvement was not as significant as in the case of Zn7.5 powder materials (Figure 4). The differences between the CP-S and HP Zn150 powder materials are significant for materials prepared under 100–300 MPa; however, only a minor influence was observed in the case of materials prepared under 400 and 500 MPa (Figure 4). This corresponds to the microstructural observations, where only a minor influence on the materials microstructure applying higher pressure (400–500 MPa) and temperature during HP was observed (Figure 3). The bonding of particles of materials prepared under 300–500 MPa was comparable for CP-S and HP materials. HP Zn150 powder materials' fracture surfaces were similar for all of the applied pressures during the preparation of materials, while differences in the CP-S materials' fracture surfaces were observed (Small pressures during materials processing resulted in intergranular failure, while the combination of inter- and trans-granular fracture correlated with higher compacting pressures (Figure 6)).

Detailed fractographic analysis of the fracture surfaces of HP processed materials revealed the mechanism of crack propagation (Figures 5f and 6f). Due to the tensile loading, microcracks through individual powder particles were observed, characteristic of the cleavage facets in Figures 5f and 6f. The final crack responsible for the sample's failure followed the powder particles' boundaries connecting cleavage facets (cracked particles).

5. Conclusions

Pure zinc materials prepared by methods of powder metallurgy were examined from the microstructural and mechanical properties point of view. Bulk materials from zinc powder of a mean particles size of 7.5 and 150 μm were prepared with three preparation methods: Cold pressing (CP), cold pressing followed by sintering (CP-S) and hot pressing (HP). The obtained results show the following:

(1) The deformability of zinc powder substantially depends on the size and shape of the powder particles. Smaller spherical particles of Zn7.5 were observed to be only slightly deformable at room temperature (CP), whereas large irregularly-shaped particles of Zn150 powder showed good deformability resulting in low porosity of samples and good material compaction. The addition of sintering (CP-S) and HP improved the deformability of the Zn7.5 powder; however, no significant influence was observed in the case of Zn150 powder.

(2) The increase of compacting pressure was connected with the increase of the bending characteristics of prepared Zn powder materials.

(3) The flexural strength of CP Zn150 powder materials was substantially higher when compared to materials prepared from Zn7.5 powder. Low deformability of small-sized powder particles led to the poor mechanical interlocking and low flexural strength of materials prepared from Zn7.5 powder compacted at room temperature.

(4) CP-S of Zn powder-based materials led to the significant improvement of the bending characteristics of prepared materials. The maximal values of the flexural strength were measured for CP-S materials prepared under 500 MPa; 76 MPa for Zn7.5 powder and 148 MPa for Zn150 powder materials.

(5) HP resulted in even more pronounced improvement of the bending characteristics of Zn powder-based materials. The highest value reached of the flexural strength were measured for Zn7.5 material prepared under 400 MPa (322 MPa) and for Zn150 material prepared under 100 and 200 MPa (203 MPa).

(6) The intercrystalline fracture mechanism was characteristic for CP materials prepared under all of the applied pressures and CP-S materials prepared under pressures from 100–400 MPa. The combination of inter- and trans-crystalline fracture mechanisms was characteristic for CP-S materials prepared under 500 MPa and for HP materials.

Acknowledgments: This work was supported by Project No. LO1211, Materials Research Centre and by Project No. LO1202, Center of New Technologies for Mechanical Engineering (NETME), plus the projects of the Ministry of Education, Youth and Sports of the Czech Republic under the "National sustainability program". We are also grateful to Pavla Puškášová and Jakub Puškáš for the professional English editing service.

Author Contributions: Pavel Doležal conceived of and designed the experiments. Michaela Krystýnová and Matěj Březina performed the experiments. Pavel Doležal, Michaela Krystýnová, Stanislava Fintová and Matěj Březina analyzed the data. Pavel Doležal, Josef Zapletal and Jaromír Wasserbauer contributed reagents/materials/analysis tools. Michaela Krystýnová and Stanislava Fintová wrote the paper.

Conflicts of Interest: The authors declare no conflict of interest.

References

1. Witte, F.; Hort, N.; Vogt, C.; Cohen, S.; Kainer, K.U.; Willumeit, R.; Feyerabend, F. Degradable biomaterials based on magnesium corrosion. *Curr. Opin. Solid State Mater. Sci.* **2008**, *12*, 63–72. [CrossRef]
2. Tan, L.; Yu, X.; Wan, P.; Yang, K. Biodegradable Materials for Bone Repairs: A Review. *J. Mater. Sci. Technol.* **2013**, *29*, 503–513. [CrossRef]
3. Staiger, M.P.; Pietak, A.M.; Huadmai, J.; Dias, G. Magnesium and its alloys as orthopedic biomaterials: A review. *Biomaterials* **2006**, *27*, 1728–1734. [CrossRef] [PubMed]
4. Heublein, B.; Rohde, R.; Kaese, V.; Niemeyer, M.; Hartung, W.; Haverich, A. Biocorrosion of magnesium alloys: A new principle in cardiovascular implant technology? *Heart* **2003**, *89*, 651–656. [CrossRef] [PubMed]

5. Waksman, R.; Pakala, R.; Kuchulakanti, P.K.; Baffour, R.; Hellinga, D.; Seabron, R.; Tio, F.O.; Wittchow, E.; Hartwig, S.; Harder, C.; et al. Safety and efficacy of bioabsorbable magnesium alloy stents in porcine coronary arteries. *Catheter. Cardiovasc. Interv.* **2006**, *68*, 607–617. [CrossRef] [PubMed]

6. Shuai, C.; Zhou, Y.; Yang, Y.; Feng, P.; Liu, L.; He, C.; Zhao, M.; Yang, S.; Gao, C.; Wu, P. Biodegradation Resistance and Bioactivity of Hydroxyapatite Enhanced Mg-Zn Composites via Selective Laser Melting. *Materials* **2017**, *10*, 307. [CrossRef] [PubMed]

7. Chen, J.; Wu, P.; Wang, Q.; Yang, Y.; Peng, S.; Zhou, Y.; Shuai, C.; Deng, Y. Influence of Alloying Treatment and Rapid Solidification on the Degradation Behavior and Mechanical Properties of Mg. *Metals* **2016**, *6*, 259. [CrossRef]

8. He, C.; Bin, S.; Wu, P.; Gao, C.; Feng, P.; Yang, Y.; Liu, L.; Zhou, Y.; Zhao, M.; Yang, S.; et al. Microstructure Evolution and Biodegradation Behavior of Laser Rapid Solidified Mg–Al–Zn Alloy. *Metals* **2017**, *7*, 105. [CrossRef]

9. Gu, X.; Zheng, Y.; Cheng, Y.; Zhong, S.; Xi, T. In vitro corrosion and biocompatibility of binary magnesium alloys. *Biomaterials* **2009**, *30*, 484–498. [CrossRef] [PubMed]

10. Ilich, J.Z.; Kerstetter, J.E. Nutrition in Bone Health Revisited: A Story beyond Calcium. *J. Am. Coll. Nutr.* **2000**, *19*, 715–737. [CrossRef] [PubMed]

11. Xu, L.; Yu, G.; Zhang, E.; Pan, F.; Yang, K. In vivo corrosion behavior of Mg-Mn-Zn alloy for bone implant application. *J. Biomed. Mater. Res. Part A* **2007**, *83*, 703–711. [CrossRef] [PubMed]

12. Hänzi, A.C.; Gerber, I.; Schinhammer, M.; Löffler, J.F.; Uggowitzer, P.J. On the in vitro and in vivo degradation performance and biological response of new biodegradable Mg–Y–Zn alloys. *Acta Biomater.* **2010**, *6*, 1824–1833. [CrossRef] [PubMed]

13. Zhang, E.; Yin, D.; Xu, L.; Yang, L.; Yang, K. Microstructure, mechanical and corrosion properties and biocompatibility of Mg–Zn–Mn alloys for biomedical application. *Mater. Sci. Eng. C* **2009**, *29*, 987–993. [CrossRef]

14. Li, Z.; Gu, X.; Lou, S.; Zheng, Y. The development of binary Mg–Ca alloys for use as biodegradable materials within bone. *Biomaterials* **2008**, *29*, 1329–1344. [CrossRef] [PubMed]

15. Gu, X.; Zheng, Y.; Zhong, S.; Xi, T.; Wang, J.; Wang, W. Corrosion of, and cellular responses to Mg–Zn–Ca bulk metallic glasses. *Biomaterials* **2010**, *31*, 1093–1103. [CrossRef] [PubMed]

16. Zhang, S.; Zhang, X.; Zhao, C.; Li, J.; Song, Y.; Xie, C.; Tao, H.; Zhang, Y.; He, Y.; Jiang, Y. Research on an Mg–Zn alloy as a degradable biomaterial. *Acta Biomater.* **2010**, *6*, 626–640. [CrossRef] [PubMed]

17. Vojtěch, D.; Kubásek, J.; Šerák, J.; Novák, P. Mechanical and corrosion properties of newly developed biodegradable Zn-based alloys for bone fixation. *Acta Biomater.* **2011**, *7*, 3515–3522. [CrossRef] [PubMed]

18. Tapiero, H.; Tew, K.D. Trace elements in human physiology and pathology: Zinc and metallothioneins. *Biomed. Pharmacother.* **2003**, *57*, 399–411. [CrossRef]

19. Aggett, P.J.; Harries, J.T. Current status of zinc in health and disease states. *Arch. Dis. Child.* **1979**, *54*, 909–917. [CrossRef] [PubMed]

20. Fosmire, G.J. Zinc toxicity. *Am. J. Clin. Nutr.* **1990**, *51*, 225–227. [PubMed]

21. Bowen, P.K.; Drelich, J.; Goldman, J. Zinc Exhibits Ideal Physiological Corrosion Behavior for Bioabsorbable Stents. *Adv. Mater.* **2013**, *25*, 2577–2582. [CrossRef] [PubMed]

22. Wang, Y.; Wei, M.; Gao, J.; Hu, J.; Zhang, Y. Corrosion process of pure magnesium in simulated body fluid. *Mater. Lett.* **2008**, *62*, 2181–2184. [CrossRef]

23. Ollig, J.; Kloubert, V.; Weßels, I.; Haase, H.; Rink, L. Parameters Influencing Zinc in Experimental Systems in Vivo and in Vitro. *Metals* **2016**, *6*, 71. [CrossRef]

24. Kirkland, N.T.; Lespagnol, J.; Birbilis, N.; Staiger, M.P. A survey of bio-corrosion rates of magnesium alloys. *Corros. Sci.* **2010**, *52*, 287–291. [CrossRef]

25. ASM International. *ASM Handbook*, 10th ed.; ASM International: Materials Park, OH, USA, 2004.

26. Brandes, E.A.; Brook, G.B.; Smithells, C.J. *Smithells Metals Reference Book*, 7th ed.; Brandes, E.A., Brook, G.B., Eds.; Butterworth-Heinemann: Boston, MA, USA, 1998; Volume 22, p. 94.

27. Kubásek, J.; Vojtěch, D.; Jablonská, E.; Pospíšilová, I.; Lipov, J.; Ruml, T. Structure, mechanical characteristics and in vitro degradation, cytotoxicity, genotoxicity and mutagenicity of novel biodegradable Zn–Mg alloys. *Mater. Sci. Eng. C* **2016**, *58*, 24–35. [CrossRef] [PubMed]

28. Sikora-Jasinska, M.; Mostaed, E.; Mostaed, A.; Beanland, R.; Mantovani, D.; Vedani, M. Fabrication, mechanical properties and in vitro degradation behavior of newly developed Zn Ag alloys for degradable implant applications. *Mater. Sci. Eng. C* **2017**, *77*, 1170–1181. [CrossRef] [PubMed]

29. Wang, X.; Lu, H.M.; Li, X.L.; Li, L.; Zheng, Y.F. Effect of cooling rate and composition on microstructures and properties of Zn-Mg alloys. *Trans. Nonferrous Met. Soc. China* **2007**, *17*, 122–125.

30. Bowen, P.K.; Seitz, J.M.; Guillory, R.J.; Braykovich, J.P.; Zhao, S.; Goldman, J.; Drelich, J.W. Evaluation of wrought Zn-Al alloys (1, 3, and 5 wt % Al) through mechanical and in vivo testing for stent applications. *J. Biomed. Mater. Res. Part B Appl. Biomater.* **2017**. [CrossRef] [PubMed]

31. Demirtas, M.; Purcek, G.; Yanar, H.; Zhang, Z.J.; Zhang, Z.F. Effect of chemical composition and grain size on RT superplasticity of Zn-Al alloys processed by ECAP. *Lett. Mater.* **2015**, *5*, 328–334. [CrossRef]

32. Shaw, C.A.; Tomljenovic, L. Aluminum in the central nervous system (CNS): Toxicity in humans and animals, vaccine adjuvants, and autoimmunity. *Immunol. Res.* **2013**, *56*, 304–316. [CrossRef] [PubMed]

33. Baba, K.; Hatada, R.; Flege, S.; Ensinger, W.; Shibata, Y.; Nakashima, J.; Sawase, T.; Morimura, T. Preparation and antibacterial properties of Ag-containing diamond-like carbon films prepared by a combination of magnetron sputtering and plasma source ion implantation. *Vacuum* **2013**, *89*, 179–184. [CrossRef]

34. Munir, Z.A. Analytical treatment of the role of surface oxide layers in the sintering of metals. *J. Mater. Sci.* **1979**, *14*, 2733–2740. [CrossRef]

35. Nassef, A.; El-Garaihy, W.; El-Hadek, M. Mechanical and Corrosion Behavior of Al-Zn-Cr Family Alloys. *Metals* **2017**, *7*, 171. [CrossRef]

36. Prasad, Y.V.R.K.; Sasidhara, S. *Hot Working Guide: A Compendium of Processing Maps*; ASM International: Materials Park, OH, USA, 1997; p. 496.

MDPI

St. Alban-Anlage 66

4052 Basel

Switzerland

Tel. +41 61 683 77 34

Fax +41 61 302 89 18

www.mdpi.com

Metals Editorial Office

E-mail: metals@mdpi.com

www.mdpi.com/journal/metals

www.ingramcontent.com/pod-product-compliance
Lightning Source LLC
Chambersburg PA
CBHW051851210326
41597CB00033B/5857